Unified SuperStandard Theory and The SuperUniverse Model: The Foundation of Science

Based Solely on Complex General Relativity and Quantum Field Theory
Internal Symmetry: $SU(2)\otimes U(1)\otimes SU(3)\otimes SU(2)\otimes U(1)]^4\otimes U(4)]^9$
Universe Expansion based on Vacuum Polarization
Exact QED Calculation of Fine Structure Constant α
Approximate Standard Model Coupling Constants
Proof: Universe Vacuum Polarization Generates Expansion
Gravity Regimes from Earth to Cosmic Large Scale Structures
Physics and Chemistry of 4D SuperUniverses

Stephen Blaha Ph. D.
Blaha Research

Pingree-Hill Publishing
MMXIX

ISBN: 978-1-7328245-9-1

Rev. 00/00/01 November 19, 2019

To Margaret

Some Other Books by Stephen Blaha

All the Megaverse! Starships Exploring the Endless Universes of the Cosmos using the Baryonic Force (Blaha Research, Auburn, NH, 2014)

SuperCivilizations: Civilizations as Superorganisms (McMann-Fisher Publishing, Auburn, NH, 2010)

All the Universe! Faster Than Light Tachyon Quark Starships & Particle Accelerators with the LHC as a Prototype Starship Drive Scientific Edition (Pingree-Hill Publishing, Auburn, NH, 2011).

Cosmos Creation: The Unified SuperStandard Theory, Volume 2, SECOND EDITION (Pingree Hill Publishing, Auburn, NH, 2018).

Immortal Eye: God Theory: Second Edition (Pingree Hill Publishing, Auburn, NH, 2018).

Calculation of: QED α = 1/137, and Other Coupling Constants of the Unified SuperStandard Theory (Pingree Hill Publishing, Auburn, NH, 2019).

Coupling Constants of the Unified SuperStandard Theory SECOND EDITION (Pingree Hill Publishing, Auburn, NH, 2019).

Unification of God Theory and Unified SuperStandard Theory THIRD EDITION (Pingree Hill Publishing, Auburn, NH, 2018).

The Exact QED Calculation of the Fine Structure Constant Implies ALL 4D Universes have the Same Physics/Life Prospects (Pingree Hill Publishing, Auburn, NH, 2019).

Available on Amazon.com, bn.com Amazon.co.uk and other international web sites as well as at better bookstores (through Ingram Distributors).

CONTENTS

FIGURES and TABLES

INTRODUCTION

This book is a companion to the primary book *Unification of God Theory and Unified SuperStandard Theory THIRD EDITION*. It provides additional topics, greater depth on many topics, and revised views in some areas.

The major new features in this book are:

1. A detailed derivation of the Unified SuperStandard Theory from Complex General Relativity and Quantum Field Theory without recourse to ad hoc assumptions. Five axioms are described that specify primary features of the derivation.

2. A more detailed view of the Generation group and the Layer groups that lead to the 192 fermion spectrum and an understanding of the proportions of normal and Dark energy and matter in the universe..

3. A calculation of the QED Fine Structure Constant and the Standard Model coupling constants from quantum field theory vacuum polarization.

4. A detailed view of the impact of Dark Energy on the expansion of the universe.

5. A calculation of the gravity potential from earth to the galaxies with a possible application to understanding the new data on large scale structures in the universe up to distant quasars.

6. A development of a SuperUniverse Model based on the author's studies of elementary particles and universe evolution. Their deep basis in the author's theories suggests all four-dimensional SuperUniverses have a common Physics, Chemistry and Biology.

1. Origin of Science

We trace the origin of Chemistry and Biology ultimately to Physics. We can further trace the basis of Physics to the Physics of Elementary Particles and of General Relativity. In this book we will establish a basis for Physics in a theory of Elementary Particles and Complex General Relativity.

We choose to start with Complex General Relativity because it leads to Complex Special Relativity which subsequently leads to the general form of the spectrum of spin ½ fermions as a set of four types of fermions: charged leptons, neutral leptons, up-type quarks and down-type quarks.[1]

Complex General Relativity is one of the parents in the Foundation of Physics. The other parent is Quantum Field Theory.[2]

One might ask why this choice of parentage was made. We argue the key factors that lead to the choice are motion and creation/destruction. Motion requires coordinates for its specification. Complex General Relativity offers a most general formulation leading to coordinates and the description of motion. Needless to say, motion requires entities that move. Particles are perhaps the simplest entities. The acts of creation and destruction seem to be a requirement of particles to avoid a static universe. The simplest method of implementing creation and destruction is Quantum Field Theory. It avoids the alternatives of interacting blobs of matter and of more complex forms of entities such as strings. Thus we have a simple rationale for the Foundation of Physics.

This Foundation leads to the known Standard Model in a fairly direct way and beyond that to our SuperStandard Theory described in earlier books. We now turn to follow the path that ultimately leads to the SuperStandard Theory. Unlike other attempts at a fundamental theory we do not posit groups for interactions but rather provide strong arguments (derivations) for the choice of the Standard Model group and the SuperStandard Theory group.

1.1 Foundation Complex General Relativity

We assume that the fundamental theory of space-time coordinates is Complex General Relativity. It is General Relativity extended to complex-valued coordinates with a complex-valued energy-momentum tensor. Complex General Relativity can be factored into a U(4) group that rotates complex coordinates and a residual Complex

---

[1] One cannot have a Complex Special Relativity without embedding it in a Complex General Relativity.

[2] We do not accept a deeper foundation for Fundamental Physics, such as Emergent Physics, due to the total absence of experimental evidence for a deeper level.

General Relativity factor. We call the U(4) group the Coordinates Species group for reasons given later.

From Complex General Relativity we proceed to consider the flat space-time case with Complex Special Relativity. Complex Special Relativity is described by the Complex Lorentz Group which has the subgroups SU(2), U(1), SU(3), and additional SU(2) and U(1) subgroups.

After showing a map from coordinate subgroup symmetries to elementary particle group symmetries we find the Coordinates Species group maps to a U(4) Particle Species group.

1.2 Quantum Field Theoretic Elementary Particle Theory: SU(11)

The group of the internal symmetry group direct product corresponding to the Complex Lorentz Group's subgroups is

$$SU(2)\otimes U(1)\otimes SU(3)\otimes SU(2)\otimes U(1) \tag{1.1}$$

Its covering group is SU(7). When one combines the SU(7) elementary particles group, and a U(4) Generation Group that generates four generations we obtain

$$SU(2)\otimes U(1)\otimes SU(3)\otimes SU(2)\otimes U(1)\otimes U(4) \tag{1.2}$$

with an SU(11) covering group.[3] It is a modest generalization of the Standard Model having a U(4) Generation Group[4] factor that gives a four generation fermion spectrum.

If we further add the Particle Species Group we obtain an SU(15) covering group with the internal symmetry subgroup

$$SU(2)\otimes U(1)\otimes SU(3)\otimes SU(2)\otimes U(1)\otimes [U(4)]^2 \tag{1.3}$$

---

[3] The SU(11) 11-dimensional fundamental representation "contains" the fundamental representations of the factors of eq. 1.1, namely 2 + 3 + 2 + 4 = 11.

[4] The appearance of the Generation Group is due to the consideration of the conserved (or almost conserved) particle number operators seen later.

Thus we obtain the symmetry of an Extended Standard Model (eq. 1.3). The Extended Standard Model group contains the Standard Model symmetries and adds an additional SU(2)⊗U(1) specifying the Weak interactions of Dark Matter and an additional U(4) Particle Species group generates fermion mass terms.

Complex General Relativity (plus the Generation group) lead to the SU(15) particle species covering group with subgroup factors given by eq. 1.3. *Thus we avoid postulating a separate, independent group structure for elementary particles. We view the Extended Standard Model as flowing directly (by analogy) from Complex General Relativity.*

The "universe" of Extended Standard Model particle interactions "mimics" the subgroup symmetries within Complex General Relativity. One wonders if there is a "particle universe' that parallels (in form) our space-time universe?

1.3 Foundation Quantum Field Theory

Particles, and particle symmetries, to which we have been alluding above, emerge from Quantum Field Theory—the other Foundation of Physics. Why do we view Quantum Field Theory as a fundamental foundation? Of all the forms a dynamics theory might take, Quantum Field Theory is the simplest form that supports the creation and annihilation of matter—a fundamental attribute of matter as we know it. Blob creation and annihilation would take us into complexity as would realistic string creation and annihilation.

Quantum Field Theory offers a simplicity that is easily seen in the representation of creation and annihilation using Feynman diagrams.

Thus we opt for Quantum Field Theory and find it, and Complex General Relativity, sufficient to describe all known features of elementary particles and their combinations into more complex forms of matter and energy. As we saw in Blaha (2018e) and our earlier books the purported problems of Quantum Field Theory are easily curable.

A direct benefit of Quantum Field Theory is the appearance of particle number operators which lead to the Generation Group and the Layer Groups See Chapter 4.

The resulting total internal symmetry, with covering group SU(64) is

$$SU(2)\otimes U(1)\otimes SU(3)\otimes SU(2)\otimes U(1)]^4\otimes U(4)]^9$$

1.4 Axioms of the Unified SuperStandard Theory

Complex General Relativity and Quantum Field Theory lead to the Unified SuperStandard Theory. The fundamental axioms that specify the basis of the derivation of the Model are listed in Fig. 1.1. Their detailed implications are:

1. Each space-time symmetry subgroup of the Complex Lorentz Group has a corresponding particle interaction symmetry group. The particle symmetry groups combine in a direct product.

   The specific subgroups of the Complex Lorentz groups with corresponding particle interaction symmetry groups is only well-defined if we further require that they correspond to the distinction between space and time. Thus the SU(3) subgroup emerges from the 3 × 3 space part of Complex Lorentz Group elements. The pair of SU(2) ⊗ U(1) subgroups emerge from the "boost" parts of Complex Lorentz Group elements.[5] Consequently the corresponding enlarged Standard Model particle interaction symmetry is

   $$SU(2)\otimes U(1)\otimes SU(3)\otimes SU(2)\otimes U(1)$$

The minimal group in which the above direct product is a subgroup is SU(7) due to the Dark SU(2)⊗U(1) factor.

2. Quantum Field Theory supports fundamental particles that form a countable set. Each particle number operator is a generator of a particle interaction group.

   All matter and energy is composed of discrete particles.

3. All quantum field theory calculations are finite.

[5] See Appendix A and appendices C through G for details.

Perturbation theory calculations have divergences in conventional quantum field theory that require renormalization. In the author's Two Tier formulation[6] there are no divergences—eliminating the need for renormalization to eliminate divergences.

4. The Quantum Field Theory of particles can be defined in any curved space-time.

   In certain curved space-times conventional second quantization leads to ambiguities in the definition of particle states. The author's generalized second quantization procedure[7] called Pseudoquantization eliminates these ambiguities. It also supports a canonical lagrangian formulation of higher derivative field theories.[8]

5. Each particle wave function has a functional[9] defining the particle state in a space without a distance measure.
   There is a space of functionals (called monads or cores) with an element for each particle in the universe.[10] The quantum entanglement of particles at a distance can be instantaneous because the functionals (which embody the state of each particle) exist in a space without a distance measure. In a sense this feature embodies the unity of creation.

The five axioms imply the detailed list of axioms in Appendix J.

---

[6] See Blaha (2005a).
[7] S. Blaha, Phys. Rev. D**17**, 994 (1978) and references therein to earlier papers by the author such as Phys. Rev. D**10**, 4268 (1974) and Il Nuovo Cimento **49A**, 35 (1979) and **49A**, 113 (1979).
[8] S. Blaha, Phys. Rev. **D10**, 4268 (1974) and **D11**, 2921 (1975) and references therein.
[9] See Blaha (2018e) for a detailed discussion. Our approach eliminates the issues of "spookiness" and instantaneous action at a distance that clouds quantum entanglement.
[10] See Blaha (2018e).

AXIOMS

1. Each space-time symmetry group of the Complex Lorentz Group has a corresponding particle interaction symmetry group. The particle symmetry groups combine in a direct product.

2. Quantum Field Theory supports fundamental particles that form a countable set. Each particle number operator is a generator in a particle interaction group.

3. All quantum field theory calculations are finite.

4. The Quantum Field Theory of particles can be defined in any curved space-time.

5. Each particle wave function has a functional defining the particle state in a space without a distance measure.

Figure 1.1 The axioms of the Unified SuperStandard Theory.

2. Complex General Relativity and its Reality Group

Complex General Relativity is the foundation of the coordinates of physical reality. Its flat space limit is Complex Special Relativity. Complex Special Relativity transformations form the Complex Lorentz Group. We shall see that the Complex Lorentz Group, plus a U(4) group[11] within Complex General Relativity, leads directly to the symmetries of an Extended Standard Model based on an SU(11) covering group.

We start by noting the a Complex General Relativistic transformation can be represented by

$$\Lambda_c = H\Lambda H^{-1} \tag{2.1}$$

where H is a localU(4) (Reality group) transformation that maps a complex coordinate system to real-valued coordinates and Λ is a real-valued General Relativistic transformation.

Appendix A develops a tetrad formalism that describes the Reality group. Chapter 3 applies its U(4) particle interaction group analogue (called the Species group) to particle equations and shows that a Higgs Mechanism for the Species group adds a term to each fermion maas. The commonality of the added term enables us to prove gravitational mass equals inertial mass.

---

[11] We call it the General Relativity Reality Group. Appendix A describes a tetrad (vierbein) formulation of Complex General Relativity that yields the General Relativity Reality Group.

3. Species Group U(4) Gauge Fields and Inertial Mass

The General Relativistic Reality Group described in appendix A has a non-abelian U(4) coordinate space group as its flat space-time limit. We call this group the *Species Group*. Its gauge fields are denoted $A_{Rk}{}^{\lambda}(y)$.

We assume[12] there are corresponding internal symmetry gauge fields $A_{Sk}{}^{\lambda}(y)$ – the U(4) Internal Symmetry Species Group gauge fields. The interaction that appears in covariant derivatives is $g_8\mathbf{A}_S{}^{\mu}(x) = g_8\mathbf{A}_{Sk}{}^{\mu}(x)\mathbf{G}_{Sk}$ where the $\mathbf{G}_{Sk}$ are U(4) generator matrices and k is summed from 1, … , 16.

Below we will see that the effect of the Internal Symmetry Species Group is to U(4) rotate the four components of each fermion's field performing a U(4) rotation of the spinors of each fermion. Since it preserves the species of each fermion we were led to call this group the *Species Group*.

We will see that the Higgs Mechanism breakdown of the Species Group endows each fermion with a mass contribution that breaks the scale invariance of the Unified SuperStandard Theory.

Since the Species Group Higgs Mechanism breakdown gives each fermion a 'gravity generated' mass, and since this mass sets the mass scale for each fermion, we conclude later that the principle of the equality of inertial and gravitational mass is a direct consequence. ***Inertial mass equals gravitational mass.***

3.1 Species Group Covariance

A Species Group transformation on a Dirac equation must be covariant. Consider the Dirac equation lagrangian term under an Internal Symmetry Species Group transformation:

---

[12] This assumption parallels the assumptions for the SU(3)⊗SU(2)⊗U(1)⊗SU(2)⊗U(1) Internal Symmetry Reality Group presented in previous chapters.

$$\bar{\psi}(x)[i\gamma_\mu(\partial/\partial x_\mu - ig_8 A_{Sk}{}^\mu(x)G_{Sk}) - m]\psi(x) \qquad (3.1)$$

with a sum over k. If we perform a Species group transformation U on lagrangian terms:

$$\bar{\psi}(x)[i\gamma_\mu(\partial/\partial x_\mu - ig_8 A_{Sk}{}^\mu(x)G_{Sk}) - m]U^{-1}U\psi(x)$$

or

$$\bar{\psi}(x)U^{-1}U[iU^{-1}U\gamma_\mu U^{-1}U(\partial/\partial x_\mu - ig_8 A_{Sk}{}^\mu(x)G_{Sk}) - m]U^{-1}U\psi(x)$$

we find

$$\bar{\psi}'(x)[i\gamma_\mu{}'U(\partial/\partial x_\mu - ig_8 A_{Sk}{}^\mu(x)G_{Sk})U^{-1} - m]\psi'(x)$$

where

$$\gamma_\mu{}'(x) = U\gamma_\mu U^{-1}$$

is locally equivalent to a Dirac matrix by Good's Theorem.[13] If we set

$$A'_S{}^\mu(x) = -(i/g_8)U[\partial U^{-1}/\partial x^\mu] + UA_S{}^\mu(x)U^{-1}$$

then the transformed lagrangian expression becomes

$$\bar{\psi}'(x)[i\gamma_\mu{}'(x)(\partial/\partial x_\mu - ig_8 A'_{Sk}{}^\mu(x)G_{Sk}) - m]\psi'(x) \qquad (3.2)$$

It has the same form as the original eq. 3.1 above and thus the expression is covariant. We note the indices of the matrices G_{Sk} are spinor indices and so $G_{Sk}\gamma_\mu$ has an implicit spinor matrix summation. But the symmetry group is U(4).

The coordinate dependence of $\gamma_\mu{}'(x)$ introduces locality into the Dirac matrix. This locality might be viewed with concern except that an inverse Species group transformation exists that removes the locality. Thus the physical impact of this 'new' locality is eliminated.

---

[13] R. H. Good, Jr., Rev. Mod. Phys., **27**, 187 (1955).

3.2 Spontaneous Symmetry Breaking of the Species Group

We begin the discussion of the Internal Symmetry Species Group symmetry breaking[14] by defining a Higgs field η which is a Species group 4-vector

$$\eta = \begin{bmatrix} \rho_1 \\ \rho_2 \\ \rho_3 \\ \rho_4 \end{bmatrix} \tag{3.3}$$

where ρ_1, ρ_2, ρ_3 and ρ_4 are real fields.[15] Then the covariant derivative of η (taking account only of the Species group) is

$$D_{\ldots\mu}\eta = \{\partial/\partial X^\mu + \ldots - \tfrac{1}{2} i g_8 \Sigma A_{Sk}{}^\mu(x) G_{Sk}\} \begin{bmatrix} \rho_1 \\ \rho_2 \\ \rho_3 \\ \rho_4 \end{bmatrix} \tag{3.3}$$

Following familiar we find (with ρ_i being the vacuum expectation value of the Higgs field):

$$\begin{aligned}(D_{\ldots\mu}\eta)^\dagger D_{\ldots}{}^{\mu}\eta = &\ \partial\rho_1/\partial X^\mu\, \partial\rho_1/\partial X_\mu + \partial\rho_2/\partial X^\mu\, \partial\rho_2/\partial X_\mu + \partial\rho_3/\partial X^\mu\, \partial\rho_3/\partial X_\mu + \\ &+ \partial\rho_4/\partial X^\mu\, \partial\rho_4/\partial X_\mu + \\ &+ \tfrac{1}{4} g_8^2\{\rho_1^2 A_{S1}^2 + \rho_2^2 A_{S2}^2 + \rho_3^2 A_{S3}^2 + \rho_4^2 A_{S4}^2 + \\ &+ (\rho_1^2 + \rho_2^2)(V_5^2 + V_6^2) + \tfrac{1}{4}(\rho_1^2 + \rho_3^2)(V_7^2 + V_8^2) + \\ &+ (\rho_1^2 + \rho_4^2)(V_9^2 + V_{10}^2) + \tfrac{1}{4}(\rho_2^2 + \rho_3^2)(V_{11}^2 + V_{12}^2) + \\ &+ (\rho_2^2 + \rho_4^2)(V_{13}^2 + V_{14}^2) + \tfrac{1}{4}(\rho_3^2 + \rho_4^2)(V_{15}^2 + V_{16}^2)\}\end{aligned} \tag{3.4}$$

---

[14] Since the Species gauge fields have been shown to have a mass it might seem redundant to introduce Higgs symmetry breaking as well. However the Higgs breaking introduces the further benefit of giving a mass term to each fermion particle – thus establishing the equality of gravitational mass and inertial mass.

[15] Each field ρ_i can be expressed as a PseudoQuantum field: $\rho_i = \varphi_{1i} + \varphi_{2i}$ where φ_{1i} has the vacuum expectation value ρ_{i0} for $i = 1, \ldots, 4$. Thus our PseudoQuantum field theory version is implemented easily. See Blaha (2018e).

up to total divergences, which generate surface terms which we discard. We also assume that all fields satisfy the gauge condition

$$\partial A_{Si}{}^{\mu}/\partial X^{\mu} = 0 \tag{3.5}$$

Eq. 3.4 shows all Species Group gauge fields have masses. Thus Species Group symmetry is completely broken. The combination of an ultra-weak coupling constant and very large gauge field masses results in extremely weak Species interactions.

We assume Species group gauge field masses are very large – of the order of the Planck mass in view of its origin in Complex General Relativity.

3.3 Species Group Higgs Mechanism Contributions to Fermion Masses

The symmetry breaking of the Species Group results in a contribution to each fermion mass of all types, species, generations, and layers. The Species Group contributions to normal and Dark fermion mass terms are

$$\mathcal{L}^{Higgs}{}_{FermionMassesSpecies} = \Sigma_{s,g,l}\bar{\psi}_{sglL}\rho_s m_{sgl}\psi_{sglR} + \Sigma_{s,g,l}\bar{\psi}_{DsglL}\rho_s m_{Dsgl}\psi_{DsglR} + \text{c.c.} \tag{3.6}$$

The η field expectation value has components labeled ρ_s.[16] The mass matrices m_{sgl} and m_{Dsgl} are the complex constant Species mass matrix contributions for normal and Dark fermion species. Each fermion acquires a mass contribution from the Species Higgs Mechanism.

3.4 Species Group Higgs Masses Shows Inertial Mass Equals Gravitational Mass

In Blaha (2016h) we showed that a Complex General Relativity transformation can be factored into the product of a complex-valued transformation and a real-valued General Coordinate transformation. The set of complex valued transformations form a

---

[16] The Higgs fields η… in our PseudoQuantum formulation are $\eta\ldots = \varphi_{1\ldots}(x) + \varphi_{2\ldots}(x)$ as described in Blaha (2018e).

U(4) group that we called the General Coordinate Reality group. The analogous Internal Symmetry Species Group has gauge fields that undergo spomtaneous symmetry breaking and generate contributions to all fermion masses.

Since fermion field masses are now sums of ElectroWeak Higgs contributions, Generation group Higgs contributions (chapter 5), Layer group Higgs contributions (chapter 6), and Species Group contributions, and since the Species Group Higgs fields mass contributions appear in all fermion masses, the equality of inertial and gravitational mass is proven. The Species Group Higgs particles' equations set the mass scale of gravitational mass, and thereby of all Higgs mass contributions. The scale of inertial masses equals the scale of gravitational masses. ***Since an expression cannot mix mass scales, the gravitational mass scale must be the same as the inertial mass scale.***

Inertial mass equals gravitational mass.

We have established the equality of inertial and gravitational mass at the short distance quantum level. In our view, this explanation is far more satisfying than basing the equality on a combination of large distance phenomena and quantum phenomena. Einstein and Weyl have pointed out that all fundamental physics phenomena should be based on a local theory.

4. Interactions Based on Particle Numbers

The particle interactions following directly from Complex General Relativity are

$$SU(2)\otimes U(1)\otimes SU(3)\otimes SU(2)\otimes U(1)\otimes U(4) \quad (1.1)$$

where the U(4) factor is for the Species Group.

They have a SU(11) covering group that contains this direct product of groups. It is described briefly in chapter 1. The groups in eq. 1.1 are the particle interaction groups of an Extended Standard Model. They are discussed in detail in appendices b through G.

Unlike other attempts to develop a formulation of the Standard Model (or generalizations) the Extended Standard Model is directly based on a theory foundation consisting of Complex General Relativity and Quantum Field Theory.

To those who might prefer to base a theory on real General Relativity we note that proofs in Quantum Field Theory *require* the Complex Lorentz Group.[17] Thus the Complex Lorentz group is unavoidable for a properly (and rigorously) formulated Quantum Field Theory. Since the formulation of the Complex Lorentz Group in flat space-time can only be as the limit of Complex General Relativity, the choice of a foundation of Complex General Relativity is required.

In chapter 1 we also specified a foundation of Quantum Field Theory. Our motivation was that it provided the simplest possible formulation of particles capable of engaging in creation and annihilation.

Since particles are countable and thus have discrete particle numbers Quantum Field Theory brings particle numbers, and particle number laws such as particle conservation laws, into consideration.

---

[17] Streater (

Appendix H shows that Complex Lorentz boosts generate four types of fermion particles that we call *particle species*. We map these four species to charged leptons (such as electrons), neutral leptons (such as neutrinos), up-type quarks (such as the u quark), and down-type quarks (such as the d quark).

4.1 Basis of the Generation Group

We define a particle number operator for quark-type particles that we call Baryon Number, and a particle number operator for lepton-type particles that we call Lepton Number.

By analogy, we assume that there are four species of Dark matter: charged Dark leptons, neutral Dark leptons, Dark up-type quarks, and Dark down-type quarks.[18] Thus we are led to Dark particle numbers: Dark Baryon Number, and Dark Lepton Number.

These four fermion particle number operators are assumed to be conserved in particle interactions modulo any experimentally found violations of their conservation.

Regardless of whether their conservation is violated or not we may use these four numbers, denoted B, L, B_D, and L_D respectively, as "diagonal" operators within a U(4) group. We will call this group the Generation Group (chapter 5) and relate it to the generations of the eight (normal and Dark) species of fermions. On this basis we assume there are four generations of each species with one generation (of large masses) as yet not found. (The gauge vector bosons of the Generation Group also have large masses.) If the conservation of the fermion particle numbers is broken then we will view it as a consequence of Generation Group symmetry breaking.

4.2 Basis of the Layer Group

The set of particle number operators can be further refined if we take account of the fourfold fermion generations. In further refining the set of particle number operators we temporarily neglect interactions that would violate conservation laws for the set.

We therefore subdivide the above particle number set into four particle numbers per generation. For the ith generation we define

---

[18] Later we specify Dark quarks to be Color singlets to prevent interactions between Dark and normal quarks. Such interactions have not been found experimentally. Normal quarks will be Color triplets.

L_{iB} – The Baryon particle number for the ith generation
L_{iDB} – The Dark Baryon particle number for the ith generation
L_{iL} – The Lepton particle number for the ith generation
L_{iDL} – The Dark Lepton particle number for the ith generation

for each generation i = 1, 2, 3, 4. Individual fermions have positive $L_{ia} = +1$ values and anti-fermions have negative $L_{ia} = -1$ values for species a = 1, 2, 3, 4 (with the three color subspecies of quarks treated as part of one species.)

At this point we have four particle number operators for each generation. We wish to define a group framework for each set of particle numbers. The simplest way is to assume that each generation consists of four layers with particles in each generation being in a U(4) fundamental representation.[19] Then each generation has a U(4) group with the generation's four number operators as its diagonal operators. We call this group the Layer Group of the generation. With four generations we obtain four U(4) Layer groups.

The consequence of this expansion of particle numbers and groups is that the set of fermions increases fourfold. We now have four layers, with each having four generations, Experimentally we know of three generations of fermions—the lowest generations of the lowest level. The remaining generation and three levels of fermions is of much higher mass and yet to be found.

We can denote the 16 number operators of the $[U(4)]^4$ total Layer Group by

L_{ijB} – The layer j Baryon particle number for the ith generation
L_{ijDB} – The layer j Dark Baryon particle number for the ith generation
L_{ijL} – The layer j Lepton particle number for the ith generation
L_{ijDL} – The layer j Dark Lepton particle number for the ith generation

See chapter 6 for a detailed discussion of the Layer Group. We note in passing that the conservation of these number operators is badly broken. Yet the underlying group structure remains.

[19] See Fig. 6.2 for a depiction of the "splitting" of fermions: first into generations, then into layers.

The definition of four layers of fermions raises the question of their vector gauge field interactions. Since we see no evidence of the higher layers in particle interactions it seems reasonable to assume that each layer has its own Extended Standard Model set of interactions and its own Generation Group. Consequently the total symmetry group (although broken) is[20]

$$[SU(2)\otimes U(1)\otimes SU(3)\otimes SU(2)\otimes U(1))\otimes U(4)]^4\otimes U(4)\otimes [U(4)]^4$$
$$= [SU(2)\otimes U(1)\otimes SU(3)\otimes SU(2)\otimes U(1)]^4\otimes U(4)]^9 \qquad (4.1)$$

The total symmetry group above has a covering group of SU(64). The fermions and symmetry group of eq. 4.1 comprise the basis of the Unified SuperStandard Theory of the author. Chapters 5 and 6 describe the fermion spectrum and the symmetry in detail. Fig. 6.3 depicts the the SuperStandard Theory fermion "periodic table." Fig. 6.4 depicts the SuperStandard Theory set of of gauge fields.

If one does not include the Layer Groups or the Species Group then the resulting total symmetry of the Extended Standard Model is

$$SU(2)\otimes U(1)\otimes SU(3)\otimes SU(2)\otimes U(1))\otimes [U(4)]^2 \qquad (4.2)$$

It has a covering group of SU(11), 48 fundamental fermions, and 32 fundamental vector bosons. SU(11) is reminiscent of 11-dimensional SuperSymmetry although it will not be pursued here in this strictly Quantum Field Theoretic development.

4.3 The Total Number of Fermions and Vactor Bosons

Given the generations and layers described above we find that there are 192 fundamental fermions (taking account of color triplet quarks.) We also find there are 192 vector gauge fields (particles). The equality of the numbers of fermions and vector bosons raises the possibility of SuperSymmetry. See Figs. 6.3 and 6.4 for the "periodic tables" for fermions and vector bosons.

---

[20] The Species group is not multiplied fourfold since all fermions in all layers experience the same gravitation and Species group.

5. Generation Group

We now turn to the derivation of the Generation Group and, in the following chapter, the Layer Groups, which are based on conserved (and broken) particle numbers.

In this chapter we consider the extension of a one generation Superstandard Model to four generations based on the U(4) Generation symmetry group.

It is based on four conservation laws for Baryon, Lepton, Dark Baryon, and Dark Lepton Numbers. The conservation laws are manifest in the fermion terms of the one generation version of The Standard Model which all have the form:

$$\mathcal{L} \sim \bar{\psi}_\alpha \gamma^\mu D_{\mu_{\alpha\beta}} \psi_\beta$$

where the covariant derivative terms preserve the four conservation laws. Symmetry breaking is brought in at a later point through Higgs fields terms. There is an evident SU(3)⊗SU(2)⊗U(1)⊗SU(2)⊗U(1) symmetry in the sum of these lagrangian terms. The next relevant question is whether the particle number conservation laws have associated gauge fields that provide interactions. This question can be partially answered in our current state of experimental knowledge.

In Blaha (2017b), and earlier books, we showed that the existence of a long range baryonic force[21] supports baryon number conservation in a manner similar to electric charge conservation due to the electromagnetic force. Lepton number conservation suggests a very weak long range force as well. By analogy we postulate two similar Dark particle number conservation laws and forces.

---

[21] See Blaha (2017b), earlier books, and later in this chapter, for a discussion of evidence for a baryonic force and conservation law. We found that gravity experiments suggested a possible baryonic potential with the (order of magnitude) coupling constant $\alpha_B = \beta^2/4\pi \simeq .118\ Gm_H^2$ where m_H is the mass of the hydrogen atom. Its section 12.1 provides details.

Having four conserved (or almost conserved) particle numbers with their attendant forces (interactions) leads naturaly to a U(4) symmetry with four 'diagonal' operators. We call the U(4) group the Generation Group.

If Baryon Number and Lepton Number are both conserved quantities then any linear combination of them is also conserved. Therefore

$$B' = aB + bL$$

is also conserved.

If we consider the Dark Matter sector of the Unified SuperStandard Theory it is reasonable to assume that *Dark Baryon Number* B_D *and Dark Lepton Number* L_D *are "conserved"* also (although there is no experimental evidence available as yet to confirm (or deny) these assumptions.)

Thus we have four conserved particle Numbers. Linear combinations of these numbers are also conserved:

$$\begin{aligned} B' &= aB + bL + cB_D + dL_D \\ L' &= eB + fL + gB_D + hL_D \\ B_D' &= iB + jL + kB_D + lL_D \\ L_D' &= mB + nL + oB_D + pL_D \end{aligned}$$

The set of 4×4 matrices form an U(4) group if we wish to perform these transformations within lagrangians of the type of the Unified SuperStandard Theory. The choice of U(4) rather than SU(4) is required since there are four independent particle numbers. U(4) has four diagonal matrices in its algebra while SU(4) only has three diagonal matrices. U(4) preserves the independence of the four independent particle numbers. It also allows complex rotations.

The U(4) Generation Group symmetry leads immediately to four fermion generations. We add an index to each fermion field ranging from 1 through four, add U(4) gauge fields to the covariant derivative, and thus Yang-Mills local gauge field terms to the lagrangian.

Then we perform the following tasks:[22]

- Define the Two-Tier Lepton and Quark Sectors
- Introduce Symmetry Breaking via the Higgs Mechanism for Fermions and U(4) Gauge Fields

Thus we now have four fermion generations, a broken Generation Group symmetry, and masses for fermions and Generation Group gauge fields.

5.1 Baryon Number Conservation and a Possible Baryonic Force

We have considered baryon number conservation and a possible baryonic force in Blaha (2014a) and (2014b). The primary forces involved in the interactions and collisions of baryons include the forces of The Standard Model, the force of gravity, and possibly a fifth force which we take to be the baryonic force, a much discussed force that depends on the baryon numbers of objects experiencing it. The Gravitation and baryonic forces (neglecting other SuperStandard Theory interactions) between two clumps of baryonic matter containing baryons and other particles: clump1 being of mass m_1 and baryon number n_1, and clump2 being of mass m_2 and baryon number n_2 is

$$F = -Gm_1m_2/r^2 + (\beta^2/4\pi)n_1n_2/r^2 \tag{5.1}$$

where G is the gravitational constant, β is the baryonic coupling constant and r is the distance between the 'widely' separated clumps. Experimentally a baryonic force between baryons has not been detected with any degree of certainty. Sakurai (1964) discusses early efforts to determine the force of gravity in detail. Eőtvős experiments on the ratio of the observed gravitational mass to the inertial mass showed that that the gravity force is constant to within one part in 100,000,000 as far back as 1922

[22] These tasks are described in detail in chapters 12 and 13 in Blaha (2017b) as well as earlier books.

indicating the baryonic force, if it exists, as we believe it does, is extremely weak compared to the gravitational force. Eőtvős et al[23] found

$$(\beta^2/4\pi)/(Gm_p^2) < 10^{-5}$$

where m_p is the proton mass.

Since then, the experiment has been redone with improved accuracy by Dicke and collaborators.[24] They have improved the accuracy to one part in 100 billion. A further analysis showed a very small discrepancy that suggested the ratio, while small, was non-zero, implying the equivalence principle might not be exact and that the discrepancy changed with the material used in the experiment – just what one might expect if a very small baryonic force was present – sometimes called the "fifth force." At present the existence and amount of the discrepancy is unclear. *Nevertheless, we will assume a fifth force – Baryonic force. We will also assume there are three additional forces corresponding to the three other conserved particle numbers.These forces are not relevant to these considerations.*

The conservation of baryon number has been repeatedly investigated by experimenters and found to be true to extremely high accuracy. For decades theorists have suggested that a baryon conservation law[25] follows from the existence of a gauge field in a manner much like electric charge conservation follows from the properties of the electromagnetic gauge field.

5.1.1 Estimate of the Baryonic Coupling Constant

The baryonic force, and coupling constant, if it exists, is known to be very small in comparison to gravity and the other known forces. However, measurements of the gravitational constant G are significantly different.[26,27] The reason(s) for these

---

[23] Eőtvős, R. V., Pekár, D., Fekete, E., Ann. d. Physik **68**, 11 (1922).
[24] P. G. Roll, R. Krotkov, R. H. Dicke, Annals of Physics, 26, 442, 1964.
[25] See Gell-Mann, M. and Levy, M. *Nuovo Cimento* 16, 705 (1960) for a proof and Sakurai (1964) for a discussion of the relation of the baryonic gauge field to gravity experimentally.
[26] T. Quinn et al, Phys. Rev. Lett. **111**, 101102 (2013).
[27] P. J. Mohr, B.N. Taylor, and D. B. Newell, Rev. Mod. Phys. 84, 1527 (2012).

discrepancies is not known. We will assume that both the 2010 and 2013 measurements of G are experimentally correct but disagree because of the baryonic force term in eq. 5.1 that would create a difference in effective G values if the experiments used different masses and thus baryon numbers. Quinn *et al* found a value for the gravitational constant of $G_1 = 6.67545\times10^{-11}$ m^{3}kg^{-1}s^{-2}. The combined 2010 CODATA value for the gravitational constant was $G_2 = 6.67384\times10^{-11}$ m^{3}kg^{-1}s^{-2}. Both values are subject to estimated uncertainties.

Suppose these values are correct and due to a difference in the chemical composition (metals) of the test masses used in the experiment. Quinn *et al* use 1.2 kg test masses composed of Cu-0.7% Te free machining alloy. The CODATA value being a composite of many experiments does not have an effective equivalent test mass value or composition specified.[28] Suppose the test mass value is $N_1^2m_1^2 + N_{1e}^2m_e^2$ for the G_1 result giving

$$-(N_1^2m_1^2 + N_{1e}^2m_e^2)G_1 = [-G(m_1^2N_1^2 + N_{1e}^2m_e^2) + (\beta^2/4\pi)N_1^2] \tag{5.2}$$

where G is the *real* value of the gravitational constant. The total test mass is $(m_1^2N_1^2 + N_{1e}^2m_e^2)$ with N_1 baryons of average mass m in each test mass and N_{1e} leptons of average mass m_e.

Suppose further the test mass value is $N_2^2m_2^2 + N_{2e}^2m_e^2$ for the G_2 result giving

$$-(N_2^2m_2^2 + N_{2e}^2m_e^2)G_2 = [-G(m_2^2N_2^2 + N_{2e}^2m_e^2) + (\beta^2/4\pi)N_2^2] \tag{5.3}$$

where G is again the *real* value of the gravitational constant. The total test mass is $(m_2^2N_2^2 + N_{2e}^2m_e^2)$ with N_2 baryons of average mass m_2 in each test mass and N_{2e} leptons of average mass m_e. Since the test masses are electrically neutral and there are approximately equal numbers of protons and neutrons in a test mass it follows approximately that

$$N_{1e} = \tfrac{1}{2}N_1 \quad \text{and} \quad N_{2e} = \tfrac{1}{2}N_2 \tag{5.4}$$

[28] The Eötvös' experiment used a 0.1 gm test mass of $RaBr_2$. R. v. Eötvös, D. Pekár, E. Fekete, Annalen der Physik (Leipzig) **68**, 11 (1922).

Subtracting eq. 5.2 from eq. 5.3 after some algebra[29] we find

$$\Delta G = -G_2 + G_1 = (\beta^2/4\pi)/(m_2^{\,2} + m_e^{\,2}/2) - (\beta^2/4\pi)/(m_1^{\,2} + m_e^{\,2}/2)$$
$$\simeq (\beta^2/4\pi)(1/m_2^{\,2} - 1/m_1^{\,2}) \quad (5.5)$$

The masses m_1 and m_2 can differ. For example, if m_H is mass of the hydrogen atom, then $m^{-1} = 1.0m_H^{-1}$ for hydrogen, for carbon $m^{-1} = 1.00782m_H^{-1}$, for copper $m^{-1} = 1.00895m_H^{-1}$, and for lead $m^{-1} = 1.00794m_H^{-1}$.[30] Thus using the Quinn *et al* results, and CODATA results, and assuming copper and lead test masses, we find the order of magnitude *estimate*:

$$\alpha_B = \beta^2/4\pi \simeq \Delta G/[(1.00895^2 - 1.00794^2)\, m_H^{\,2}]$$
$$\simeq \Delta G/G\ G\, m_H^{\,2}/.002037$$
$$\simeq (0.000241/0.002037)Gm_H^{\,2}$$
$$\simeq 0.118\ Gm_H^{\,2} \quad (5.6)$$

indicating a very weak baryonic force consistent with our general view of the universe. The baryon fine structure constant is minute in comparison to the electromagnetic fine structure constant $\alpha \simeq 1/137$.

Due to our assumptions in the calculation of α_B, which makes it merely an order of magnitude estimate at best, we suggest that an experimental group measure G with differing test masses in the same apparatus to obtain a better value for α_B.

5.2 Four Generation Unified SuperStandard Theory

Previously we derived the form of a one generation Unified SuperStandard Theory that included the known parts of the Standard Model (excepting the Higgs sector) and an SU(2)⊗U(1) part for Dark Matter.

---

[29] The reduction of the calculation to algebra reminds the author of Nobelist Hans Bethe's remark that he only felt he understood a physical phenomenon when he could reduce it to algebra. This was quite evident when the author collaborated with Professor Bethe on a study of pion condensation in neutron stars some years ago.

[30] "One Hundred Years of the Eötvös Experiment", I. Bod, E. Fischbach, G. Marx and Maria Náray-Ziegler, August, 1990.

In this section we generalize to the four generation SuperStandard Theory that results.[31] Covariant derivatives acquire another interaction term with 16 U(4) fields U_i^{μ}. In addition we add another index to each fermion field specifying its generation. Lastly a set of initially massless gauge field dynamics terms is added to the Unified SuperStandard Theory lagrangian to specify U(4) gauge field evolution and interactions.

5.2.1 Two-Tier Lepton Sector

We begin with the definition of a quadruplet of leptons – a pair of doublets, one normal and one Dark, instead of a single doublet. We define left and right lepton quadruplets with[32]

$$\Psi_{L,Ra}(X) = \begin{bmatrix} \psi_{DL,Ra}(X) \\ \psi_{NL,Ra}(X) \end{bmatrix} \tag{5.7}$$

where a is a generation index ranging from 1 to 4, $\psi_{NL,R}(X)$ is a "normal" ElectroWeak-like lepton doublet, and where $\psi_{DL,R}(X)$ is a similar Dark ElectroWeak-like lepton doublet consisting of a Dark electron-like fermion and a Dark neutrino-like fermion.

We define covariant derivative terms which we express in matrix form are

$$D_{L,R}(X) = \begin{bmatrix} \gamma^{\mu} D_{DL,R\mu} & 0 \\ 0 & \gamma^{\mu} D_{NL,R\mu} \end{bmatrix} \tag{5.8}$$

where the normal matter left-handed covariant derivative is

---

[31] It is based on the three principles based on Ockham's Razor ("The simplest choice is often the best."): 1) The only connecting interaction is a weak interaction, 2) The form of ElectroWeak theory remains unchanged, and 3) Dark Matter parallels normal matter in its general characteristics: four generations, SU(3) singlets, an SU(2)⊗U(1) symmetry analogous to ElectroWeak symmetry, SU(2)⊗U(1) dark lepton and dark quark doublets.

[32] The X's are Two-Tier coordinates.

$$D_{NL\mu} = \partial/\partial X^{\mu} - ½ig\boldsymbol{\sigma}\cdot\mathbf{W}_{\mu} + ½ig'B_{\mu} - ½ig_G\mathbf{G}\cdot\mathbf{U}_{\mu} \quad (5.9)$$

where g_G is an ultra-weak generational coupling constant, $\mathbf{G}\cdot\mathbf{U}_{\mu}$ is the sum of the inner product of 16 U(4) generators G_i and gauge fields $U_i(X)$, and where the Dark matter left-handed covariant derivative is

$$D_{DL\mu} = \partial/\partial X^{\mu} - ½ig_D\boldsymbol{\sigma}\cdot\mathbf{W}'_{\mu} + ½ig_D'B'_{\mu} - ½ig_G\mathbf{G}\cdot\mathbf{U}_{\mu} \quad (5.10)$$

with $\boldsymbol{\sigma}$ a vector composed of the Pauli matrices. The right-handed covariant derivatives have a simpler form. The normal matter right-handed covariant derivative is

$$D_{NR\mu} = \partial/\partial X^{\mu} + ½ig'B_{\mu} - ½ig_G\mathbf{G}\cdot\mathbf{U}_{\mu} \quad (5.11)$$

and the Dark matter right-handed covariant derivative is

$$D_{DR\mu} = \partial/\partial X^{\mu} + ½ig_D'B'_{\mu} - ½ig_G\mathbf{G}\cdot\mathbf{U}_{\mu} \quad (5.12)$$

The normal and Dark electroweak fields above are functions of a Two-Tier X. The Faddeev-Popov mechanism operative for these types of fields is described in appendix 19-A of Blaha (2011c).

5.2.2 Quark Sector

In the *quark* sector we define left and right quark quadruplets with

$$\Psi_{qL,Ra}(X_c) = \begin{bmatrix} \psi_{DqL,Ra}(X_c) \\ \psi_{NqL,Ra}(X_c) \end{bmatrix} \quad (5.13)$$

where $\psi_{NqL,Ra}(X_c)$ is a "normal" ElectroWeak-like quark doublet, and where $\psi_{DqL,Ra}(X_c)$ is a Dark ElectroWeak-like quark doublet consisting of a SU(3) singlet Dark up-quark of unit Dark charge and a SU(3) singlet Dark down-quark of zero Dark charge in the ath generation.

The covariant derivative terms are contained in $D_q(X_c)$ which we express in matrix form as

$$D_{qL,R}(X_c) = \begin{bmatrix} \gamma^\mu D_{qDL,R\mu}(X_c) & 0 \\ 0 & \gamma^\mu D_{qNL,R\mu}(X_c) \end{bmatrix} \qquad (5.14)$$

where the normal quark matter left-handed covariant derivative is

$$D_{qNL\mu} = \partial/\partial X_c{}^\mu - \tfrac{1}{2}ig\boldsymbol{\sigma}\cdot\mathbf{W}_\mu - ig'B_\mu/6 - \tfrac{1}{2}ig_G\mathbf{G}\cdot\mathbf{U}_\mu + ig_C\boldsymbol{\tau}\cdot A_{C\mu} \qquad (5.15)$$

and where the Dark quark left-handed covariant derivative is

$$D_{qDL\mu} = \partial/\partial X_c{}^\mu - \tfrac{1}{2}ig_D\boldsymbol{\sigma}\cdot\mathbf{W}'_\mu + \tfrac{1}{2}ig_D'B'_\mu - \tfrac{1}{2}ig_G\mathbf{G}\cdot\mathbf{U}_\mu \qquad (5.16)$$

since Dark quarks are SU(3) singlets with unit or zero Dark charge. The right-handed quark covariant derivatives have a simpler form. The normal quark right-handed covariant derivative is

$$D_{qNR\mu} = \partial/\partial X_c{}^\mu + \tfrac{1}{2}ig'B_\mu/3 - \tfrac{1}{2}ig_G\mathbf{G}\cdot\mathbf{U}_\mu + ig_C\boldsymbol{\tau}\cdot A_{C\mu} \qquad (5.17)$$

and the Dark quark right-handed covariant derivative is

$$D_{qDR\mu} = \partial/\partial X_c{}^\mu + \tfrac{1}{2}ig_D'B'_\mu - \tfrac{1}{2}ig_G\mathbf{G}\cdot\mathbf{U}_\mu \qquad (5.18)$$

The normal and Dark gauge boson fields are functions of X_c. = $(X_{r_\mu}(y_r), X_{i_\mu}(y_i))$. The Faddeev-Popov mechanism is operative for gauge boson fields and is described later in Appendix 19-A of Blaha (2011c).[33] The *complexon* quark SuperStandard Theory ElectroWeak Sector covariant derivatives in quadruplet matrix form are

[33] Those who might be concerned about the propagator term $<W_i(X), W_j(X_c)>$ and similar propagators where one field is a function of X and the other field is a function of X_c should note that such terms are to very good

$$D_{qL,R}(X_c) = \begin{bmatrix} \gamma^{\mu}D_{qDL,R\mu} & 0 \\ 0 & \gamma^{\mu}D_{qNL,R\mu} \end{bmatrix} \tag{5.19}$$

The remaining parts of the complexon Standard Model are described in chapter 23 of Blaha (2011) and summarized below. The addition of singlet Dark quark Higgs terms is also required.

The lagrangian density and action is

$$\mathcal{L}_{CSM} = \Psi_{La}{}^{\dagger}\gamma^0 i\gamma^{\mu}D_{L\mu}\Psi_{La} - \Psi_{Ra}{}^{\dagger}\gamma^0 i\gamma^{\mu}D_{R\mu}\Psi_{3Ra} + \Psi_{CLa}{}^{\dagger}\gamma^0 i\gamma^{\mu}\mathcal{D}_{qL\mu}\Psi_{CLa} + + \Psi_{CRa}{}^{\dagger}\gamma^0 i\gamma^{\mu}\mathcal{D}_{qR\mu}\Psi_{CRa} - \mathcal{L}_{BareMasses} + \mathcal{L}_{Gauge} + \mathcal{L}_{Mass} + \mathcal{L}_{Ufields} \tag{5.20}$$

where a is the generation index. $\mathcal{L}_{BareMasses}$ contains the fermion bare mass terms. Also,

$$\mathcal{L}_{Gauge} = \mathcal{L}_{GaugeEW} + \mathcal{L}_{GaugeC} + \mathcal{L}_{GaugeEWD} \tag{5.21}$$

with

$$\mathcal{L}_{GaugeEW} = -\tfrac{1}{4}\, F_W{}^{a\mu\nu}F_W{}^{a}{}_{\mu\nu} - \tfrac{1}{4}\, F_B{}^{\mu\nu}F_{B\mu\nu} + \mathcal{L}_{EW}{}^{ghost} \tag{5.22}$$

$$\mathcal{L}_{GaugeEWD} = -\tfrac{1}{4}\, F'_W{}^{a\mu\nu}F'_W{}^{a}{}_{\mu\nu} - \tfrac{1}{4}\, F_{B'}{}^{\mu\nu}F_{B'\mu\nu} + \mathcal{L}_{W'}{}^{ghost} \tag{5.23}$$

and

$$\mathcal{L}_{GaugeC} = \mathcal{L}_{CCG} + \mathcal{L}_C{}^{ghost} + \mathcal{L}_{CC}{}^{ghost} \tag{5.24}$$

$$\mathcal{L}_{Ufields} = -\tfrac{1}{4}\, F_U{}^{a\mu\nu}F_{U\mu\nu\nu} + \mathcal{L}_U{}^{ghost} + \mathcal{L}_U{}^{UHiggs} \tag{5.25}$$

---

approximation equal to <W $_i$(X), W $_j$(X)> for energies much less than M_c (which could be as large as the Planck energy.)

where $\mathcal{L}_U{}^{UHiggs}$ is discussed later. The ElectroWeak gauge bosons $W_\mu{}^a$, B_μ and B'_μ field tensors are:

$$F_{W}{}^{a}{}_{\mu\nu} = \partial W^a{}_\mu/\partial X^\nu - \partial W^a{}_\nu/\partial X^\mu + g_2 f^{abc} W^b{}_\mu W^c{}_\nu \qquad (5.26)$$

$$F_{B_{\mu\nu}} = \partial B_\mu/\partial X^\nu - \partial B_\nu/\partial X^\mu \qquad (5.27)$$

and the Dark ElectroWeak gauge bosons $W'_\mu{}^a$ and B'_μ field tensors are:

$$F_{B'_{\mu\nu}} = \partial B'_\mu/\partial X^\nu - \partial B'_\nu/\partial X^\mu$$

$$F'_{W}{}^{a}{}_{\mu\nu} = \partial W'^a{}_\mu/\partial X^\nu - \partial W'^a{}_\nu/\partial X^\mu + g_2 f^{abc} W'^b{}_\mu W'^c{}_\nu \qquad (5.28)$$

The U fields' tensor is:

$$F_{U}{}^{a}{}_{\mu\nu} = \partial U^a{}_\mu/\partial X^\nu - \partial U^a{}_\nu/\partial X^\mu + g_G f_4{}^{abc} U^b{}_\mu U^c{}_\nu \qquad (5.29)$$

where $f_4{}^{abc}$ are the U(4) algebra commutator constants.

$\mathcal{L}_{EW}{}^{ghost}$ contains the Faddeev-Popov ghost terms for the ElectroWeak $W_\mu{}^a$ gauge bosons. The complexon color gluon lagrangian $\mathcal{L}_{CCG}$ is defined by

$$\mathcal{L}_{CCG} = -\tfrac{1}{4} F_{CC}{}^{a\mu\nu}(X) F_{CC}{}^{a}{}_{\mu\nu}(X) \qquad (5.30)$$

where

$$F_{CC}{}^{a}{}_{\mu\nu} = \partial/\partial X_c{}^\nu\, A_C{}^a{}_\mu - \partial/\partial X_c{}^\mu\, A_C{}^a{}_\nu + g f_{su(3)}{}^{abc} A_C{}^b{}_\mu A_C{}^c{}_\nu \qquad (5.31)$$

where $A_C{}^a{}_\nu$ is the color gluon gauge field, g is the color coupling constant, and the $f_{su(3)}{}^{abc}$ are the SU(3) structure constants.

In addition $\mathcal{L}_C^{ghost}$ is the color SU(3) Faddeev-Popov ghost terms.[34] The mass sector $\mathcal{L}_{Mass}$ is presumably based on the Higgs Mechanism.which creates the fermion and ElectroWeak vector boson masses, and generation mixing.

The lagrangian is supplemented with the following condition on all complexon fields $\Phi_{...}$:[35]

$$\nabla_r \cdot \nabla_i \Phi ... = 0 \tag{5.32}$$

The subsidiary condition:

$$[\nabla_r \cdot \nabla_i - (\nabla_r^2 \nabla_i^2)^{1/2}]\Omega ... = 0 \tag{5.33}$$

would guarantee a particle's real momentum is parallel to its imaginary momentum. The Faddeev-Popov Method can be used to implement the eq. 5.32 constraint.

5.3 U(4) Gauge Symmetry Breaking and Long Range Forces

Above we showed that there was good experimental evidence for a conserved Baryon Number B and we proceeded to develop a simple U(1) gauge theory that would imply Baryon Number conservation in a manner analogous to QED's implying electric charge conservation. In section 5.2 we used a new symmetry group local U(4) to generalize the one generation Unified SuperStandard Theory to a four generation Unified SuperStandard Theory based on four conserved particle numbers: B, L, B_D, and L_D.[36]

We now assume in our construction that the four generation Unified SuperStandard Theory has a local U(4) symmetry that is broken by mass terms gewnerated by the Higgs Mechanism.

Further, we will assume that the Higgs breakdown yields two massless (long range) fields which we associate with Baryon Number B and Dark Baryon Number B_D. The remaining fields acquire masses and generate short range forces.

---

[34] Faddeev-Popov gauge fixing is described in appendix 19-A of Blaha (2011c), for the complexon Lorentz gauge. $\mathcal{L}_{CC}^{ghost}$ is the complexon color SU(3) constraint ghost terms defined through the Faddeev-Popov mechanism.

[35] These conditions implement the orthogonality of the real and imaginary parts of complexon 3-momentum.

[36] Charge, although a conserved number, is a part of the ElectroWeak sector, account of which has already been taken.

We use the following U(4) diagonal matrices:

$$\begin{aligned} G_1 &= \mathrm{diag}(1, 1, 1, 1) \\ G_2 &= \mathrm{diag}(0, 1, 0, 0) \\ G_3 &= \mathrm{diag}(0, 0, 1, 0) \\ G_4 &= \mathrm{diag}(0, 0, 0, 1) \end{aligned} \tag{5.34}$$

The U(4) algebra has 16 hermitean matrices that satisfy

$$G_i^{\dagger} = G_i \tag{5.35}$$

The particle numbers can be expressed in terms of the diagonal generators as

$$\begin{aligned} B &= G_1 - G_2 - G_3 - G_4 \\ B_D &= G_2 \\ L &= G_3 \\ L_D &= G_4 \end{aligned} \tag{5.36}$$

The covariant derivatives have the general form:

$$D_{\ldots\mu} = \partial/\partial X^{\mu} + \ldots - \tfrac{1}{2} i g_G \mathbf{G}\cdot\mathbf{U}_{\mu} \tag{5.37}$$

where the ellipses indicate the other details of the particular covariant derivative. We now wish to express the four gauge fields $U_i(X)$ for i = 1, 2, 3, 4 corresponding to the diagonal generators in terms of the fields of the four particle number gauge fields: B_μ, L_μ, $B_{D\mu}$, and $L_{D\mu}$.

$$U_{i\mu} = A_{ik} N_{k\mu} \tag{5.38}$$

where A_{ik} are the elements of a matrix of constants and

$$N_\mu = \begin{bmatrix} B_\mu(X) \\ L_\mu(X) \\ B_{D\mu}(X) \\ L_{D\mu}(X) \end{bmatrix} \quad (5.39)$$

is a column vector consisting of the gauge fields corresponding to each of the conserved particle numbers.

The matrix A must have non-zero determinant so that eq. 5.38 can be inverted to express the particle number fields in terms of the four $U_i(X)$ gauge fields:

$$N_\mu = A^{-1}U_\mu \quad (5.40)$$

resulting in

$$\begin{aligned} B_\mu(X) &= U_{1\mu} \\ L_\mu(X) &= U_{1\mu} + U_{2\mu} \\ B_{D\mu}(X) &= U_{1\mu} + U_{3\mu} \\ L_{D\mu}(X) &= U_{1\mu} + U_{4\mu} \end{aligned} \quad (5.41)$$

Then

$$D_{\ldots\mu} = \partial/\partial X^\mu + \ldots - \tfrac{1}{2} i g_G [\sum_{i=5}^{16} \mathbf{G}_i \mathbf{U}_{i\mu} + BB_\mu(X) + LL_\mu(X) + B_D B_{D\mu}(X) + L_D L_{D\mu}(X)] \quad (5.42)$$

where the particle numbers, which are analogous to the charges Q and Q' in ElectroWeak theory, are B, L, B_D, and L_D. They are expressed in terms of U(4) generators by eqs. 5.36.

In section 15.2 we described the Higgs mass mechanism for Generation group gauge fields using the above notation.

6. Layer Group

This chapter[37] describes the four Internal Symmetry Layer Groups. The transformations of the Complex Lorentz group transformations lead to the internal symmetry Reality group: R = SU(2)⊗U(1)⊗SU(3)⊗SU(2)⊗U(1), which is the symmetry group of the 'original' Standard Model plus Dark SU(2)⊗U(1). The Internal Symmetry U(4) Generation group was shown to follow from the four particle number conservation laws,[38] and increased the Standard Model group to SU(2)⊗U(1)⊗SU(3)⊗SU(2)⊗U(1)⊗U(4).

6.1 New Conserved Particle Numbers

The four generations of enhanced form of the model lead to a new set of particle conservation laws.

If we examine the form of the SU(2)⊗U(1)⊗SU(3)⊗SU(2)⊗U(1)⊗U(4) free fermion lagrangian terms of the Unified SuperStandard Theory

$$\mathcal{L} \sim \bar{\psi}_\alpha \gamma^\mu D_{\mu_{\alpha\beta}} \psi_\beta$$

we find that the SU(2)⊗U(1)⊗SU(3)⊗SU(2)⊗U(1)⊗U(4) symmetry of the terms not only yields the B, L, B_D, and L_D particle conservation laws but it also yields four quartets of conservation laws – one quartet for each of the four generations – totaling 16 conserved particle numbers.

The four 'conserved' layer numbers per generation are:

L_{iB} – The Baryon layer particle number for the ith generation
L_{iDB} – The Dark Baryon particle layer number for the ith generation

[37] Much of this chapter appears in Blaha (2017c).
[38] Baryon number, Lepton number, and their Dark analogues.

L_{iL} – The Lepton layer particle number for the ith generation
L_{iDL} – The Dark Lepton particle layer number for the ith generation

for each generation i = 1, 2, 3, 4. Fermions have positive L_{ia} = +1 values and anti-fermions have negative L_{ia} = –1 values for species a = 1, 2, 3, 4 (with the three color subspecies treated as part of one species.)

6.2 Four Layers of Fermions

These four sets of particle numbers lead to four U(4) Layer groups. The rationale for the new Layer groups is similar to that of the Generation group which was based on the four particle numbers B, L, B_D, and L_D. The Generation group was based on U(4) rotations related to the four number operators B, L, B_D, and L_D in the one generation SuperStandard Theory.

Similarly we base four Layer groups on the four quartets of conserved layer numbers. Each generation has a Layer group. Consequently, for each fermion species, *we assume the fermion in each generation acquires a Layer index and 'becomes' a set of four fermions in a Layer U(4)* $\underline{4}$ *representation.* Thus the set of four generations becomes four layers of four generations as shown in Figs. 6.1 and 6.2.

6.3 Layer Group Rotations

A U(4) rotation of the ith generation transforms the four i^{th} generation fermions of the four layers amongst each other (Fig. 6.3). The vertical rotations for generation k can be symbolized by:

G_k ↑	Layer 4
G_k	Layer 3
G_k	Layer 2
G_k ↓	Layer 1

We assume layer 1 is the known layer of fermions.

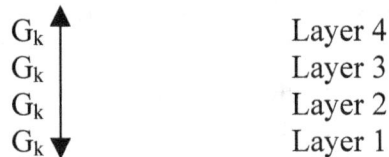

Figure 6.1 The four layers of fermions. Each layer (oval) has four generations of fermions. Layer 1 is the layer that we have found experimentally. The 4th generation of layer 1, the Dark part of layer 1, and the remaining three more massive layers constitute Dark matter.

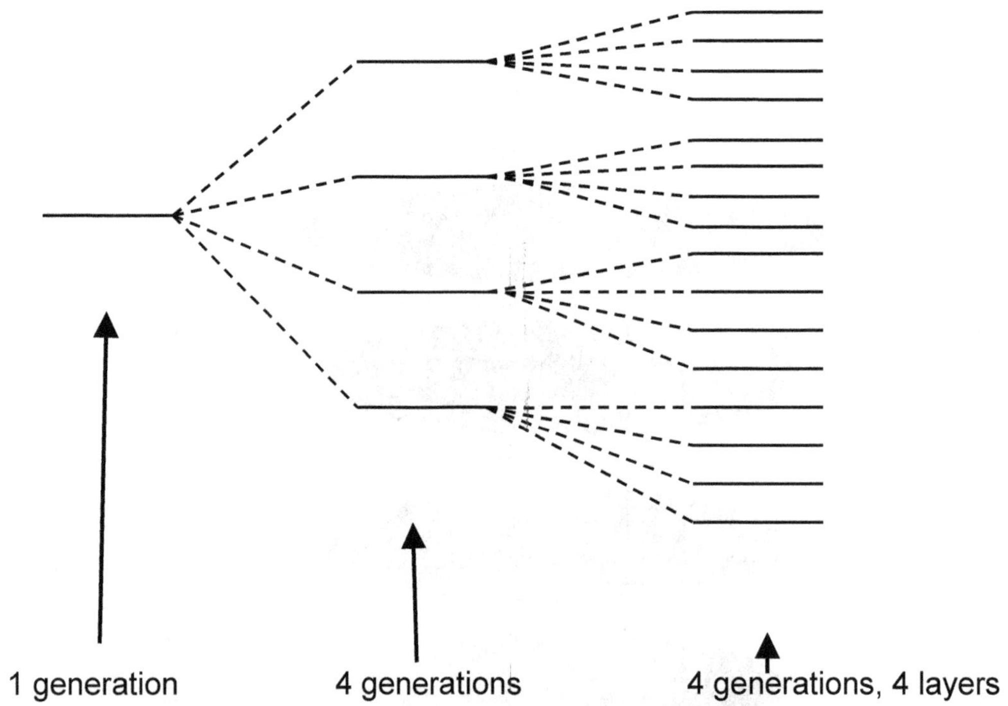

Figure 6.2 The "splitting" of a single generation fermion into four generations and then into four layers.

6.4 Layer Groups

We assume that each of the four Layer groups has associated gauge fields and particle interactions. Each generation in the four layers has a U(4) Layer symmetry group mixing the fermions in its generation across all four layers.(Fig. 6.3) Since the coefficients in Layer group transformations can be local functions, the new Layer groups are implemented with Yang-Mills fields.

It is important to note that the Layer particle numbers are independent of the baryon and lepton particle numbers that form the basis of the Generation group, and so the physics embodied in the Generation group is not the same as the physics of the Layer groups

Layer numbers are conserved under strong and electromagnetic interactions but broken by the Electromagnetic and Weak interactions as well as their Dark counterparts.

Further the gauge fields for SU(2)⊗U(1)⊗SU(3)⊗SU(2)⊗U(1)⊗U(4) must now be different for each layer.Thus interactions of these types between fermions in different layers is prevented. As a result the SuperStandard Theory symmetry becomes (subject to the introduction of other groups such as the Species Group) now is

$$[SU(3)\otimes SU(2)\otimes U(1)\otimes SU(2)\otimes U(1)\otimes U(4)\otimes U(4)]^4$$

where, for compactness, we place the four Layer groups within the quartic expression.

6.5 Steps to Introduce the Layer Groups in the SuperStandard Theory

To implement this new Layer symmetry we expand the SuperStandard Theory lagrangian with the following steps:

1, All covariant derivatives must expand to incude four U(4) Layer gauge fields terms $V^{\mu}{}_{i}$. for i = 1, 2, 3, 4. The new terms are for four layers of SU(2)⊗U(1)⊗SU(3)⊗ SU(2)⊗U(1)⊗U(4) fields. Each gauge field then has an additionl index specifying its layer. The rationale for the choice of these sets of groups to be duplicated is that they, and only they, all play a necessary role in determining the structure of the Periodic Table of Fermions.

2. The new Layer groups are the same as in the Unified SuperStandard Theory of Blaha (2017b). Each fermion particle field has an index labeling the layer of the particle making four layers of four generations of fermions.

THE FERMION PERIODIC TABLE

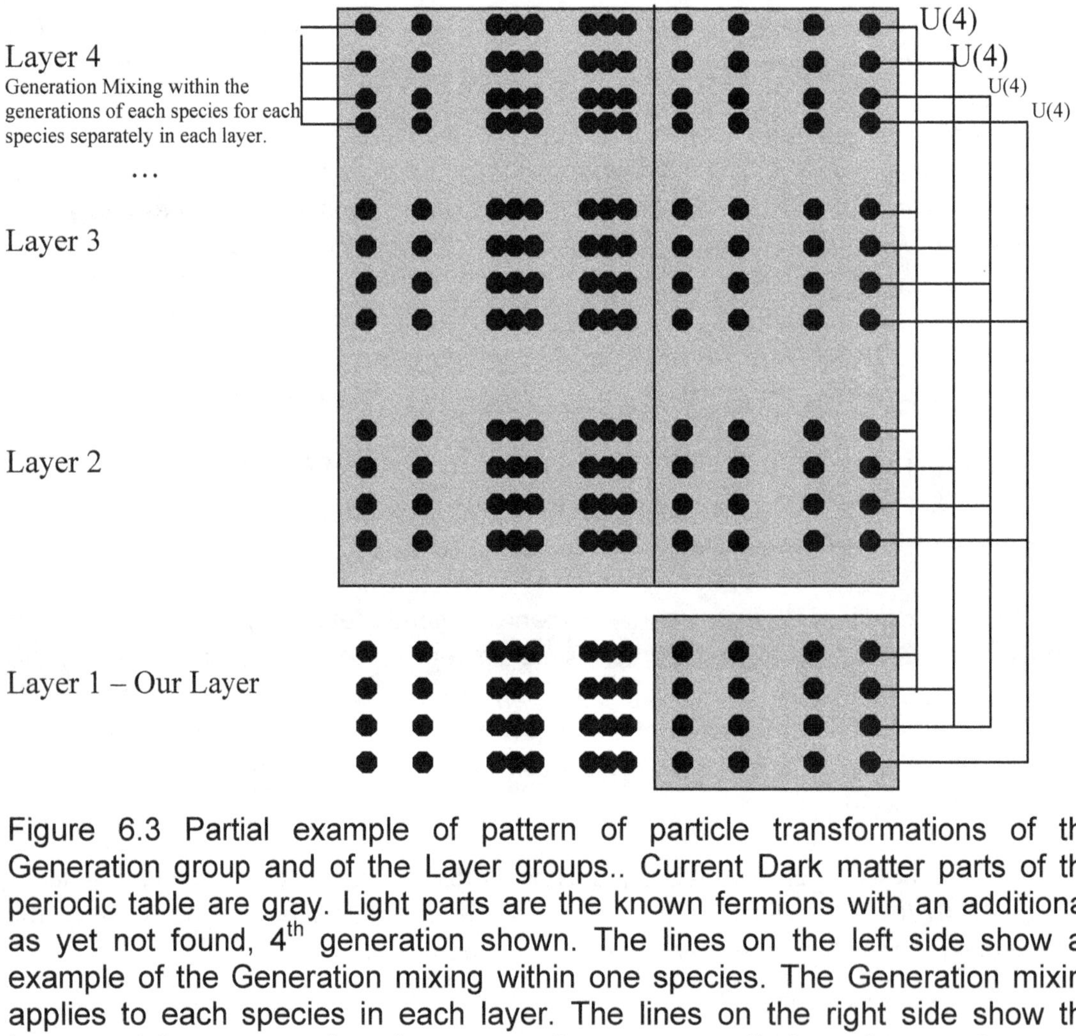

Figure 6.3 Partial example of pattern of particle transformations of the Generation group and of the Layer groups.. Current Dark matter parts of the periodic table are gray. Light parts are the known fermions with an additional, as yet not found, 4th generation shown. The lines on the left side show an example of the Generation mixing within one species. The Generation mixing applies to each species in each layer. The lines on the right side show the Layer mixing generation by generation.among all four layers for each generation individually.

THE VECTOR BOSON PERIODIC TABLE

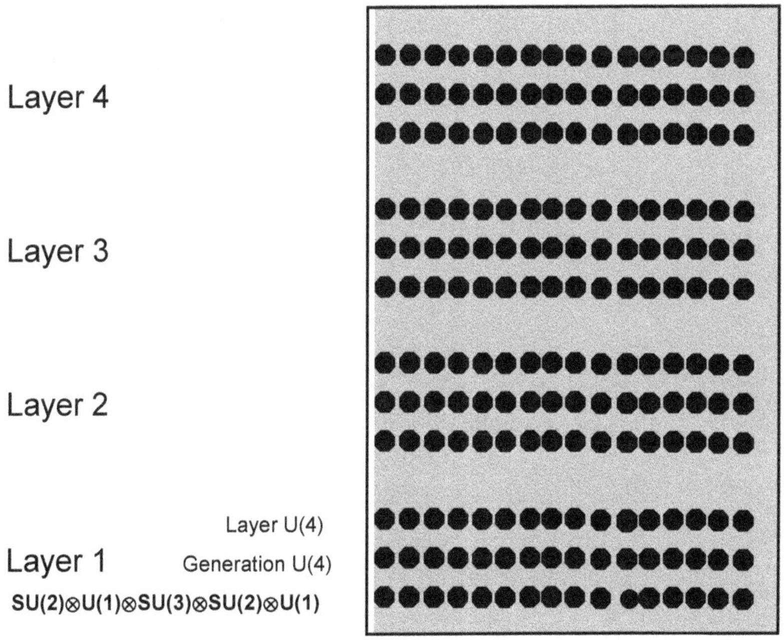

Figure 6.4 The known vector bosons are in the lowest row. The Layer groups are distributed by layer symbolically although they each straddle all four layers.

3. Each layer should have its own set of Higgs particles (modulo mixing) contributing to fermion masses. A layer index number must be added to each Higgs field. One expects that the masses of fermions should be substantially larger for the three ‘upper’ layers beyond our layer.[39] Otherwise we would have found particles from these upper layers.

---

[39] The interplayed mixing of the particles in each generation between different layers may partly explain the vast increases seen in fermion masses as one goes from generation to generation in each species. The Higgs particles in different layers are different and a possible source of the growth of fermion masses.

The form of the "periodic tables" of fermion and vector bosons that results appears in Figs. 6.4 and 6.5. The anticipated[40] large breakdown of Layer groups symmetries causes the fermion masses of each generation of each species to be significantly different. *This difference may explain the large increase in masses in each species as one goes up from the lowest mass fermion in the species (such as the e mass in the charged lepton species) to the highest (known) mass in the species (such as the τ mass in the charged lepton species.) It also explains why simple models of the fermion generations mass spectrums do not work.*

---

[40] Experimentally the top three layers of fermions have not been found. Therefore we expect that they have extremely large masses.

7. 192 SuperStandard Fundamental Fermions

We have seen that our development of the fermion particle spectrum has led to four species of fermions, four generations of fermions and four layers of fermions. The resulting 192 fermions need a form of labeling to distinguish accurately amongst them. In this chapter we provide a labeling method based on species, generations, and layers.

7.1 New Labeling of Fermion Periodic Table Particles

Fig. 7.1 contains the Periodic Table of Fermions with rows and columns labeled with quadruplets of numbers: (T, S, L, G) ≡ f_{TSLG}. In the case of normal quark species, which each consist of a color triplet, a fifth integer would be needed to identify each color quark within a triplet: (T, S, L, G, C) ≡ f_{TSLGC} where

T = type of matter: 1 for normal, 2 for Dark matter
S = species (1 for charged lepton, 2 for neutral lepton, 3 for each of three up-type quarks, 4 for each of three down-type quarks, 5 for each Dark charged lepton, 6 for each Dark neutral lepton, 7 for each Dark up-type quark species, and 8 for each Dark down-type quark
L = Layer group number (1, 2, 3, 4 where 1 is the known layer)
G = Generation group number (1, 2, 3, 4 where 1, 2, 3 are the known generations)
C = 1, 2, 3 identifies the color ("red, white or blue")

We will call the quadruplet (or quintet) of a fermion its *ID number*. See Fig. 7.1. For example, the second generation, fourth layer, normal up-type red quark is (1, 3, 4, 2, 1) where we treat 'red' as having the value 1.

Since the Periodic table of Fermions is 2-dimensional – like the Chemical Periodic Table of Elements one might ask: Why not use two numbers to identify fermions in a manner similar to the Chemistry table? The answer: our numbering better identifies the structure within the fermion periodic table.

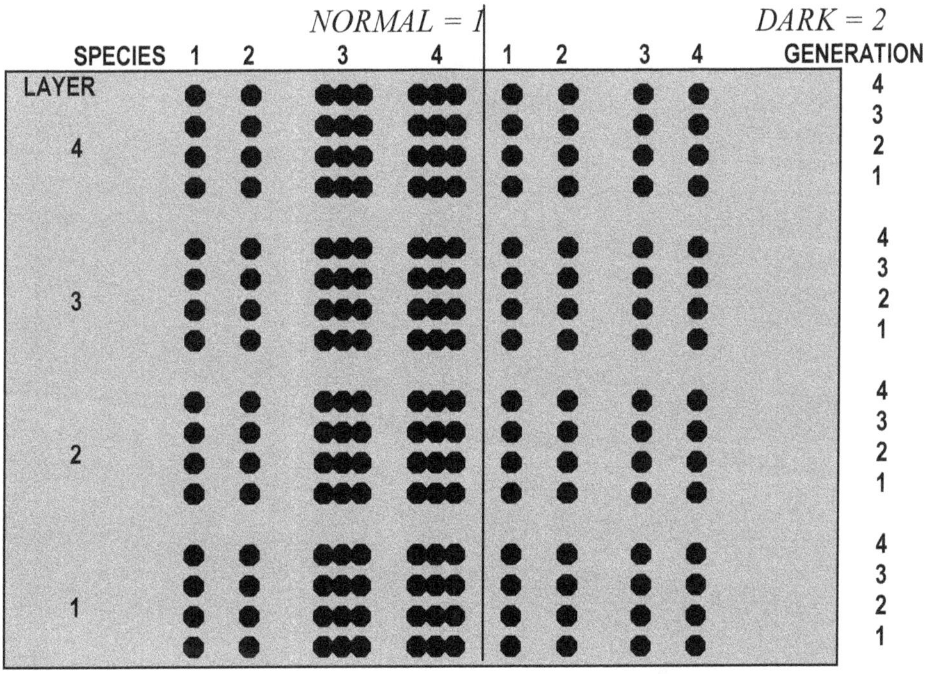

Figure 7.1 The Periodic Table of Fermions with each fermion identified by a quadruplet of integers: (T, S, L, G) where T identifies Normal or Dark, S identifies the species, G identifies the generation, and L identifies the layer. For example the electron has the triplet (1, 1, 1, 1), and the second generation, fourth layer, Dark up-type quark is (2, 3, 4, 2).

8. Equipartition Principle for Fermions and Gauge Fields

We now[41] will suggest a rationale for the abundances of "normal" and Dark mass-energy in the universe. Interestingly it directly relates the distribution of fermions and vector bosons found in the previous chapters to experimentally known proportions.

8.1 Equipartition Principle for Particle Degrees of Freedom

In a closed system at equilibrium the thermal energy of a system is equally partitioned (distributed) among its degrees of freedom. This Equipartition Principle is well known. The application of this principle to the beginning of the universe *when all particles were massless* and all symmetries were unbroken suggests that the distribution of mass-energy should be the same for all degrees of freedom at that time. Thus there should be approximately equal numbers and energies of 192 fermions and 192 vector bosons with the same fraction of the total thermal energy.

We now estimate the relative proportion of Normal and Dark matter in the universe at its beginning based on this Equipartition Principle.

8.2 Proportion of Dark Mass-Energy in the Universe

First we note that 8 of the 12 fermion species (counting quarks of each color as a separate species) in layer 1 – the layer with which we are familiar are Normal matter fermions. (Our discussion is based on our Unified SuperStandard Theory.) Four of the 12 first layer species are Dark.

The other three fermion layers are all Dark from our point of view since they have not been detected. Thus we find that 40 of the 48 species are Dark yielding a *percentage of Dark Matter equal to 40/48 = 83.33%.* (The same counting

[41] The material in this chapter appeared in Blaha (2018e).

could have been done by counting individual fermions with the same results.)

Recent studies of the proportion of Dark Matter in the universe have yielded two estimates: 84.5% by Aghanim *et al* in Astronomy and Astrophysics 1303;5062 and 81.5% from a NASA fit to various models.

Thus our estimate based on our fermion Equipartition Principle is midway between these experimental estimates.

Two possibilities emerge with respect to the present proportion of Dark Matter:

1. The percentage has not changed from the Beginning and the approximate estimates are slightly off. The lack of change could be due to the extremely small decay rates of the fermions in the higher layers.

2. The percentage of matter in the upper layers has decreased due to decay and so the current proportion may be somewhat below 83.33%.

8.3 Proportion of Dark Mass-Energy in the Universe

We know of 12 of the 192 vector bosons in the SuperStandard Theory and 24 fermions. Thus we find a total of 348 out of 384 particles are Dark yielding a Dark mass-energy of 91% of the universes mass-energy at the beginning of the universe.

The sum of Dark energy in the universe currently has been estimated to be 68% of the total energy. The energy of Dark Matter is estimated to be 26.8%. The total is 95% - a value that compares favorably with our above approximate estimate of 91%. The dfference in these values can be attributed to various factors. One distinct possibility is the decay of Dark mass-energy from higher layers to the known layer in the 13.8 billion years since the Big Bang.

9. Extended Standard Model

The Extended Standard Model omits the Species Group and the Layer Groups. Its symmetry is

$$SU(2)\otimes U(1)\otimes SU(3)\otimes SU(2)\otimes U(1)\otimes [U(4)]^2 \qquad (1.3)$$

It has an SU(15) covering group that contains it as a subgroup.

The four generation fermion spectrum for the Extended Standard Model is

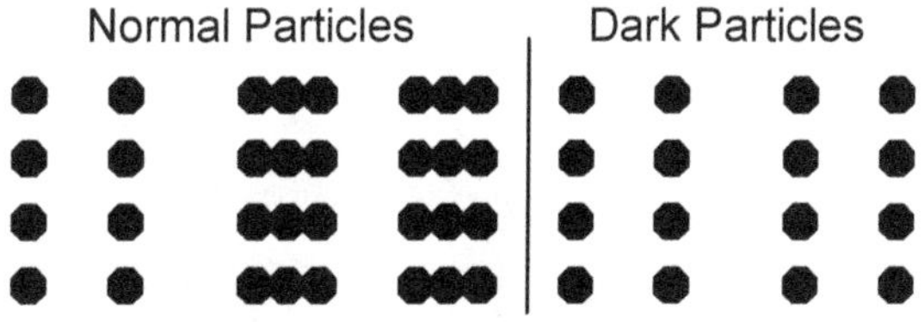

where the three low generations (excepting the Dark particles) are known. Each generation is ordered as Charged Lepton, Neutral Lepton, Up-type Quark, and Down-type quark.for both normal and Dark sectors. There are 48 fermions in the spectrum.

The gauge vector boson spectrum for the Extended Standard Model is

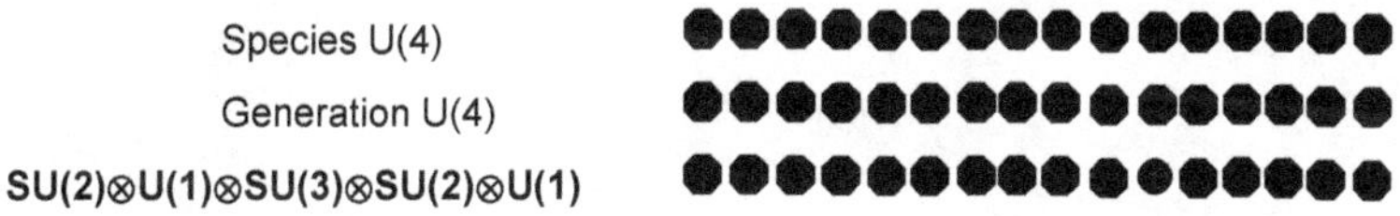

The Generation Group and Species Group vector bosons have not yet been found. The four vector bosons for Dark ElectroWeak interactions have also not been found as yet. There are 48 vector bosons in this Extended Standard Model.

10. Unified SuperStandard Theory Interactions

The Unified SuperStandard Theory symmetry is

$$[SU(2)\otimes U(1)\otimes SU(3)\otimes SU(2)\otimes U(1))\otimes U(4)]^4\otimes[U(4)]^4\otimes U(4)$$
$$= [SU(2)\otimes U(1)\otimes SU(3)\otimes SU(2)\otimes U(1)]^4\otimes U(4)]^9 \qquad (4.1)$$

It includes the interactions found by analogy with the Complex Lorentz Group, the Generation Group, and the Layer Groups—each repeated four times due to four layers:

$$[SU(2)\otimes U(1)\otimes SU(3)\otimes SU(2)\otimes U(1))\otimes U(4)]^4$$

plus the Species Group.

The total symmetry group above has a covering group of SU(64). Chapters 5 and 6 described the fermion spectrum and the symmetry in detail. Fig. 6.3 depicts the the SuperStandard Theory fermion "periodic table." Fig. 6.4 depicts the SuperStandard Theory set of of gauge fields. There are 192 fundamental fermions and 192 fundamental gauge vector bosons.

In addition to the above particles and interactions the Unified SuperStandard Theory includes real General Relativity and the General Relativistic Reality Group.

The next chapter describes its interactions in detail

11. The Riemann-Christoffel Curvature Tensor and Vector Boson Part of the Unified SuperStandard Theory Lagrangian

In view of our goal of defining a unified theory of elementary particles and General Relativity we begin by defining a Riemann-Christoffel curvature tensor which we will use to construct a lagrangian for the theory.

11.1 The Covariant Derivative

The covariant derivative[42] which appears in fermion and gravitation equations uses the vector boson 7-vector:

$$ {}^{a}\mathbf{A}_{I}{}^{\mu}(x) = ({}^{a}g_{1}{}^{a}\mathbf{A}_{SU(3)}{}^{\mu}(x_{C}),\ {}^{a}g_{2}{}^{a}\mathbf{W}^{\mu}(x)\ ,\ {}^{a}g_{3}{}^{a}\mathbf{A}_{E}{}^{\mu}(x),\ {}^{a}g_{4}{}^{a}\mathbf{W}_{D}{}^{\mu}(x),\ {}^{a}g_{5}{}^{a}\mathbf{A}_{DE}{}^{\mu}(x),\ {}^{a}g_{6}{}^{a}\mathbf{U}^{\mu}(x),\ {}^{a}g_{7}{}^{a}\mathbf{V}^{\mu}(x)) \qquad (11.1)$$

where a labels the layer, a = 1, 2, 3, 4. We we label the respective coupling constants in each layer a as ${}^{a}g_{1}$, ${}^{a}g_{2}$, …, ${}^{a}\ g_{7}$. In the equation above: the subscript 'D' labels Dark matter interactions, 'W' labels Weak fields, 'E' labels Electromagnetic fields, $U^{\mu}(x)$ labels U(4) Generation group fields, and 'V' labels U(4) Layer group fields. A_S labels the U(4) Species Group fields.

We define the sum over a and the the components of the vector ${}^{a}\mathbf{A}_{I}{}^{\mu}(x)$ labeled with i by

$$ \mathbf{C}_{I}{}^{\mu}(x) = \Sigma_{a,i}\ {}^{a}\mathbf{A}_{Ii}{}^{\mu}(x) + g_{8}\mathbf{A}_{S}{}^{\mu}(x) + g_{\Theta}\mathbf{A}_{\Theta}{}^{\mu}(x) \qquad (11.2)$$

---

[42] This section has equations obtained from Blaha (2017d).

The symmetry is $[SU(3)\otimes SU(2)\otimes U(1)\otimes SU(2)\otimes U(1)\otimes U(4)\otimes U(4)]^4\otimes U(4)\otimes U(192)$ for particles. The number of gauge fields for each of the elements of the vectors in each layer is 8. 3, 1, 3, 1, 16, and 16, totalling 48 components. For four layers there are 4×48 = 192 elements plus 16 Species gauge fields and 192^2 Θ-Symmetry fields. The Θ-Symmetry, which we also call the Interaction Rotations Group in Blaha (2018e), rotates the interactions of eq. 11.1.

We begin by considering the case of one layer of vector bosons below omitting the ^{a} superscript. The generalization to four layers is straightforward.

Using the above definitions the *PseudoQuantum* covariant derivative of a 4-vector Z_μ is

$$\begin{aligned} D_\nu Z_\mu &= (\partial_\nu + iF_\nu)Z_\mu - H^\sigma{}_{\nu\mu}Z_\sigma \\ &= [g^\sigma{}_\mu\partial_\nu + ig^\sigma{}_\mu F_\nu - H^\sigma{}_{\nu\mu}]Z_\sigma \\ &= [g^\sigma{}_\mu\partial_\nu + iD^\sigma{}_{\mu\nu}]Z_\sigma \end{aligned} \tag{11.3}$$

where[43]

$$F^\mu = C_I{}^{1\mu}(x) + \mathbf{C}_I{}^{2\mu}(x) + A_R{}^{1\mu} + A_R{}^{2\mu} + B^{1\mu} + B^{2\mu} \tag{11.4}$$

and

$$\begin{aligned} H^\sigma{}_{\nu\mu} &= \Gamma_{GR}{}^{\sigma}{}_{\nu\mu} + \Gamma_{GR}{}^{2\sigma}{}_{\nu\mu} \\ D^\sigma{}_{\mu\nu} &= g^\sigma{}_\mu F_\nu + iH^\sigma{}_{\nu\mu} \end{aligned} \tag{11.5}$$

where we have abstracted the complex part of the complex affine connection into the U(4) gauge field $A_S{}^\mu$. $H^\sigma{}_{\nu\mu}$ is the real-valued part of the complex affine connection.

Commutators of the vector fields in F_μ are implicit when the covariant derivative is applied to vectors and tensors such as Z_σ.

11.2 The Curvature Tensor

The curvature tensor applied to a 4-vector Z_β is

$$R'^\beta{}_{\sigma\nu\mu}Z_\beta = g^\alpha{}_\mu(\partial_\nu + iF_\nu)g^\beta{}_\sigma(\partial_\alpha + iF_\alpha)Z_\beta - H^\alpha{}_{\mu\nu}g^\beta{}_\sigma(\partial_\alpha + iF_\alpha)Z_\beta + \tag{11.6}$$

[43] We will omit the insertion of the spinor coupling constants of the spinor connection $B^{1\mu}$ and $B^{2\mu}$ in eq. 11.2 in the interests of simplifying expressions.

$$+ H^{\alpha}{}_{\mu\nu}H^{\beta}{}_{\sigma\alpha}Z_\beta - g^{\alpha}{}_{\mu}(\partial_\nu + iF_\nu)H^{\beta}{}_{\sigma\alpha}Z_\beta - H^{\gamma}{}_{\nu\sigma}\{g^{\alpha}{}_{\gamma}(\partial_\mu + iF_\mu)Z_\alpha - H^{\alpha}{}_{\gamma\mu}Z_\alpha\} - \{\mu\leftrightarrow\nu\}$$

$$= ig^{\beta}{}_{\sigma}(\partial_\nu F_\mu - \partial_\mu F_\nu - i[F_\nu, F_\mu])Z_\beta + (\partial_\mu H^{\beta}{}_{\sigma\nu} - \partial_\nu H^{\beta}{}_{\sigma\mu} + H^{\gamma}{}_{\nu\sigma}H^{\beta}{}_{\gamma\mu} - H^{\gamma}{}_{\mu\sigma}H^{\beta}{}_{\gamma\nu})Z_\beta$$

$$= ig^{\beta}{}_{\sigma}(F_E{}^1{}_{\nu\mu} + F_E{}^2{}_{\nu\mu} + F_W{}^1{}_{\nu\mu} + F_W{}^2{}_{\nu\mu} + F_{DE}{}^1{}_{\nu\mu} + F_{DE}{}^2{}_{\nu\mu} + F_{DW}{}^1{}_{\nu\mu} + F_{DW}{}^2{}_{\nu\mu} + F_{SU(3)}{}^1{}_{\nu\mu} + F_{SU(3)}{}^2{}_{\nu\mu} + F_U{}^1{}_{\nu\mu} + F_U{}^2{}_{\nu\mu} + F_V{}^1{}_{\nu\mu} + F_V{}^2{}_{\nu\mu} + F_S{}^1{}_{\nu\mu} + F_S{}^2{}_{\nu\mu} + F_\Theta{}^1{}_{\nu\mu} + F_\Theta{}^2{}_{\nu\mu} + F_B{}^1{}_{\nu\mu} + F_B{}^2{}_{\nu\mu} + F_A{}^1{}_{\nu\mu} + F_A{}^2{}_{\nu\mu})Z_\beta + (\partial_\mu H^{\beta}{}_{\sigma\nu} - \partial_\nu H^{\beta}{}_{\sigma\mu} + H^{\gamma}{}_{\nu\sigma}H^{\beta}{}_{\gamma\mu} - H^{\gamma}{}_{\mu\sigma}H^{\beta}{}_{\gamma\nu})Z_\beta$$

$$= R'_E{}^{\beta}{}_{\sigma\nu\mu}Z_\beta + R'_{SU(2)}{}^{\beta}{}_{\sigma\nu\mu}Z_\beta + R'_{DE}{}^{\beta}{}_{\sigma\nu\mu}Z_\beta + R'_{DSU(2)}{}^{\beta}{}_{\sigma\nu\mu}Z_\beta + R'_{SU(3)}{}^{\beta}{}_{\sigma\nu\mu}Z_\beta + R'_U{}^{\beta}{}_{\sigma\nu\mu}Z_\beta + R'_V{}^{\beta}{}_{\sigma\nu\mu}Z_\beta + R'_S{}^{\beta}{}_{\sigma\nu}Z_\beta + R'_\Theta{}^{\beta}{}_{\sigma\nu}Z_\beta + R'_A{}^{\beta}{}_{\sigma\nu}Z_\beta + R'_B{}^{\beta}{}_{\sigma\nu}Z_\beta + R'_G{}^{\beta}{}_{\sigma\nu\mu}Z_\beta$$

where all $F_{\dots}{}^1{}_{\nu\mu}$ and $F_{\dots}{}^2{}_{\nu\mu}$ terms have summations over the four layers (see below) except the terms[44] $F_S{}^1{}_{\nu\mu} + F_S{}^2{}_{\nu\mu} + F_\Theta{}^1{}_{\nu\mu} + F_\Theta{}^2{}_{\nu\mu} + F_A{}^1{}_{\nu\mu} + F_A{}^2{}_{\nu\mu} + F_B{}^1{}_{\nu\mu} + F_B{}^2{}_{\nu\mu}$, and where[45]

$$\begin{aligned}
R'_{SU(3)}{}^{\beta}{}_{\sigma\nu\mu} &= ig^{\beta}{}_{\sigma}(F_{SU(3)}{}^1{}_{\nu\mu} + F_{SU(3)}{}^2{}_{\nu\mu}) \qquad (11.7)\\
R'_{SU(2)}{}^{\beta}{}_{\sigma\nu\mu} &= ig^{\beta}{}_{\sigma}(F_W{}^1{}_{\nu\mu} + F_W{}^2{}_{\nu\mu})\\
R'_E{}^{\beta}{}_{\sigma\nu\mu} &= ig^{\beta}{}_{\sigma}(F_E{}^1{}_{\nu\mu} + F_E{}^2{}_{\nu\mu})\\
R'_U{}^{\beta}{}_{\sigma\nu\mu} &= ig^{\beta}{}_{\sigma}(F_U{}^1{}_{\nu\mu} + F_U{}^2{}_{\nu\mu})\\
R'_V{}^{\beta}{}_{\sigma\nu\mu} &= ig^{\beta}{}_{\sigma}(F_V{}^1{}_{\nu\mu} + F_V{}^2{}_{\nu\mu})\\
R'_{DSU(2)}{}^{\beta}{}_{\sigma\nu\mu} &= ig^{\beta}{}_{\sigma}(F_{DW}{}^1{}_{\nu\mu} + F_{DW}{}^2{}_{\nu\mu})\\
R'_{DE}{}^{\beta}{}_{\sigma\nu\mu} &= ig^{\beta}{}_{\sigma}(F_{DE}{}^1{}_{\nu\mu} + F_{DE}{}^2{}_{\nu\mu})\\
R'_S{}^{\beta}{}_{\sigma\nu\mu} &= ig^{\beta}{}_{\sigma}(F_S{}^1{}_{\nu\mu} + F_S{}^2{}_{\nu\mu})\\
R'_\Theta{}^{\beta}{}_{\sigma\nu\mu} &= ig^{\beta}{}_{\sigma}(F_\Theta{}^1{}_{\nu\mu} + F_\Theta{}^2{}_{\nu\mu})\\
R'_B{}^{\beta}{}_{\sigma\nu\mu} &= ig^{\beta}{}_{\sigma}(F_B{}^1{}_{\nu\mu} + F_B{}^2{}_{\nu\mu})\\
R'_A{}^{\beta}{}_{\sigma\nu\mu} &= ig^{\beta}{}_{\sigma}(F_A{}^1{}_{\nu\mu} + F_A{}^2{}_{\nu\mu})
\end{aligned}$$

and

---

44 The U(4) General Relativity Reality Group fields $A_R{}^{\beta}$ have $F_A{}^1{}_{\nu\mu} + F_A{}^2{}_{\nu\mu}$.

45 The B field is the General Relativistic spinor connection. Its effects are miniscule in physical situations except for extreme cases that have not as yet been encountered experimentally.

$$R'_G{}^{\beta}{}_{\sigma\nu\mu} = \partial_\mu H^{1\beta}{}_{\sigma\nu} - \partial_\nu H^{1\beta}{}_{\sigma\mu} + H^{1\gamma}{}_{\nu\sigma}H^{1\beta}{}_{\gamma\mu} - H^{1\gamma}{}_{\mu\sigma}H^{1\beta}{}_{\gamma\nu} + \partial_\mu H^{2\beta}{}_{\sigma\nu} - \partial_\nu H^{2\beta}{}_{\sigma\mu} + \quad (11.8)$$
$$+ H^{2\gamma}{}_{\nu\sigma}H^{2\beta}{}_{\gamma\mu} - H^{2\gamma}{}_{\mu\sigma}H^{2\beta}{}_{\gamma\nu} + H^{1\gamma}{}_{\nu\sigma}H^{2\beta}{}_{\gamma\mu} - H^{1\gamma}{}_{\mu\sigma}H^{2\beta}{}_{\gamma\nu} + H^{2\gamma}{}_{\nu\sigma}H^{1\beta}{}_{\gamma\mu} - \Gamma^{2\gamma}{}_{\mu\sigma}\Gamma^{\beta}{}_{\gamma\nu}$$
$$= R^{1\beta}{}_{\sigma\nu\mu} + R^{2\beta}{}_{\sigma\nu\mu}$$

with

$$H^{\beta}{}_{\sigma\nu\mu} = \partial_\mu H^{\beta}{}_{\sigma\nu} - \partial_\nu H^{\beta}{}_{\sigma\mu} + H^{\gamma}{}_{\nu\sigma}H^{\beta}{}_{\gamma\mu} - H^{\gamma}{}_{\mu\sigma}H^{\beta}{}_{\gamma\nu} \quad (11.9)$$
$$R^{1\beta}{}_{\sigma\nu\mu} = \partial_\mu H^{1\beta}{}_{\sigma\nu} - \partial_\nu H^{1\beta}{}_{\sigma\mu} + H^{1\gamma}{}_{\nu\sigma}H^{1\beta}{}_{\gamma\mu} - H^{1\gamma}{}_{\mu\sigma}H^{1\beta}{}_{\gamma\nu}$$
$$R^{2\beta}{}_{\sigma\nu\mu p} = \partial_\mu H^{2\beta}{}_{\sigma\nu} - \partial_\nu H^{2\beta}{}_{\sigma\mu} + H^{2\gamma}{}_{\nu\sigma}H^{2\beta}{}_{\gamma\mu} - H^{2\gamma}{}_{\mu\sigma}H^{2\beta}{}_{\gamma\nu} +$$
$$+ H^{1\gamma}{}_{\nu\sigma}H^{2\beta}{}_{\gamma\mu} - H^{1\gamma}{}_{\mu\sigma}H^{2\beta}{}_{\gamma\nu} + H^{2\gamma}{}_{\nu\sigma}H^{1\beta}{}_{\gamma\mu} - H^{2\gamma}{}_{\mu\sigma}H^{1\beta}{}_{\gamma\nu}$$

and

$$H^{1\sigma}{}_{\nu\mu} = \Gamma_{GR}{}^{\sigma}{}_{\nu\mu} \quad (11.10)$$
$$H^{2\sigma}{}_{\nu\mu} = \Gamma_{GR}{}^{2\sigma}{}_{\nu\mu}$$

and with summations over four layers indicated by Σ (Layer numbers on fields are not shown to avoid clutter.) As a result we have

$$F_{SU(3)}{}^1{}_{\varkappa\mu} = \Sigma\ \{\partial A_{SU(3)}{}^1{}_\mu/\partial x^\varkappa - \partial A_{SU(3)}{}^1{}_\varkappa/\partial x^\mu + ig_1[A_{SU(3)}{}^1{}_\varkappa,\ A_{U(3)}{}^1{}_\mu]\ \} \quad (11.11)$$
$$F_W{}^1{}_{\varkappa\mu} = \Sigma\ \{\partial W^1{}_\mu/\partial x^\varkappa - \partial W^1{}_\varkappa/\partial x^\mu + ig_2[W^1{}_\varkappa,\ W^1{}_\mu]\ \}$$
$$F_E{}^1{}_{\varkappa\mu} = \Sigma\ \{\partial A_E{}^1{}_\mu/\partial x^\varkappa - \partial A_E{}^1{}_\varkappa/\partial x^\mu\ \}$$
$$F_{DW}{}^1{}_{\varkappa\mu} = \Sigma\ \{\partial W_D{}^1{}_\mu/\partial x^\varkappa - \partial W_D{}^1{}_\varkappa/\partial x^\mu + ig_4[W_D{}^1{}_\varkappa,\ W_D{}^1{}_\mu]\ \}$$
$$F_{DE}{}^1{}_{\varkappa\mu} = \Sigma\ \{\partial A_{DE}{}^1{}_\mu/\partial x^\varkappa - \partial A_{DE}{}^1{}_\varkappa/\partial x^\mu\}$$
$$F_U{}^1{}_{\varkappa\mu} = \Sigma\ \{\partial U^1{}_\mu/\partial x^\varkappa - \partial U^1{}_\varkappa/\partial x^\mu + ig_6[U^1{}_\varkappa,\ U^1{}_\mu]\ \}$$
$$F_V{}^1{}_{\varkappa\mu} = \Sigma\ \{\partial V^1{}_\mu/\partial x^\varkappa - \partial V^1{}_\varkappa/\partial x^\mu + ig_7[V^1{}_\varkappa,\ V^1{}_\mu]\ \}$$
$$F_S{}^1{}_{\varkappa\mu} = \partial A_S{}^1{}_\mu/\partial x^\varkappa - \partial A_S{}^1{}_\varkappa/\partial x^\mu + ig_8[A_S{}^1{}_\varkappa,\ A_S{}^1{}_\mu]$$
$$F_\Theta{}^1{}_{\varkappa\mu} = \partial A_\Theta{}^1{}_\mu/\partial x^\varkappa - \partial A_\Theta{}^1{}_\varkappa/\partial x^\mu + ig_\Theta[A_\Theta{}^1{}_\varkappa,\ A_\Theta{}^1{}_\mu]$$
$$F_B{}^1{}_{\varkappa\mu} = \partial B^1{}_\mu/\partial x^\varkappa - \partial B^1{}_\varkappa/\partial x^\mu\ + i[B^1{}_\varkappa,\ B^1{}_\mu]$$
$$F_A{}^1{}_{\varkappa\mu} = \partial A_R{}^1{}_\mu/\partial x^\varkappa - \partial\ A_R{}^1{}_\varkappa/\partial x^\mu\ + i[A_R{}^1{}_\varkappa,\ A_R{}^1{}_\mu]$$

$$F_{SU(3)}{}^2{}_{\varkappa\mu} = \Sigma\ \{\partial A_{SU(3)}{}^2{}_\mu/\partial x^\varkappa - \partial A_{SU(3)}{}^2{}_\varkappa/\partial x^\mu\ + ig_1[A_{SU(3)}{}^2{}_\varkappa,\ A_{SU(3)}{}^2{}_\mu] + \quad (11.12)$$

$$+ ig_1[A_{SU(3)}{}^1{}_\varkappa, A_{SU(3)}{}^2{}_\mu] + ig_1[A_{SU(3)}{}^2{}_\varkappa, A_{SU(3)}{}^1{}_\mu]\}$$

$$F_W{}^2{}_{\varkappa\mu} = \Sigma\ \{\partial W^2{}_\mu/\partial x^\varkappa - \partial W^2{}_\varkappa/\partial x^\mu + ig_2[W^2{}_\varkappa, W^2{}_\mu] + ig_2[W^1{}_\varkappa, W^2{}_\mu] + ig_2[W^2{}_\varkappa, W^1{}_\mu]\}$$

$$F_E{}^2{}_{\varkappa\mu} = \Sigma\ \{\partial A_E{}^2{}_\mu/\partial x^\varkappa - \partial A_E{}^2{}_\varkappa/\partial x^\mu\}$$

$$F_{DW}{}^2{}_{\varkappa\mu} = \Sigma\ \{\partial W_D{}^2{}_\mu/\partial x^\varkappa - \partial W_D{}^2{}_\varkappa/\partial x^\mu + ig_4[W_D{}^2{}_\varkappa, W_D{}^2{}_\mu] + ig_4[W_D{}^1{}_\varkappa, W_D{}^2{}_\mu] + $$
$$+ ig_4[W_D{}^2{}_\varkappa, W_D{}^1{}_\mu]\}$$

$$F_{DE}{}^2{}_{\varkappa\mu} = \Sigma\ \{\partial A_{DE}{}^2{}_\mu/\partial x^\varkappa - \partial A_{DE}{}^2{}_\varkappa/\partial x^\mu\}$$

$$F_U{}^2{}_{\varkappa\mu} = \Sigma\ \{\partial U^2{}_\mu/\partial x^\varkappa - \partial U^2{}_\varkappa/\partial x^\mu + ig_6[U^2{}_\varkappa, U^2{}_\mu] + ig_6[U^1{}_\varkappa, U^2{}_\mu] + ig_6[U^2{}_\varkappa, U^1{}_\mu]\}$$

$$F_V{}^2{}_{\varkappa\mu} = \Sigma\ \{\partial V^2{}_\mu/\partial x^\varkappa - \partial V^2{}_\varkappa/\partial x^\mu + ig_7[V^2{}_\varkappa, V^2{}_\mu] + ig_7[V^1{}_\varkappa, V^2{}_\mu] + ig_7[V^2{}_\varkappa, V^1{}_\mu]\}$$

$$F_S{}^2{}_{\kappa\mu} = \partial A_S{}^2{}_\mu/\partial x^\kappa - \partial A_S{}^2{}_\kappa/\partial x^\mu + ig_8[A_S{}^2{}_\kappa, A_S{}^2{}_\mu] + ig_8[A_S{}^1{}_\kappa, A_S{}^2{}_\mu] + ig_8[A_S{}^2{}_\kappa, A_S{}^1{}_\mu]$$

$$F_\Theta{}^2{}_{\kappa\mu} = \partial A_\Theta{}^2{}_\mu/\partial x^\kappa - \partial A_\Theta{}^2{}_\kappa/\partial x^\mu + ig_\Theta[A_\Theta{}^2{}_\kappa, A_\Theta{}^2{}_\mu] + ig_\Theta[A_\Theta{}^1{}_\kappa, A_\Theta{}^2{}_\mu] + ig_\Theta[A_\Theta{}^2{}_\kappa, A_\Theta{}^1{}_\mu]$$

$$F_B{}^2{}_{\varkappa\mu} = \partial B^2{}_\mu/\partial x^\varkappa - \partial B^2{}_\varkappa/\partial x^\mu + i[B^2{}_\mu, B^2{}_\varkappa] + i[B^1{}_\mu, B^2{}_\varkappa] + i[B^2{}_\mu, B^1{}_\varkappa]$$

$$F_A{}^2{}_{\varkappa\mu} = \partial\ A_R{}^2{}_\mu/\partial x^\varkappa - \partial\ A_R{}^2{}_\varkappa/\partial x^\mu + i[A_R{}^2{}_\mu, A_R{}^2{}_\varkappa] + i[A_R{}^1{}_\mu, A_R{}^2{}_\varkappa] + i[A_R{}^2{}_\mu, A_R{}^1{}_\varkappa]$$

Note that $R'^\beta{}_{\sigma\nu\mu}$ factorizes into $[U(1)\otimes SU(2)\otimes U(1)\otimes SU(2)\otimes SU(3)\otimes U(4)\otimes U(4)]^4\otimes U(4)\otimes U(192)$ parts and a Riemann-Christoffel Gravitational curvature tensor part. For later use in defining a lagrangian we define

$$R'^\beta{}_{\sigma\nu\mu} = R'_E{}^{1\beta}{}_{\sigma\nu\mu} + R'_E{}^{2\beta}{}_{\sigma\nu\mu} + R'_{SU(2)}{}^{1\beta}{}_{\sigma\nu\mu} + R'_{SU(2)}{}^{2\beta}{}_{\sigma\nu\mu} + R'_{DE}{}^{1\beta}{}_{\sigma\nu\mu} + R'_{DE}{}^{2\beta}{}_{\sigma\nu\mu} +$$
$$+ R'_{DSU(2)}{}^{1\beta}{}_{\sigma\nu\mu} + R'_{DSU(2)}{}^{2\beta}{}_{\sigma\nu\mu} + R'_{SU(3)}{}^{1\beta}{}_{\sigma\nu\mu} + R'_{SU(3)}{}^{2\beta}{}_{\sigma\nu\mu} + R'_U{}^{1\beta}{}_{\sigma\nu\mu} +$$
$$+ R'_U{}^{2\beta}{}_{\sigma\nu\mu} + R'_V{}^{1\beta}{}_{\sigma\nu\mu} + R'_V{}^{2\beta}{}_{\sigma\nu\mu} + R'_S{}^{1\beta}{}_{\sigma\nu\mu} + R'_S{}^{2\beta}{}_{\sigma\nu\mu} + R'_\Theta{}^{1\beta}{}_{\sigma\nu\mu} +$$
$$+ R'_\Theta{}^{2\beta}{}_{\sigma\nu\mu} + R'_B{}^{1\beta}{}_{\sigma\nu\mu} + R'_B{}^{2\beta}{}_{\sigma\nu\mu} + R'_A{}^{1\beta}{}_{\sigma\nu\mu} + R'_A{}^{2\beta}{}_{\sigma\nu\mu} + R^{1\beta}{}_{\sigma\nu\mu} + R^{2\beta}{}_{\sigma\nu\mu} \qquad (11.13)$$

where

$$R'_E{}^{1\beta}{}_{\sigma\nu\mu} = ig^\beta{}_\sigma F_E{}^1{}_{\nu\mu} \tag{11.14}$$
$$R'_E{}^{2\beta}{}_{\sigma\nu\mu} = ig^\beta{}_\sigma F_{DE}{}^2{}_{\nu\mu}$$

$$R'_{DE}{}^{1\beta}{}_{\sigma\nu\mu} = ig^\beta{}_\sigma F_E{}^1{}_{\nu\mu}$$
$$R'_{DE}{}^{2\beta}{}_{\sigma\nu\mu} = ig^\beta{}_\sigma F_{DE}{}^2{}_{\nu\mu}$$
$$R'_{SU(2)}{}^{1\beta}{}_{\sigma\nu\mu} = ig^\beta{}_\sigma F_W{}^1{}_{\nu\mu}$$
$$R'_{SU(2)}{}^{2\beta}{}_{\sigma\nu\mu} = ig^\beta{}_\sigma F_{DW}{}^2{}_{\nu\mu}$$

$$R'_{DSU(2)}{}^{1\beta}{}_{\sigma\nu\mu} = ig^\beta{}_\sigma F_W{}^1{}_{\nu\mu}$$
$$R'_{DSU(2)}{}^{2\beta}{}_{\sigma\nu\mu} = ig^\beta{}_\sigma F_{DW}{}^2{}_{\nu\mu}$$

$$R'_{SU(3)}{}^{1\beta}{}_{\sigma\nu\mu} = ig^\beta{}_\sigma F_{SU(3)}{}^1{}_{\nu\mu}$$
$$R'_{SU(3)}{}^{2\beta}{}_{\sigma\nu\mu} = ig^\beta{}_\sigma F_{SU(3)}{}^2{}_{\nu\mu}$$

$$R'_U{}^{1\beta}{}_{\sigma\nu\mu} = ig^\beta{}_\sigma F_U{}^1{}_{\nu\mu}$$
$$R'_U{}^{2\beta}{}_{\sigma\nu\mu} = ig^\beta{}_\sigma F_U{}^2{}_{\nu\mu}$$

$$R'_V{}^{1\beta}{}_{\sigma\nu\mu} = ig^\beta{}_\sigma F_V{}^1{}_{\nu\mu}$$
$$R'_V{}^{2\beta}{}_{\sigma\nu\mu} = ig^\beta{}_\sigma F_V{}^2{}_{\nu\mu}$$

$$R'_S{}^{1\beta}{}_{\sigma\nu\mu} = ig^\beta{}_\sigma F_S{}^1{}_{\nu\mu}$$
$$R'_S{}^{2\beta}{}_{\sigma\nu\mu} = ig^\beta{}_\sigma F_S{}^2{}_{\nu\mu}$$

$$R'_\Theta{}^{1\beta}{}_{\sigma\nu\mu} = ig^\beta{}_\sigma F_\Theta{}^1{}_{\nu\mu}$$
$$R'_\Theta{}^{2\beta}{}_{\sigma\nu\mu} = ig^\beta{}_\sigma F_\Theta{}^2{}_{\nu\mu}$$

$$R'_B{}^{1\beta}{}_{\sigma\nu\mu} = ig^\beta{}_\sigma B^1{}_{\nu\mu}$$
$$R'_B{}^{2\beta}{}_{\sigma\nu\mu} = ig^\beta{}_\sigma B^2{}_{\nu\mu}$$

$$R'_A{}^{1\beta}{}_{\sigma\nu\mu} = ig^\beta{}_\sigma A_R{}^1{}_{\nu\mu}$$
$$R'_A{}^{2\beta}{}_{\sigma\nu\mu} = ig^\beta{}_\sigma A_R{}^2{}_{\nu\mu}$$

The total Ricci tensor is

$$R'_{\sigma\mu} = R'^\beta{}_{\sigma\beta\mu} \tag{11.15}$$

$$\begin{aligned} = & iF_E{}^1{}_{\sigma\mu} + iF_E{}^2{}_{\sigma\mu} + iF_W{}^1{}_{\sigma\mu} + iF_W{}^2{}_{\sigma\mu} + iF_{DE}{}^1{}_{\sigma\mu} + iF_{DE}{}^2{}_{\sigma\mu} + iF_{DW}{}^1{}_{\sigma\mu} + iF_{DW}{}^2{}_{\sigma\mu} + iF_{SU(3)}{}^1{}_{\sigma\mu} + iF_{SU(3)}{}^2{}_{\sigma\mu} + \\ & + iF_U{}^1{}_{\sigma\mu} + iF_U{}^2{}_{\sigma\mu} + iF_V{}^1{}_{\sigma\mu} + iF_V{}^2{}_{\sigma\mu} + iF_S{}^1{}_{\sigma\mu} + iF_S{}^2{}_{\sigma\mu} + iF_\Theta{}^1{}_{\sigma\mu} + iF_\Theta{}^2{}_{\sigma\mu} + iF_B{}^1{}_{\sigma\mu} + iF_B{}^2{}_{\sigma\mu} + \\ & + iF_A{}^1{}_{\sigma\mu} + iF_A{}^2{}_{\sigma\mu} + \partial_\mu H^{1\beta}{}_{\sigma\beta} - \partial_\beta H^{1\beta}{}_{\sigma\mu} + H^{1\gamma}{}_{\beta\sigma} H^{1\beta}{}_{\gamma\mu} - H^{1\gamma}{}_{\mu\sigma} H^{1\beta}{}_{\gamma\beta} + \\ & + \partial_\mu H^{2\beta}{}_{\sigma\beta} - \partial_\beta H^{2\beta}{}_{\sigma\mu} + H^{2\gamma}{}_{\beta\sigma} H^{2\beta}{}_{\gamma\mu} - H^{2\gamma}{}_{\mu\sigma} H^{2\beta}{}_{\gamma\beta} + H^{1\gamma}{}_{\beta\sigma} H^{2\beta}{}_{\gamma\mu} - H^{1\gamma}{}_{\mu\sigma} H^{2\beta}{}_{\gamma\beta} + \\ & + H^{2\gamma}{}_{\beta\sigma} H^{1\beta}{}_{\gamma\mu} - H^{2\gamma}{}_{\mu\sigma} H^{1\beta}{}_{\gamma\beta} \end{aligned}$$

$$\begin{aligned} = & R'_E{}^1{}_{\sigma\mu} + R'_E{}^2{}_{\sigma\mu} + R'_{SU(2)}{}^1{}_{\sigma\mu} + R'_{SU(2)}{}^2{}_{\sigma\mu} + R'_{DE}{}^1{}_{\sigma\mu} + R'_{DE}{}^2{}_{\sigma\mu} + R'_{DSU(2)}{}^1{}_{\sigma\mu} + R'_{DSU(2)}{}^2{}_{\sigma\mu} + \\ & + R'_{SU(3)}{}^1{}_{\sigma\mu} + R'_{SU(3)}{}^2{}_{\sigma\mu} + R'_U{}^1{}_{\sigma\mu} + R'_U{}^2{}_{\sigma\mu} + R'_V{}^1{}_{\sigma\mu} + R'_V{}^2{}_{\sigma\mu} + R'_S{}^1{}_{\sigma\mu} + R'_S{}^2{}_{\sigma\mu} + R'_\Theta{}^1{}_{\sigma\mu} + \\ & + R'_\Theta{}^2{}_{\sigma\mu} + R'_A{}^{1\beta}{}_{\sigma\beta\mu} + R'_A{}^{2\beta}{}_{\sigma\beta\mu} + R'_B{}^{1\beta}{}_{\sigma\beta\mu} + R'_B{}^{2\beta}{}_{\sigma\beta\mu} + R^1{}_{\sigma\mu} + R^2{}_{\sigma\mu} \\ = & R'^1{}_{\sigma\mu} + R'^2{}_{\sigma\mu} \end{aligned}$$

where

$$\begin{aligned} R'^1{}_{\sigma\mu} = & R'_E{}^1{}_{\sigma\mu} + R'_{SU(2)}{}^1{}_{\sigma\mu} + R'_{DE}{}^1{}_{\sigma\mu} + R'_{DSU(2)}{}^1{}_{\sigma\mu} + R'_{SU(3)}{}^1{}_{\sigma\mu} + R'_U{}^1{}_{\sigma\mu} + R'_V{}^1{}_{\sigma\mu} + \\ & + R'_S{}^1{}_{\sigma\mu} + R'_\Theta{}^1{}_{\sigma\mu} + R'_A{}^{1\beta}{}_{\sigma\beta\mu} + R'_B{}^{1\beta}{}_{\sigma\beta\mu} + R^1{}_{\sigma\mu} \end{aligned} \tag{11.16}$$

$$\begin{aligned} R'^2{}_{\sigma\mu} = & R'_E{}^2{}_{\sigma\mu} + R'_{SU(2)}{}^2{}_{\sigma\mu} + R'_{DE}{}^2{}_{\sigma\mu} + R'_{DSU(2)}{}^2{}_{\sigma\mu} + R'_{SU(3)}{}^2{}_{\sigma\mu} + R'_U{}^2{}_{\sigma\mu} + R'_V{}^2{}_{\sigma\mu} + \\ & + R'_S{}^2{}_{\sigma\mu} + R'_\Theta{}^2{}_{\sigma\mu} + R'_A{}^{2\beta}{}_{\sigma\beta\mu} + R'_B{}^{2\beta}{}_{\sigma\beta\mu} + R^2{}_{\sigma\mu} \end{aligned} \tag{11.17}$$

with

$$R'_E{}^1{}_{\sigma\mu} = iF_E{}^1{}_{\sigma\mu} \tag{11.18}$$
$$R'_E{}^2{}_{\sigma\mu} = iF_E{}^2{}_{\sigma\mu}$$
$$R'_{SU(2)}{}^1{}_{\sigma\mu} = iF_W{}^1{}_{\sigma\mu}$$
$$R'_{SU(2)}{}^2{}_{\sigma\mu} = iF_W{}^2{}_{\sigma\mu}$$
$$R'_{DE}{}^1{}_{\sigma\mu} = iF_{DE}{}^1{}_{\sigma\mu}$$

$$R'_{DE}{}^{2}{}_{\sigma\mu} = iF_{DE}{}^{2}{}_{\sigma\mu}$$
$$R'_{DSU(2)}{}^{1}{}_{\sigma\mu} = iF_{DW}{}^{1}{}_{\sigma\mu}$$
$$R'_{DSU(2)}{}^{2}{}_{\sigma\mu} = iF_{DW}{}^{2}{}_{\sigma\mu}$$
$$R'_{SU(3)}{}^{1}{}_{\sigma\mu} = iF_{SU(3)}{}^{1}{}_{\sigma\mu}$$
$$R'_{SU(3)}{}^{2}{}_{\sigma\mu} = iF_{SU(3)}{}^{2}{}_{\sigma\mu}$$
$$R'_{U}{}^{1}{}_{\sigma\mu} = iF_{U}{}^{1}{}_{\sigma\mu}$$
$$R'_{U}{}^{2}{}_{\sigma\mu} = iF_{U}{}^{2}{}_{\sigma\mu}$$
$$R'_{V}{}^{1}{}_{\sigma\mu} = iF_{V}{}^{1}{}_{\sigma\mu}$$
$$R'_{V}{}^{2}{}_{\sigma\mu} = iF_{V}{}^{2}{}_{\sigma\mu}$$
$$R'_{S}{}^{1}{}_{\sigma\mu} = iF_{S}{}^{1}{}_{\sigma\mu}$$
$$R'_{S}{}^{2}{}_{\sigma\mu} = iF_{S}{}^{2}{}_{\sigma\mu}$$
$$R'_{\Theta}{}^{1}{}_{\sigma\mu} = iF_{\Theta}{}^{1}{}_{\sigma\mu}$$
$$R'_{\Theta}{}^{2}{}_{\sigma\mu} = iF_{\Theta}{}^{2}{}_{\sigma\mu}$$
$$R'_{A}{}^{1}{}_{\sigma\mu} = iF_{B}{}^{1}{}_{\sigma\mu}$$
$$R'_{A}{}^{2}{}_{\sigma\mu} = iF_{B}{}^{2}{}_{\sigma\mu}$$
$$R'_{B}{}^{1}{}_{\sigma\mu} = iF_{B}{}^{1}{}_{\sigma\mu}$$
$$R'_{B}{}^{2}{}_{\sigma\mu} = iF_{B}{}^{2}{}_{\sigma\mu}$$

with the further definition of $R''^{1}{}_{\sigma\mu}$ and $R''^{2}{}_{\sigma\mu}$:

$$\begin{aligned} R''^{1}{}_{\sigma\mu} &= R'_{SU(3)}{}^{1}{}_{\sigma\mu} + R^{1}{}_{\sigma\mu} \\ R''^{2}{}_{\sigma\mu} &= R'_{SU(3)}{}^{2}{}_{\sigma\mu} + R^{2}{}_{\sigma\mu} \end{aligned} \tag{11.19}$$

$R'^{1}{}_{\sigma\mu}$ is the Ricci tensor. An additional Ricci-like tensor is

$$H_{\sigma_\mu} = H^{\beta}{}_{\sigma\beta\mu} \tag{11.20}$$

The curvature scalar is

$$\begin{aligned} R' = g^{\sigma\mu}R'_{\sigma\mu} = &+ \partial^{\sigma}H^{1\beta}{}_{\sigma\beta} - \partial_{\beta}H^{1\beta}{}_{\sigma}{}^{\sigma} + H^{1\gamma}{}_{\beta\sigma}H^{1\beta}{}_{\gamma}{}^{\sigma} - H^{1\gamma}{}_{\mu\sigma}H^{1\beta}{}_{\gamma\beta} + \partial^{\sigma}H^{2\beta}{}_{\sigma\beta} - \partial_{\beta}H^{2\beta}{}_{\sigma}{}^{\sigma} + \\ &+ H^{2\gamma}{}_{\beta\sigma}H^{2\beta}{}_{\gamma}{}^{\sigma} - H^{2\gamma\sigma}{}_{\sigma}H^{2\beta}{}_{\gamma\beta} + H^{1\gamma}{}_{\beta\sigma}H^{2\beta}{}_{\gamma}{}^{\sigma} - H^{1\gamma\sigma}{}_{\sigma}H^{2\beta}{}_{\gamma\beta} + H^{2\gamma}{}_{\beta\sigma}H^{1\beta}{}_{\gamma}{}^{\sigma} - \end{aligned}$$

$$- H^{2\gamma\sigma}{}_{\sigma}H^{1\beta}{}_{\gamma\beta} \tag{11.21}$$

$$= g^{\sigma\mu}(R^{1\beta}{}_{\sigma\beta\mu} + R^{2\beta}{}_{\sigma\beta\mu})$$

11.3 Vector Boson and Graviton Lagrangian Terms

We choose the vector boson and gravitational part of the lagrangian[46] of the Unified SuperStandard (with the Higgs sector and the Faddeev-Popov terms gauge sector not displayed here) to be:

$$\mathcal{L} = \mathrm{Tr}\ \sqrt{g}[MD_\nu R''^{1}{}_{\sigma\mu}D^\nu R''^{2\sigma\mu} + aR'^{1}{}_{\sigma\mu}R'^{2\sigma\mu} + bR' + cg^{\sigma\mu}g^{2}{}_{\sigma\mu} + c'g^{2\sigma\mu}g^{2}{}_{\sigma\mu} - \\ - dA_{SU(3)}{}^{2}{}_{\mu}A_{SU(3)}{}^{2\mu}] \tag{11.22}$$

where M, a, b, c, c', and d are constants, and $R''^{i}{}_{\sigma\mu}$ for i = 1, 2 determined above.[47]

---

[46] The rationale for this choice is 1) to obtain the known Stanard Model interactions, 2) to obtain a canonical formulation for this higher derivative theory, and 3) to introduce higher derivative terms that yield quark confinement and a theory of gravity that accounts for known deviations from Newtonian gravity such as described by MoND. See Blaha (2018e) and earlier books for details on these points.

[47] One may ask why $R''^{1}{}_{\sigma\mu}$ and $R''^{2}{}_{\sigma\mu}$ appear in the first term of the lagrangian, and not other interaction terms. We believe the primary reason is: "The extended vierbein $l^{\mu ai}(x)$ can be viewed as located at a point in a higher dimensional complex-valued space.

$$l^{\mu ai}(x) = (\partial \xi_X{}^{ai}(x)/\partial x_\mu)_{X=h(x)}$$

where $\xi_X{}^{ai}$ is a set of locally inertial coordinates located at point X, and x = h(x) is a 4-dimensional point in a tangent subspace of the higher dimensional space:

$$X = h(x)$$

The relation between complex 4-dimensional coordinates x and the higher dimensional coordinates X is an embedding of a 4-dimensional surface within the higher dimensional complex space when account is taken of the range of possible x values. We have considered such embeddings in Blaha (2015a), and in earlier books, and developed a theory of a higher dimensional complex-valued space (the *Megaverse*) that contains our universe and probably many other universes." Thus SU(3) and Gravitation have a special role in our particle dynamics based on geometry. The second reason is the common feature of color SU(3) and real-valued General Relativity is that they are the only interactions that do not participate in 'rotations of interactions' as described earlier and in chapter 31 of Blaha (2017b). The third, practical reason is the experimental reality that the Strong Interaction and Gravitation are known to have 'anomalous' features that will be seen to be remedied by these insertions while the other interactions are 'conventional.'

This higher derivative lagrangian maintains the locality of the theory but does entail a modest modification in the derivation of the Euler-Lagrange equations of motion. It also requires the use of principal value propagators rather than ordinary Feynman propagators for gluon and graviton interactions. Thus the Strong Interaction sector, and the Gravitation sector are Action-at-a-Distance theories that are similar in spirit to Wheeler-Feynman Electrodynamics. The two U(1) Electromagnetic sectors, the Generation group U(4) gauge field sector, the Layer group U(4) gauge field sector, the two SU2) Weak sectors, the U(4) A_S gauge field sector, the spinor connection sector, and the Θ-interaction sector may, or may not, be Action-at-a-Distance fields. They are not constrained to be Action-at-a-Distance by the present considerations.

Since we wish to apply our theory cosmologically, and within hadrons, where the gravitational spinor connections are negligible due to the smallness of the gravitational constant G and the 'smallness' of Gravitational B spinor connection effects on the cosmological scale, we set $F^1{}_{\nu\mu} = F^2{}_{\nu\mu} = 0$ and find[48]

$$\mathcal{L} = \mathrm{Tr}\ \sqrt{g}[MD_\nu(R'^1{}_{SU(3)\sigma\mu} + R^1{}_{\sigma\mu})D^\nu(R'_{SU(3)}{}^{2\sigma\mu} + R^{2\sigma\mu}) + \\ + aR'^1{}_{\sigma\mu}R'^{2\sigma\mu} + bR' + cg^{\sigma\mu}g^2{}_{\sigma\mu} + c'g^{2\sigma\mu}g^2{}_{\sigma\mu} - dA_{SU(3)}{}^2{}_\mu A_{SU(3)}{}^{2\mu}] \tag{11.23}$$

Since there are no strong interaction fields in 'empty' space and gravity is negligible within hadrons,[49] we can drop the interaction terms between the Strong interaction and the Gravity interaction. However, we cannot drop the interaction terms amongst Electromagnetism, the Weak interaction, the Strong Interaction, the Generation group U(4) interaction, the U(4) Layer groups interactions, the U(4) Species group interaction, and the U(192) Θ-interaction – within, and between, hadrons. The interaction terms between Electromagnetism and Gravitation are important cosmologically.

The above lagrangian terms can therefore be expressed as:[50]

---

[48] The constants have the dimensions: M has the dimension of inverse mass squared, b has dimension mass squared, a is dimensionless, c and c' have dimension mass, and d has dimension mass squared.

[49] We show gravity weakens at very short distances using our Two-Tier Quantum Field Theory formalism. See Appendix A, and Blaha (2003) and (2005a) among other books by the author.

[50] We only consider the gauge field lagrangian terms.

$$\mathcal{L} = \mathcal{L}_E + \mathcal{L}_{SU(2)} + \mathcal{L}_{DE} + \mathcal{L}_{DSU(2)} + \mathcal{L}_{SU(3)} + \mathcal{L}_U + \mathcal{L}_V + \mathcal{L}_S + \mathcal{L}_\Theta + \mathcal{L}_G + \mathcal{L}_{int} \quad (11.24)$$

where taking traces of $\mathcal{L}$s terms is understood, and with coupling constants not displayed below to avoid clutter,

$$\mathcal{L}_E = \text{Tr}\,\surd g\{M\{[\partial_\nu + i(A_E{}^1{}_\nu + A_E{}^2{}_\nu)]F^1{}_{E\sigma\mu}[\partial^\nu + i(A_E{}^{1\nu} + A_E{}^{2\nu})]F^2{}_E{}^{\sigma\mu}\} + aF_E{}^1{}_{\sigma\mu}F_E{}^{2\sigma\mu}\} \quad (11.25)$$

$$\mathcal{L}_{SU(2)} = \text{Tr}\,\surd g[aF_W{}^1{}_{\sigma\mu}F_W{}^{2\sigma\mu}]$$

$$\mathcal{L}_{DE} = \text{Tr}\,\surd g\{M\{[\partial_\nu + i(A_{DE}{}^1{}_\nu + A_{DE}{}^2{}_\nu)]F^1{}_{DE\sigma\mu}[\partial^\nu + i(A_{DE}{}^{1\nu} + A_{DE}{}^{2\nu})]F_{DE}{}^{2\sigma\mu}\} + {} + aF_{DE}{}^1{}_{\sigma\mu}F_{DE}{}^{2\sigma\mu}\}$$

$$\mathcal{L}_{DSU(2)} = \text{Tr}\,\surd g[aF_W{}^1{}_{\sigma\mu}F_W{}^{2\sigma\mu}]$$

$$\mathcal{L}_{SU(3)} = \text{Tr}\,\surd g\{M[\partial_\nu + i(A_{SU(3)}{}^1{}_\nu + A_{SU(3)}{}^2{}_\nu)]F_{SU(3)}{}^1{}_{\sigma\mu}[\partial^\nu + i(A_{SU(3)}{}^{1\nu} + {} + A_{SU(3)}{}^{2\nu})]F_{SU(3)}{}^{2\sigma\mu} + aF_{SU(3)}{}^1{}_{\sigma\mu}F_{SU(3)}{}^{2\sigma\mu} - dA_{SU(3)}{}^2{}_\mu A_{SU(3)}{}^{2\mu}\}$$

$$\mathcal{L}_U = \text{Tr}\,\surd g[aF_U{}^1{}_{\sigma\mu}F_U{}^{2\sigma\mu}]$$

$$\mathcal{L}_V = \text{Tr}\,\surd g[aF_V{}^1{}_{\sigma\mu}F_V{}^{2\sigma\mu}]$$

$$\mathcal{L}_S = \text{Tr}\,\surd g[aF_S{}^1{}_{\sigma\mu}F_S{}^{2\sigma\mu}]$$

$$\mathcal{L}_\Theta = \text{Tr}\,\surd g[aF_\Theta{}^1{}_{\sigma\mu}F_\Theta{}^{2\sigma\mu}]$$

$$\mathcal{L}_G = \text{Tr}\,\surd g[MD_\nu R^1{}_{\sigma\mu}D^\nu R^{2\sigma\mu} + aR^1{}_{\sigma\mu}R^{2\sigma\mu} + bg^{\sigma\mu}(R^{1\beta}{}_{\sigma\beta\mu} + R^{2\beta}{}_{\sigma\beta\mu}) + cg^{\sigma\mu}g^2{}_{\sigma\mu} + c'g^{2\sigma\mu}g^2{}_{\sigma\mu}]$$
$$= \text{Tr}\,\surd g[MD_\nu R^1{}_{\sigma\mu}D^\nu R^{2\sigma\mu} + aR^1{}_{\sigma\mu}R^{2\sigma\mu} + bH + cg^{\sigma\mu}g^2{}_{\sigma\mu} + c'g^{2\sigma\mu}g^2{}_{\sigma\mu}]$$

$$\mathcal{L}_{int} = \mathcal{L} - (\mathcal{L}_E + \mathcal{L}_{SU(2)} + \mathcal{L}_{DE} + \mathcal{L}_{DSU(2)} + \mathcal{L}_{SU(3)} + \mathcal{L}_U + \mathcal{L}_V + \mathcal{L}_S + \mathcal{L}_\Theta + \mathcal{L}_G) \quad (11.26)$$

with appropriate sums over layers and gravitational B spinor connection terms omitted. Thus $\mathcal{L}_{SU(3)}$, $\mathcal{L}_{SU(2)}$, $\mathcal{L}_E$, $\mathcal{L}_{DE}$, $\mathcal{L}_{DSU(2)}$, $\mathcal{L}_U$,, $\mathcal{L}_V$, $\mathcal{L}_S$, $\mathcal{L}_\Theta$, and parts of $\mathcal{L}_{int}$ are the dominant interactions within hadrons, and $\mathcal{L}_G$, $\mathcal{L}_E$ and parts of $\mathcal{L}_{int}$ are the dominant interactions in space within the framework of this discussion.

The $D_\nu R^1{}_{\sigma\mu}$ and $D^\nu R^{2\sigma\mu}$ terms have the form:

$$D_\nu R^i{}_{\sigma\mu} = + \partial_\nu R^i{}_{\sigma\mu} - H^{1\beta}{}_{\sigma\nu}R^i{}_{\beta\mu} - H^{1\beta}{}_{\nu\mu}R^i{}_{\sigma\beta} \quad (11.27)$$

for i = 1, 2.

11.4 New Vector Boson Interactions

The above lagrangian can be broken up into pieces in the following manner:

$$\mathcal{L}_E = \mathrm{Tr}\ \sqrt{g}\{M\{[\partial_\nu + i(A_E{}^1{}_\nu + A_E{}^2{}_\nu)]F^1{}_{E\sigma\mu}[\partial^\nu + i(A_E{}^{1\nu} + A_E{}^{2\nu})]F^2{}_E{}^{\sigma\mu}\} + aF_E{}^1{}_{\sigma\mu}F_E{}^{2\sigma\mu}\} \tag{11.28}$$

$$\mathcal{L}_{SU(2)} = \mathrm{Tr}\ \sqrt{g}[aF_W{}^1{}_{\sigma\mu}F_W{}^{2\sigma\mu}]$$
$$\mathcal{L}_{DE} = \mathrm{Tr}\ \sqrt{g}\{M\{[\partial_\nu + i(A_{DE}{}^1{}_\nu + A_{DE}{}^2{}_\nu)]F^1{}_{DE\sigma\mu}[\partial^\nu + i(A_{DE}{}^{1\nu} + A_{DE}{}^{2\nu})]F_{DE}{}^{2\sigma\mu}\} + $$
$$+ aF_{DE}{}^1{}_{\sigma\mu}F_{DE}{}^{2\sigma\mu}\}$$
$$\mathcal{L}_{DSU(2)} = \mathrm{Tr}\ \sqrt{g}[aF_W{}^1{}_{\sigma\mu}F_W{}^{2\sigma\mu}]$$
$$\mathcal{L}_{SU(3)} = \mathrm{Tr}\ \sqrt{g}\{M[\partial_\nu + i(A_{SU(3)}{}^1{}_\nu + A_{SU(3)}{}^2{}_\nu)]F_{SU(3)}{}^1{}_{\sigma\mu}[\partial^\nu + i(A_{SU(3)}{}^{1\nu} +$$
$$+ A_{SU(3)}{}^{2\nu})]F_{SU(3)}{}^{2\sigma\mu} + aF_{SU(3)}{}^1{}_{\sigma\mu}F_{SU(3)}{}^{2\sigma\mu} - dA_{SU(3)}{}^2{}_\mu A_{SU(3)}{}^{2\mu}\} \tag{11.29}$$

$$\mathcal{L}_U = \mathrm{Tr}\ \sqrt{g}[aF_U{}^1{}_{\sigma\mu}F_U{}^{2\sigma\mu}]$$
$$\mathcal{L}_V = \mathrm{Tr}\ \sqrt{g}[aF_V{}^1{}_{\sigma\mu}F_V{}^{2\sigma\mu}]$$
$$\mathcal{L}_S = \mathrm{Tr}\ \sqrt{g}[aF_S{}^1{}_{\sigma\mu}F_S{}^{2\sigma\mu}]$$
$$\mathcal{L}_\Theta = \mathrm{Tr}\ \sqrt{g}[aF_\Theta{}^1{}_{\sigma\mu}F_\Theta{}^{2\sigma\mu}]$$
$$\mathcal{L}_G = \mathrm{Tr}\ \sqrt{g}[MD_\nu R^1{}_{\sigma\mu}D^\nu R^{2\sigma\mu} + aR^1{}_{\sigma\mu}R^{2\sigma\mu} + bg^{\sigma\mu}(R^{1\beta}{}_{\sigma\beta\mu} + R^{2\beta}{}_{\sigma\beta\mu}) + cg^{\sigma\mu}g^2{}_{\sigma\mu} + c'g^{2\sigma\mu}g^2{}_{\sigma\mu}]$$
$$= \mathrm{Tr}\ \sqrt{g}[MD_\nu R^1{}_{\sigma\mu}D^\nu R^{2\sigma\mu} + aR^1{}_{\sigma\mu}R^{2\sigma\mu} + bH + cg^{\sigma\mu}g^2{}_{\sigma\mu} + c'g^{2\sigma\mu}g^2{}_{\sigma\mu}]$$
$$\mathcal{L}_{int} = \mathcal{L} - (\mathcal{L}_E + \mathcal{L}_{SU(2)} + \mathcal{L}_{DE} + \mathcal{L}_{DSU(2)} + \mathcal{L}_{SU(3)} + \mathcal{L}_U + \mathcal{L}_V + \mathcal{L}_S + \mathcal{L}_\Theta + \mathcal{L}_G) \tag{11.30}$$

again with appropriate sums over layers and with coupling constants not displayed to avoid clutter. Thus $\mathcal{L}_{SU(3)}$, $\mathcal{L}_{SU(2)}$, $\mathcal{L}_E$, $\mathcal{L}_{DE}$, $\mathcal{L}_{DSU(2)}$, $\mathcal{L}_U$,, $\mathcal{L}_V$, $\mathcal{L}_S$, $\mathcal{L}_\Theta$, and parts of $\mathcal{L}_{int}$ are the dominant interactions within hadrons, and $\mathcal{L}_G$, $\mathcal{L}_E$ and parts of $\mathcal{L}_{int}$ are the dominant interactions in space within the framework of this discussion. The terms of $\mathcal{L}_{int}$ have 'new' interactions between gauge fields that are described in some detail in Blaha (2017b) and other books. These interactions are not in the conventional Standard Model. They lead to modifications of gravity, the Strong Interactions, spin dynamics and so on.

12. Gravitational Potential on the Three Distance Scales

This chapter[51] and the next chapter describe some of the possible results of the unified lagrangian terms in eq. 11.22 and 11.30. In this chapter we put together results on the gravitational potential that were previously found in Blaha (2016g), (2016h) and (2017a). We note that a new experimental study of 33,000 galaxies indicate that the gravitational potential at inter-galactic distances deviates significantly from the Newtonian potential G/r. In Blaha (2017a) we showed that such a deviation occurs in our theory. This experimental result was not known at the time of its writing.[52,53]

In this chapter we see that the theory has higher derivative dynamic equations but, unlike the Strong Interaction sector (next chapter) which yields color (quark-gluon) confinement, the gravitation dynamic equations do not have confinement. They do yield a modified form of gravity at various intermediate distances of the order of the average galactic radius, and beyond.

The modification of gravity implied by our theory is consistent with the need for Dark Matter described in our Theory of Everything books (Blaha (2015a) and (2016c)).[54] It is also consistent with a MoND theory predictions[55] with the addition of

---

[51] Most of this chapter appears in Blaha (2017a) and earlier books by the author.

[52] The gravitational potential was found to be greater than G/r at inter-galactic distances in a survey of 33,000 galaxies by M. Brouwer and colleagues at the Leiden Observatory (The Netherlands) in an announcement on December 18, 2016.

[53] There are other higher derivative theories – some with two metrics, and some with a metric plus vector plus scalar field formulation. The present work is based on a unification of Strong and Gravity sectors and a totally different formalism. Some significant references are: M. Milgrom, Phys. Rev. **D80**, 123536 (2009); C. Skordis et al, Phys. Rev. Lett. **96**, 011301 (2006); R. H. Sanders, Astrophysical Journal **480**, 492 (1997); and references therein; J. D. Bekenstein, Phys. Rev. **D70**, 083509 (2004) and references therein; J-P. Bruneton, Phys. Rev. **D76**, 124012 (2007) with a higher derivative gravity and metric, vector, scalar fields. See also references within these articles.

[54] I. Ferreras et al, Phys. Rev. Lett., **96**, 011301 (2006) shows the need for both Dark Matter and MOND based on studies of astrophysical data.

sixth order derivatives in the Gravitation sector lagrangian in a manner consistent with the higher order terms appearing in the Strong Interaction sector of the unified theory.

Thus our approach to MoND does not require a major change in Mechanics, quantum theory, or General Relativity (modulo higher order derivatives). The consistency between the need for higher order derivatives in both the Strong Interaction and Gravitation sectors is encouraging.

The developments in this chapter are a generalization with higher order derivative terms of the unified theory presented in chapter 6 of Blaha (2016d).

12.1 Gravitation Sector Dynamic Equations

Our gravity sector has two metric fields, $g_{\mu\nu}$ and $g^2{}_{\mu\nu}$ derived from the unified formalism described earlier. Some of the relevant gravitation equations found in sections 11.2 and 11.3 are:

$$H^\sigma{}_{\nu\mu} = \Gamma^\sigma{}_{\nu\mu} + \Gamma^{2\sigma}{}_{\nu\mu}$$

$$H^\beta{}_{\sigma\nu\mu} = \partial_\mu H^\beta{}_{\sigma\nu} - \partial_\nu H^\beta{}_{\sigma\mu} + H^\gamma{}_{\nu\sigma} H^\beta{}_{\gamma\mu} - H^\gamma{}_{\mu\sigma} H^\beta{}_{\gamma\nu}$$

$$H_{\sigma\mu} = H^\beta{}_{\sigma\beta\mu}$$

$$H = g^{\sigma\mu} H_{\sigma\mu}$$

$$\mathcal{H} = R'^1 + R'^2$$

We use the gravitational sector lagrangian:

$$\mathcal{L}_G = \sqrt{g}[M D_\nu R'^1{}_{G\sigma\mu} D^\nu R'^2{}_G{}^{\sigma\mu} + a R'^1{}_{G\sigma\mu} R'^2{}_G{}^{\sigma\mu} + b g^{\sigma\mu}(R'^1{}_G{}^\beta{}_{\sigma\beta\mu} + R'^2{}_G{}^\beta{}_{\sigma\beta\mu}) + c g^{\sigma\mu} g^2{}_{\sigma\mu} + \\ + e g^{2\sigma\mu} g^2{}_{\sigma\mu}]$$

$$\mathcal{L}_G = \sqrt{g}[M D_\nu R'^1{}_{G\sigma\mu} D^\nu R'^2{}_G{}^{\sigma\mu} + a R'^1{}_{G\sigma\mu} R'^2{}_G{}^{\sigma\mu} + bH + c g^{\sigma\mu} g^2{}_{\sigma\mu} + e g^{2\sigma\mu} g^2{}_{\sigma\mu}]$$

where

---

[55] Our gravity theory has aspects that are very similar to the MOND theories described in A. Balakin et al, Phys. Rev. **D70**, 064027 (2004); H-S Zhao et al, Phys. Rev. **D82**, 103001 (2010); and references therein. However our approach is very different.

$$D_\nu Z_\mu = (\partial_\nu + iF_\nu) Z_\mu - H^\sigma{}_{\nu\mu} Z_\sigma$$

$$= [g^\sigma{}_\mu \partial_\nu + ig^\sigma{}_\mu F_\nu - H^\sigma{}_{\nu\mu}] Z_\sigma$$
$$= [g^\sigma{}_\mu \partial_\nu + iD^\sigma{}_{\mu\nu}] Z_\sigma$$

where M, a, b, c, and e are constants determined later. We determine a in chapter 13[56]

$$a = 1/(2f) = 7.47 \times 10^{181} \quad (12.0)$$

The remaining values are determined later in this chapter in section 12.8.

The lagrangian dynamic equations that result are difficult. We consequently will examine the weak gravitation limiting case where we can approximate the two metrics with

$$g_{\mu\nu} \cong \eta_{\mu\nu} + h_{\mu\nu} \quad (12.1)$$
$$g^2{}_{\mu\nu} \cong \eta_{\mu\nu} + h^2{}_{\mu\nu} \quad (12.2)$$

where

$$|h_{\mu\nu}| << 1 \quad (12.3)$$
$$|h^2{}_{\mu\nu}| << 1$$

Using the relations

$$\partial_\mu h^\mu{}_\nu = \tfrac{1}{2}\partial_\nu h^\mu{}_\mu \quad (12.4)$$
$$\partial_\mu h^{2\mu}{}_\nu = \tfrac{1}{2}\partial_\nu h^{2\mu}{}_\mu \quad (12.5)$$

and neglecting higher order terms in $h_{\mu\nu}$ and $h^2{}_{\mu\nu}$ we find

$$R'^1{}_{\mu\nu} = \tfrac{1}{2}[\Box h_{\mu\nu} - \partial_\mu\partial_\lambda h^\lambda{}_\nu - \partial_\nu\partial_\lambda h^\lambda{}_\mu + \partial_\nu\partial_\mu h^\lambda{}_\lambda]$$
$$\cong \tfrac{1}{2}\Box h_{\mu\nu} \quad (12.6)$$
$$R'^1 \cong \tfrac{1}{2}\Box h_\mu{}^\mu$$

$$R'^2{}_{\mu\nu} = \tfrac{1}{2}[\Box h^2{}_{\mu\nu} - \partial_\mu\partial_\lambda h^{2\lambda}{}_\nu - \partial_\nu\partial_\lambda h^{2\lambda}{}_\mu + \partial_\nu\partial_\mu h^{2\lambda}{}_\lambda]$$
$$\cong \tfrac{1}{2}\Box h^2{}_{\mu\nu} \quad (12.7)$$

---

[56] The value of a is obtained from the Charmonium calculation in chapter 13.

$$R'^2 \cong ½\Box h^2{}_\mu{}^\mu$$

with

$$D_\nu = \partial_\nu$$

upon neglecting higher order terms.

Substituting we find the *effective quadratic* part of the lagrangian (in $h_{\mu\nu}$ and $h^2{}_{\mu\nu}$) is

$$\begin{aligned}\mathcal{L}_G = \sqrt{g}[&M\partial_\nu R'^1{}_{G\sigma\mu}\partial^\nu R'^2{}_G{}^{\sigma\mu} + a\Box h_{\sigma\mu}\Box h^{2\sigma\mu}/4 + ½b(\partial_\alpha h^{\sigma\mu}\partial^\alpha h_{\sigma\mu} + \partial_\alpha h^{2\sigma\mu}\partial^\alpha h^2{}_{\sigma\mu}) + \\ &+ c(4 + \eta^{\sigma\mu}h^2{}_{\sigma\mu} + h^{\sigma\mu}\eta_{\sigma\mu} + h^{\sigma\mu}h^2{}_{\sigma\mu}) + \\ &+ e(2\eta^{\sigma\mu}h^2{}_{\sigma\mu} + h^{2\sigma\mu}h^2{}_{\sigma\mu}) + 1/4(h_{\mu\nu} + h^2{}_{\mu\nu})T^{\mu\nu}]\end{aligned} \quad (12.8)$$

Using partial integrations, we find the standard technique for determining the equations of motion from a lagrangian for independent variations with respect to $h_{\mu\nu}$ and $h^2{}_{\mu\nu}$ yields

$$-M\Box^3h^{2\mu\nu}/4 + a\Box^2h^{2\mu\nu}/4 + ½b\Box h^{\mu\nu} + c(\eta^{\mu\nu} + h^{2\mu\nu}) + 1/4T^{\mu\nu} = 0 \quad (12.9)$$

$$-M\Box^3h^{\mu\nu}/4 + a\Box^2h^{\mu\nu}/4 + ½b\Box h^{2\mu\nu} + c(\eta^{\mu\nu} + h^{\mu\nu}) + 2e(\eta^{\mu\nu} + h^{2\mu\nu}) + 1/4T^{\mu\nu} = 0 \quad (12.10)$$

The term $c\eta^{\mu\nu}$ can be viewed as part of the total energy-momentum tensor $T'^{\mu\nu}$:

$$T'^{\mu\nu} = T^{\mu\nu} + 2c\eta^{\mu\nu} \quad (12.10a)$$

It plays a role similar to the Cosmological Constant. Subtracting the equations we find

$$M(\Box^3h^{\mu\nu}/4 - \Box^3h^{2\mu\nu}/4) + a\Box^2h^{2\mu\nu}/4 - a\Box^2h^{\mu\nu}/4 + ½b\Box h^{\mu\nu} - ½b\Box h^{2\mu\nu} + c(h^{2\mu\nu} - h^{\mu\nu}) - 2e(\eta^{\mu\nu} + h^{2\mu\nu}) = 0 \quad (12.11)$$

and thus

$$[-M\Box^3 + a\Box^2 - ½b\Box + (4c - 8e)]h^{2\mu\nu} = 8e\eta^{\mu\nu} - M\Box^3h^{\mu\nu} + a\Box^2h^{\mu\nu} - ½b\Box h^{\mu\nu} + 4ch^{\mu\nu} \quad (12.12)$$

Therefore we determine the metric equation for $h^{2\mu\nu}$ to be

$$h^{2\mu\nu} = [-M\Box^3 + a\Box^2 - \tfrac{1}{2}b\Box + (4c - 8e)]^{-1}[8e\eta^{\mu\nu} - M\Box^3h^{\mu\nu} + a\Box^2h^{\mu\nu} - \tfrac{1}{2}b\Box h^{\mu\nu} + 4ch^{\mu\nu}] \tag{12.13}$$

Substituting in eq. 12.9 we obtain

$$[-M\Box^3/4 + a\Box^2/4 + c][-M\Box^3 + a\Box^2 - \tfrac{1}{2}b\Box + (4c - 8e)]^{-1}[8e\eta^{\mu\nu} - M\Box^3 + a\Box^2h^{\mu\nu} - \tfrac{1}{2}b\Box h^{\mu\nu} + 4ch^{\mu\nu}] + \\ + \tfrac{1}{2}b\Box h^{\mu\nu} + 1/4T'^{\mu\nu} = 0 \tag{12.14}$$

We now redefine the energy-momentum tensor with the result eq. 12.14 becomes

$$\{[-M\Box^3/4 + a\Box^2/4 + c][-M\Box^3 + a\Box^2 - \tfrac{1}{2}b\Box + (4c - 8e)]^{-1}[-M\Box^3 + a\Box^2 - \tfrac{1}{2}b\Box + 4c] + \\ + \tfrac{1}{2}b\Box\}h^{\mu\nu} = -1/4T'^{\mu\nu} \tag{12.15}$$

12.2 Real General Relativity Gravity Potential

Assuming that we are dealing with non-relativistic matter we can calculate the gravity potential contribution from eq. 12.15

$$V_{G1}(\mathbf{x}) = -\int d^3k \exp(i\mathbf{k}\cdot\mathbf{x})V_{G1}(\mathbf{k})/(2\pi)^3 \tag{12.16}$$

where

$$V_{G1}(\mathbf{k}) = \{[M\mathbf{k}^6/4 + a\mathbf{k}^4/4 + c][M\mathbf{k}^6 + a\mathbf{k}^4 + \tfrac{1}{2}b\mathbf{k}^2 + (4c - 8e)]^{-1}[M\mathbf{k}^6 + a\mathbf{k}^4 + \tfrac{1}{2}b\mathbf{k}^2 + 4c] - \\ - \tfrac{1}{2}b\mathbf{k}^2\}^{-1} \\ = [M\mathbf{k}^6 + a\mathbf{k}^4 + \tfrac{1}{2}b\mathbf{k}^2 + (4c - 8e)]/\{[M\mathbf{k}^6/4 + a\mathbf{k}^4/4 - \tfrac{1}{2}b\mathbf{k}^2 + c][M\mathbf{k}^6 + a\mathbf{k}^4 + \tfrac{1}{2}b\mathbf{k}^2 + \\ + 4c] + 4eb\mathbf{k}^2\} \tag{12.17}$$

Similarly eq. 12.13 implies the other contribution to the total gravity potential is

$$V_{G2}(\mathbf{x}) = -\int d^3k \exp(i\mathbf{k}\cdot\mathbf{x})V_{G2}(\mathbf{k})/(2\pi)^3 \tag{12.18}$$

where

$$V_{G2}(\mathbf{k}) = \{[Mk^6 + ak^4 + ½bk^2 + 4c]^{-1}[Mk^6 + a\,\mathbf{k}^4 + ½b\mathbf{k}^2 + 4c - 8e]\}V_{G1}(\mathbf{k}) \quad (12.19)$$

The total gravity potential energy *for real-valued General Relativity* is thus

$$V^{tot}_{RG}(\mathbf{x}) = V_{G1}(\mathbf{x}) + V_{G2}(\mathbf{x}) \quad (12.20)$$

12.3 Real-valued General Relativistic Gravity Solution

Eqs. 12.16 - 12.17 can generate a massless graviton, which seems a requirement based on cosmological considerations, and also generate a pair of massive gravitons of very low mass. The massive gravitons generate a MoND-like potential that might explicate the anomalous gravitation effects seen at distances of the order of galactic dimensions.

If we set the "Cosmological Constants" c = e = 0, then eq. 12.17 becomes

$$V_{G1}(\mathbf{k}) = 4/\{M\mathbf{k}^2[\mathbf{k}^4 + a\mathbf{k}^2/M - 2b/M]\} \quad (12.21)$$

The denominator of eq. 12.21 can be factored into the form

$$V_{G1}(\mathbf{k}) = 4/\{M\mathbf{k}^2[\mathbf{k}^2 + ½m_A^2 + ½(m_A^4 + 8bm_A^2/a)^{½}][\mathbf{k}^2 + ½m_A^2 - ½(m_A^4 + 8bm_A^2/a)^{½}]\} \quad (12.22)$$

where

$$m_A = (a/M)^{½}$$

Assuming $m_A^2 << 8b/a$, or $a^2/8 < 1/8 << bM = M(2\pi G)^{-1}$, which is reasonable since a is approximately one and M is enormous, we see

$$V_{G1}(\mathbf{k}) \cong 4/\{M\mathbf{k}^2[\mathbf{k}^2 + (2bm_A^2/a)^{½}][\mathbf{k}^2 - (2bm_A^2/a)^{½}]\} \quad (12.23)$$
$$= (1/b)\{-2/\mathbf{k}^2 + 1/[\mathbf{k}^2 + (2bm_A^2/a)^{½}] + 1/[\mathbf{k}^2 - (2bm_A^2/a)^{½}]\}$$

up to negligible terms.

From eq. 12.19 we see $V_{G2}(\mathbf{k}) = V_{G1}(\mathbf{k})$ if c = e = 0. Thus the total gravitational potential due to eq. 12.8 *for **real-valued** General Relativity* (in momentum space) is

$$V^{tot}{}_{RG}(\mathbf{k}) = ½(V_{G1}(\mathbf{k}) + V_{G2}(\mathbf{k})) \quad (12.24)$$
$$= (1/b)\{-2/\mathbf{k}^2 + 1/[\mathbf{k}^2 + (2bm_A{}^2/a)^{½}] + 1/[\mathbf{k}^2 - (2bm_A{}^2/a)^{½}]\}$$
$$\cong \pi G\{-2/\mathbf{k}^2 + 1/[\mathbf{k}^2 + (2bm_A{}^2/a)^{½}] + 1/[\mathbf{k}^2 - (2bm_A{}^2/a)^{½}]\}$$

with b set to

$$b = (2\pi G)^{-1} = 2.364 \times 10^{55}\ eV^2 \quad (12.24a)$$

introducing the connection of b to G, Newton's gravitational constant and using J.1d.

The coordinate space potential is[57,58]

$$V^{tot}{}_{RG}(\mathbf{r}) = -G/r + a_1Ge^{-m_Gr}/r + a_2G\cos(m_Gr)/r \quad (12.25)$$

where

$$m_G{}^2 = (2bm_A{}^2/a)^{½} = (2b/M)^{½} \quad (12.25a)$$
$$a_1 = ½$$
$$a_2 = ½$$

where m_G is an extremely small mass.

12.4 U(4) General Relativistic Reality Group Gravity Contribution

The Reality group of Complex General Relativity generates a gravitational interaction. Its role is to rotate 4-vectors of coordinates to complex-values. Appendix B discusses it in some detail.

Assuming that we are dealing with non-relativistic matter we can calculate the gravity potential contribution from the General Relativistic Reality Group using eqs. B.22 and B.26:

$$V_{GA1}(\mathbf{x}) = -\ 1/(32M) \int d^3k\ \exp(i\mathbf{k}\cdot\mathbf{x})V_{GA1}(\mathbf{k})/(2\pi)^3 \quad (12.26)$$

where

$$V_{GA1}(\mathbf{k}) = (\mathbf{k}^4 + (a/M)\mathbf{k}^2)^{-1} \quad (12.27)$$

---

[57] Since the theory has higher order drivatives that could lead to unitarity problems Feynman propagators must be taken in Principal order. Since potentials are a part of Feynman propagators the potentials real value must be used.

[58] The third term in eq.12.25 is an oscillating Yukawa term that, because of the smallness of m_G, is slowly varying towards the end of a galaxy and thus could be well within observational error bounds. It appears that the real part of the third term is the contribution to the total gravitational potential using Principal value propagators.

up to a U(4) matrix factor with matrix elements of order 1.

The eq. 12.29 denominator can be separated into two terms:

$$V_{GA1}(\mathbf{k}) = (M/a)[1/k^2 - 1/(\mathbf{k}^2 + a/M)] \quad (12.28)$$

which upon integration yield

$$V_{GA1}(\mathbf{r}) = -[1/(128\pi a)][1/r - e^{-m_A r}/r] \quad (12.29)$$

We use:[59]

$$M \sim 1.61 \times 10^{163}\ eV^{-2} \quad (12.30)$$

and thus

$$m_A = (a/M)^{½} = 1.94 \times 10^{-101}\ eV \quad (12.31)$$

since $a = 7.47 \times 10^{181}$ by eq. 12.0. The coupling constant is

$$\alpha_R = 1/(128\pi a) \cong 3.3 \times 10^{-185} \quad (12.32)$$

In comparison the electromagnetic fine structure constant is

$$\alpha_{QED} \cong 0.0073$$

Thus the General Relativistic Reality Group fields (A_R) coupling constant is negligible in comparison.

12.5 Total Gravity

We now turn to combine the General Relativistic Reality group gravity contribution with the gravitational potential of real-valued General Relativity using:

$$V^{tot}_{RG}(\mathbf{r}) = -G/r + a_1 G e^{-m_G r}/r + a_2 G\cos(m_G r)/r \quad (12.25)$$

[59] M is determined later in section 12.9.

The total combined gravitational potential of Complex General Relativity (modulo Higgs corrections)[60]

$$V_{TOT}(\mathbf{r}) = -G/r - a_1Ge^{-m_G r}/r + a_2G\cos(m_G r)/r - [1/(128\pi a)][1/r - e^{-m_A r}/r] \quad (12.33)$$

with the mass constant calculated from eq. 12.25a above:

$$m_G = (2b/M)^{1/4} = 1.31 \times 10^{-27} \text{ eV} \quad (12.34)$$

The *other* graviton mass m_A for the Species gravity potential part is given by

$$m_A = (a/M)^{1/2} \cong 1.94 \times 10^{-101} \text{ eV} \quad (12.35)$$

The input constants are[61]

$$b = (2\pi G)^{-1} = 2.364 \times 10^{55} \text{ eV}^2 \quad (12.36)$$

$$M \sim 1.61 \times 10^{163} \text{ eV}^{-2} \quad (12.37)$$

We now examine $V_{TOT}(\mathbf{r})$ at short distances (within the solar system), distances of tens of thousands of light years (intra-galactic distances), and distances between galaxies (hundreds of thousands to millions of light years and beyond).

12.6 Influence of Gravitational Gauge Field on Gravitation

The gravitational gauge field potential $V_{TOT}(\mathbf{r})$ in eq. 12.35 has a relatively large coupling constant that makes it competitive with the known force of gravity at large distances of the scale of galactic distances. This force, which is negligible at short distances of the order of planetary distances, may be part of the MoND phenomena that affects the motion of stars.

---

[60] The third term in eq. 12.25 is an oscillating Yukawa term that, because of the smallness of m_G, is slowly varying towards the end of a galaxy and thus could be well within observational error bounds. It appears that the real part of the third term is the contribution to the total gravitational potential using Principal value propagators.

[61] M is determined later in section 12.9.

12.7 Intra-Solar System Distance Scale

Since m_G and m_A are extremely small the gravitational potential at distances of perhaps up to at least several light years is

$$V_{TOT}(\mathbf{r}) \cong -G/r \qquad (12.38)$$

to well within feasible experimental limits since the factor Reality Group contribution is negligible due to the extremely small values of m_G and m_A causing the terms

$$-a_1Ge^{-m_Gr}/r + a_2G\cos(m_Gr)/r - [1/(128\pi a)][1/r - e^{-m_Ar}/r] \cong 0 \qquad (12.39)$$

in eq. 12.33 to be zero to very good approximation.

12.8 Galactic Distance Scale

At distances of several tens of thousands of light years up to perhaps 100,000's of light years we find the m_G terms in eq. 12.33 to be the major cause of the deviations from Newton's law since

$$m_G >> m_A \qquad (12.40)$$

by eqs. 12.34 and 12.25. Thus

$$\begin{aligned} V_{TOT}(\mathbf{r}) &\cong -G/r - 0.5Ge^{-m_Gr}/r + 0.5G\cos(m_Gr)/r \\ &\cong -G/r + 0.5Gm_G - 0.5Gm_G^2r + Gm_G^3r^2/12 \end{aligned} \qquad (12.41)$$

for $r \approx m_G^{-1}$. The resultant force is

$$\mathbf{F} = -\nabla V_{TOT}(\mathbf{r})(\mathbf{r}) \sim -G\mathbf{r}/r^3 + 0.5Gm_G^2\mathbf{r}/r - Gm_G^3r\,\mathbf{r}/6 + \ldots \qquad (12.42)$$

12.9 Value of M

We determine M by requiring it to influence galactic motions at radii of the order of 100,000 light years (the radius of the Andromeda galaxy).[62]

Transforming all distances to electron volts (eV) using

$$1 \text{ eV}^{-1} = 1.2398 \times 10^{-6} \text{ m} \qquad (12.43)$$

we find 100,000 ly (light years) ≡ 7.63×10^{26} eV^{-1} implying one of the graviton masses is approximately (order of magnitude):

$$m_G \approx 1.31 \times 10^{-27} \text{ eV} \qquad (12.44)$$

if the gravitation potential (force) terms are to modify gravity at galactic distances of the order of a hundred thousand light years.

Solving for M from

$$m_G = (2b/M)^{1/4} \qquad (12.36)$$

we find

$$M = 2b/m_G^4 = 1.61 \times 10^{163} \text{ eV}^{-2} \qquad (12.45)$$

Thus a *zero mass graviton determines gravity at short and ultra-long distances. A graviton of mass m_G affects gravity at intermediate distances.*

12.10 Intergalactic Distance Scale

We can estimate the gravitational potential of $V_{TOT}(\mathbf{r})$ in eq. 12.33 for large distances $r \approx m_A^{-1}$ of the order of many hundreds of thousands of light years, and beyond. We find

$$V_{TOT}(\mathbf{r}) \cong -G/r + [1/(128\pi a)](m_A^2 r/2 - m_A) \qquad (12.46)$$

for $r \approx m_A^{-1}$. The resultant force is

---

[62] All coupling constant values are based on data from K. A. Olive et al (Particle Data Group), Chinese Physics **C38**, 090001 (2014).

$$\mathbf{F} = -\nabla V_{TOT}(\mathbf{r})(\mathbf{r}) \sim -G\mathbf{r}/r^3 - [1/(256\pi M)]\mathbf{r}/r \qquad (12.47)$$

to good approximation since m_A sets a distance scale of the order of tens of thousands of light years causing the oscillating term to 'wash out,' and causing the $a_1 Ge^{-m_G r}/r$ term to be negligible.

Consequently we find a deeper potential, and thus a larger attractive gravitational force at inter-galactic distances, in agreement with the 33,000 galaxy survey of M. Brouwer and colleagues.

12.11 Qualitative Agreement With Gravitational Data at all Known Distances

Our results are to be compared to the MoND force of A. Balakin et al, Phys. Rev. **D70**, 064027 (2004):[63]

$$F = -\lambda Gm[M/r^2 - |\,\Pi_c\,|\, r/c^2]$$

and the vector form suggested by H-S Zhao et al, Phys. Rev. **D82**, 103001 (2010):

$$\partial\Phi/\partial\mathbf{r} = Gm\mathbf{r}/r^3 + (Gm)^{1/2}\mathbf{r}/r^2$$

Recent studies of 153 galaxies confirm the MoND discrepancy from Newtonian gravitation.[64]

The resemblance of our results in eqs. 12.42 and 12.47 to MOND estimates is clear.

We chose the value of M such that the gravity potential at distances of the order of a 100,000 light years (the radius of the Andromeda galaxy) is increased by the Yukawa like terms due to the small value of m_G.

We conclude our unified theory agrees with known gravitational data at the three distance scales.

---

[63] The constant c above is the speed of light, and M is the mass (not the M used in our lagrangian equations.)

[64] S. S. McGaugh, F. Lelli, and J. M. Schombert, arXiv: 1609.0591 (2016).

12.12 Coherence of Large Scale Structures in the Universe

We have chosen M to be such that m_G is of the order of the inverse radius of a galaxy. If we chose m_G^{-1} to be of the order of the distances associated with the recently observed[65] coherence of large scale structures, such as galaxies separated by distances of the order of 20 million light years, then M would be of the order of 25.8×10^{191} eV^{-2} and could account for large scale coherence through gravitational effects.

Then m_A would be many tens of billions of light years (parsecs) accounting, perhaps, for the alignments of supermassive black holes inside quasars over distances of billions of parsecs.[66]

12.13 Unity with the Strong Interaction

The value of a and M are also relevant to our Strong Interaction sector. In chapter 13 we show that they are consistent with chatmonium data.

12.14 Three Gravitons

The sixth order graviton dynamic equations lead to three gravitons: one with mass zero and the other two being m_G and m_A. The three gravitons generate variations from Newtonian gravity on different distance scales giving three regions of gravity.

---

[65] See J. h. Lee *et al*, The Astrophysical Journal 884, Number 2 (2019) for a study of large scale structures in the universe.

[66] D. Hutsemékers *et al*, Astronomy and Astrophysics Manuscript Number aa24631 (Septrmber, 2014) arXiv:1409.6098 (2014).

13. The Strong Interaction Sector

In this chapter[67] we describe some of the implications of the Strong Interaction sector of our theory. We will show a linear quark potential and quark confinement follows and determine the parameter a used in chapter 12.

13.1 Strong Interaction Lagragian Terms

The flat space-time Strong Interaction gauge field lagrangian terms is

$$\mathcal{L}_{SU(3)} = \text{Tr}\, \sqrt{g}\{M[\partial_\nu + ig_1(A_{SU(3)}{}^1{}_\nu + A_{SU(3)}{}^2{}_\nu)]F_{SU(3)}{}^1{}_{\sigma\mu}[\partial^\nu + i\, g_1(A_{SU(3)}{}^{1\nu} + A_{SU(3)}{}^{2\nu})]F_{SU(3)}{}^{2\sigma\mu} + aF_{SU(3)}{}^1{}_{\sigma\mu}F_{SU(3)}{}^{2\sigma\mu} - dA_{SU(3)}{}^2{}_\mu A_{SU(3)}{}^{2\mu}\} \tag{11.29}$$

Dropping the subscript $_{SU(3)}$ for added clarity and adding color fermion terms we obtain[68]

$$\mathcal{L}_{SU(3)} = \text{Tr}\,\{MD_\nu F^1{}_{\sigma\mu}D^\nu F^{2\sigma\mu} + aF^1{}_{\sigma\mu}F^{2\sigma\mu} - dA^2{}_\mu A^{2\mu}\} + \bar{\psi}[i\nabla\!\!\!/ + g_1(A\!\!\!/^1 + A\!\!\!/^2) - m]\psi \tag{13.1}$$

where (for j = 1, 2)

$$D_\nu F^j{}_{\sigma\mu} = \partial_\nu F^j{}_{\sigma\mu} + ig_1[A^1{}_\nu, F^j{}_{\sigma\mu}] + ig_1[A^2{}_\nu, F^j{}_{\sigma\mu}] \tag{13.2}$$

We should start with eq. 13.1. However, with a view towards perturbation theory which appears reasonable in view of the smallness of the strong interaction coupling constant $f^2/4\pi = 0.024$ seen below, we will abstract a quadratic expresssion in the fields from eq.

---

[67] Most of this chapter appears in Blaha (2016h) and earlier books by the author.

[68] We note the constant a, that appears in this chapter is NOT the Charmonium constant denoted $a_{cornell}$.

13.1 and then proceed to develop gluon propagators and the strong interaction potential. The 'free' Strong Interaction lagrangian that we use is

$$\mathcal{L}_{SU(3)F} = \mathrm{Tr}\{M D_{F\nu} F_F{}^{1a}{}_{\sigma\mu} D_F{}^{\nu} F_F{}^{2a\sigma\mu} + aF_F{}^{1a}{}_{\sigma\mu} F_F{}^{2a\sigma\mu} - dA^{2a}{}_{\mu} A^{2a\mu}\} + \\ + \bar{\psi}[i\partial\!\!\!/ + f(A\!\!\!/^1 + A\!\!\!/^2) - m]\psi \qquad (13.3)$$

where

$$F^{1a}{}_{\mu\varkappa} = \partial A^{1a}{}_{\mu}/\partial x^{\varkappa} - \partial A^{1a}{}_{\varkappa}/\partial x^{\mu} \qquad (13.4)$$
$$F^{2a}{}_{\mu\varkappa} = \partial A^{2a}{}_{\mu}/\partial x^{\varkappa} - \partial A^{2a}{}_{\varkappa}/\partial x^{\mu}$$

and

$$D_{F\nu} = \partial_\nu \qquad (13.5)$$

The conjugate momenta to $A^{1a}{}_{\mu}$ and $A^{2a}{}_{\mu}$ are respectively

$$\pi^{1a}{}_{\mu} = \partial\mathcal{L}_{SU(3)F}/(\partial A^{1a}{}_{\mu}/\partial t) = aF_F{}^{2a\mu t} \qquad (13.6)$$
$$\pi^{2a}{}_{\mu} = \partial\mathcal{L}_{SU(3)F}/(\partial A^{2a}{}_{\mu}/\partial t) = aF_F{}^{1a\mu t}$$

The non-zero, equal time commutation relations are

$$[\pi^{ia}{}_{\mu}(\mathbf{x}, t), A^{jb}{}_{\nu}(\mathbf{y}, t)] = i(1 - \delta^{ij})\delta^{ab}\delta^{G(\mu\nu)}(\mathbf{x} - \mathbf{y}) \qquad (13.7)$$

where i and j label the fields, and $G(\mu\nu)$ indicates the gauge[69] G and the associated index expressions, with

$$\delta^{G(\mu\nu)}(\mathbf{x} - \mathbf{y}) = \int d^4k \exp(-ik\cdot x) b^G{}_{\mu\nu}(k)/(2\pi)^4 \qquad (13.8)$$

where $b^G{}_{\mu\nu}(k)$ is a polynomial in k with a δ-function factor restricting the integration over k.

---

[69] Not the gravitational coupling constant.

13.1.1 Dynamical Equations

After performing partial integrations on the $MD_{F\nu}F_F{}^1{}_{\sigma\mu}D_F{}^\nu F_F{}^{2\sigma\mu}$ term (and discarding surface terms at 'infinity') the Euler-Lagrange dynamical equations (in the Landau gauge) due to independent variations with respect to $A^{1a}{}_\mu$ is

$$2M\Box^2 A^{2a}{}_\mu - 2a\Box A^{2a}{}_\mu = -g_1\bar{\psi}T^a\gamma_\mu\psi \tag{13.9}$$

and, with respect to $A^{2a}{}_\mu$, is

$$2M\Box^2 A^{1a}{}_\mu - 2a\Box A^{1a}{}_\mu - 2dA^{2a}{}_\mu = -g_1\bar{\psi}T^a\gamma_\mu\psi \tag{13.10}$$

where T^a is an SU(3) generator. Subtracting the equations we find

$$2M\Box^2 A^{1a}{}_\mu - 2a\Box A^{1a}{}_\mu - 2M\Box^2 A^{2a}{}_\mu + 2a\Box A^{2a}{}_\mu - 2dA^{2a}{}_\mu = 0$$

or

$$A^{2a}{}_\mu = [2M\Box^2 - 2a\Box + 2d]^{-1}[2M\Box^2 A^{1a}{}_\mu - 2a\Box A^{1a}{}_\mu] \tag{13.11}$$

with the result

$$\{2M\Box^2 - 2a\Box - 2d[2M\Box^2 - 2a\Box + 2d]^{-1}[2M\Box^2 - 2a\Box]\}A^{1a}{}_\mu = -g_1\bar{\psi}T^a\gamma_\mu\psi$$

or

$$\{2M\Box^2 - 2a\Box - 2d[2M\Box^2 - 2a\Box + 2d]^{-1}[2M\Box^2 - 2a\Box]\}A^{1a}{}_\mu = -g_1\bar{\psi}T^a\gamma_\mu\psi$$

$$\{2M\Box^2 - 2a\Box - 2d + 4d^2[2M\Box^2 - 2a\Box + 2d]^{-1}\}A^{1a}{}_\mu = -g_1\bar{\psi}T^a\gamma_\mu\psi \tag{13.12}$$

Eq. 13.12 leads to the Principal Value (Feynman) propagator:

$$\begin{aligned} D^{11}{}_{\mu\nu}(x-y) &= P\, -i\langle 0|T(A^1{}_\mu(x), A^1{}_\nu(y)|0\rangle \\ &= P\int d^4k\, \exp(-ik\cdot x) b_{\mu\nu}(k) D_1(k)/(2\pi)^4 \end{aligned} \tag{13.13}$$

where $b_{\mu\nu}(k)$ is a Landau gauge polynomial in k, and

$$D_1(k) = \{2Mk^4 - 2ak^2 - 2d + 4d^2[2Mk^4 - 2ak^2 + 2d]^{-1}\}^{-1}$$
$$= [2Mk^4 - 2ak^2 + 2d](2Mk^4 - 2ak^2)^{-2}$$

Thus

$$D^{11}{}_{\mu\nu}(x-y) = P\int d^4k \exp(-ik\cdot x)b_{\mu\nu}(k)[2Mk^4 - 2ak^2 + 2d]/[(2\pi)^4(2Mk^4 - 2ak^2)^2]$$
$$= P\int d^4k \exp(-ik\cdot x)b_{\mu\nu}(k)[2Mk^4 - 2ak^2 + 2d]/[(2\pi)^4k^4(2Mk^2 - 2a)^2] \quad (13.14)$$

indicating a linear potential r term as well as terms of lower powers in r and Yukawa-like terms with a mass of $(a/M)^{1/2}$. We will describe the resulting effective Strong Interaction potential in more detail later.

Eq. 13.9 leads to the other propagator:

$$D^{12}{}_{\mu\nu}(x-y) = P -i\langle 0|T(A^1{}_\mu(x), A^2{}_\nu(y)|0\rangle$$
$$= P\int d^4k \exp(-ik\cdot x)b_{\mu\nu}(k)D_2(k)/(2\pi)^4 \quad (13.15)$$

where

$$D_2(k) = [2Mk^4 - 2ak^2]^{-1} \quad (13.16)$$

Thus

$$D^{12}{}_{\mu\nu}(x-y) = P\int d^4k \exp(-ik\cdot x)b_{\mu\nu}(k)/[(2\pi)^4k^2(2Mk^2 - 2a)] \quad (13.17)$$

indicating a 1/r potential term plus a Yukawa term with a mass of $(a/M)^{1/2}$.

Due to the form of the interaction with quarks the total effective gluon interaction between quarks is

$$D^{tot}{}_{\mu\nu}(x-y) = D^{11}{}_{\mu\nu}(x-y) + 2D^{12}{}_{\mu\nu}(x-y)$$
$$= P\int d^4k \exp(-ik\cdot x)b_{\mu\nu}(k)\{[3Mk^4 - 3ak^2 + 2d]/[(2\pi)^4k^4(2Mk^2 - 2a)^2]\} \quad (13.18)$$

13.2 Strong Interaction Potential

Eq. 13.18 leads to the form of the total Strong Interaction potential. We note that the $\mu = \nu = 0$ part of the Feynman propagator for transverse gluons has the form:

$$D^{tot}{}_{00}(x - y) = \ldots \; - \int d^4k \, V_{SI}(\mathbf{k}) \exp(-ik\cdot(x - y))/(2\pi)^4 = \ldots + V_{SI}(\mathbf{x} - \mathbf{y})\delta(x_0 - y_0) \tag{13.19}$$

where

$$V_{SI}(x) = - \int d^3k \exp(ik\cdot x) V_{SI}(k)/(2\pi)^3$$

with

$$V_{SI}(\mathbf{k}) = [3M\mathbf{k}^4 + 3a\mathbf{k}^2 + 2d]/[\mathbf{k}^4(2M\mathbf{k}^2 + 2a)^2] \tag{13.20}$$
$$= (2M)^{-2}\{2d(M/a)^2/\,\mathbf{k}^4 + [3a(M/a)^2 - 4d(M/a)^3]/\mathbf{k}^2$$
$$+ 2d(M/a)^2/(\mathbf{k}^2 + a/M)^2 + [-3a(M/a)^2 + 4d(M/a)^3]/(\mathbf{k}^2 + a/M)\}$$

Letting

$$m_{SI} = (a/M)^{1/2} \equiv m_A \tag{13.21}$$

by eq. 12.35 we find

$$V_{SI}(\mathbf{k}) = (2a)^{-2}\{2d/\mathbf{k}^4 + [3a - 4dm_{SI}{}^{-2}]/\mathbf{k}^2 + 2d/(\mathbf{k}^2 + m_{SI}{}^2)^2 + [4dm_{SI}{}^{-2} - 3a]/(\mathbf{k}^2 + m_{SI}{}^2)\} \tag{13.22}$$

The constant, a, is dimensionless and of order 1. The constant M has the dimension of inverse mass squared. We anticipate M will be extremely large resulting in a very small gluon mass m_{SI}.

There are massless gluon terms that generate color confinement reducing the impact of the massive gluon terms to a negligible effect outside hadronic regions.

We also note that the value of the inverse of the large distanve graviton mass is of the order of the average galactic radius (the average galactic radius is large) and thus generate a Modified Newtonian potential (MoND) as seen in chapter 12.

Substituting eq. 13.26 we obtain a sum of massless and Yukawa-like potentials. A Yukawa potential has the expansion:

$$V_Y(\mathbf{r}) = \int d^3k \exp(i\mathbf{k}\cdot\mathbf{r})/[(2\pi)^3(\mathbf{k}^2 + m^2)] = \exp(-mr)/[4\pi r] \tag{13.23}$$

Thus we obtain

$$V_{SI}(\mathbf{r}) = -(2a)^{-2}\{2d(dV_Y(\mathbf{r})/dm^2)|_{m=0} + [3a - 4dm_{SI}^{-2}]/(4\pi r) - 2d(dV_Y(\mathbf{r})/dm^2|_{m=m_{SI}}) + [4dm_{SI}^{-2} - 3a]V_Y(\mathbf{r})|_{m=m_{SI}}\} \tag{13.24}$$

with the form

$$V_{SI}(\mathbf{r}) = \alpha_1 r + \alpha_2/r + \alpha_3 e^{-m_{SI}r}/(4\pi m_{SI}) + \alpha_4 e^{-m_{SI}r}/(4\pi r) \tag{13.25}$$

where the constants α_i are:

$$\alpha_1 = -(2a)^{-2}d/(8\pi) \quad \text{(up to an infrared divergent constant)} \tag{13.26}$$
$$\alpha_2 = (2a)^{-2}[3a - 4dm_{SI}^{-2}]/(4\pi)$$
$$\alpha_3 = -(2a)^{-2}d$$
$$\alpha_4 = (2a)^{-2}[4dm_{SI}^{-2} - 3a]$$

Thus we find the form of the potential of eq. 13.24 is 1/r and linear terms plus Yukawa-like terms with very small mass m_{SI} – perhaps near zero. As a result the relevant form of the effective potential is

$$V(r) = -2g^2/r + g^2\lambda^2 r \tag{13.27}$$

where

$$-2g^2 = \alpha_2 = (2a)^{-2}[3a - 4dm_{SI}^{-2}]/(4\pi) \tag{13.28}$$
$$g^2\lambda^2 = \alpha_1 = (2a)^{-2}d/(16\pi)$$

13.3 Charmonium and the Strong Interaction

In 1974 a bound state of a charmed and an anti-charmed quark was discovered by two experimental groups. Since charmed quarks are quite massive theoretical attempts were made to understand the charmed quark bound states within the framework of non-relativistic quantum mechanics. The "Cornell group" developed a fairly satisfactory[70] charmed quark bound state spectrum in 1974-5 using a combination

[70] As did a Harvard group.

of a linear and inverse 1/r potential as the strong interaction. In a recent fit[71] they gave the potential energy:

$$V(r) = -\kappa/r + r/a_{cornell}^2 \tag{13.29}$$

where $\kappa = 0.61$, $a_{cornell} = 2.38\ \text{GeV}^{-1}$ and the charmed quark mass was 1.84 GeV. Based on eq. 13.29 we find

$$g^2 = \kappa/2$$
$$\lambda = (2/(\kappa a_{cornell}^{\ 2}))^{1/2}$$

giving

$$g = 0.552$$
$$\lambda = 0.761\ \text{GeV}^2$$

The constants m_G and b, and thus M, are known from chapter 12. The mass $m_A = m_{SI}$ is given by eq. 13.21. Solving for a and d:

$$-2g^2 = (2a)^{-2}(3a - 4dm_{SI}^{\ -2})/(4\pi) = (3a - 4dm_{SI}^{\ -2})/(16\pi a^2) = -\kappa = -0.61 \tag{13.30}$$

$$g^2\lambda^2 = d/(64\pi a^2) = 1/a_{cornell}^{\ 2} = 1/2.38^2\ \text{GeV}^2 = 0.177\ \text{GeV}^2$$

we find

$$a = [16(0.177 \times 10^{18})\pi M - 3/16]/(0.61\pi)$$
$$= 7.47 \times 10^{181}$$
$$d = 1.99 \times 10^{383}\ \text{eV}^2$$

using

$$M \sim 1.61 \times 10^{163}\ \text{eV}^{-2} \tag{12.37}$$

Consequently the masses of chapter 12 are

---

[71] E. J. Eichten, K. Lane, and C. Quigg, arXiv:hep-ph/ 0206018 (2002). See this paper for references to earlier work by the "Cornell group" and the "Harvard group" as well as papers by other researchers. Their results appear in the Particle Data Group (PDG) tables.

$$m_G = (2b/M)^{1/4} = 1.31 \times 10^{-27}\ \text{eV} \tag{12.34}$$

and

$$m_A = m_{SI} = (a/M)^{1/2} \cong 1.94 \times 10^{-101}\ \text{eV} \tag{13.31}$$

13.4 The Origin of the Linear Potential

The linear potential appears to have originated in a suggestion of Feynman in the Spring of 1974. This author proposed[72] a non-Abelian gauge quantum field theory, which yielded a linear potential. These papers, which had 4th order dynamic equations for the gauge fields, showed how to avoid the problems previously associated with higher derivative theories by using principal-value gauge field propagators that were similar in concept to the action-at-a-distance propagators used by Feynman and Wheeler in the late 1940's to formulate action-at-a-distance Quantum Electrodynamics.

Thus a non-Abelian quantum field theory of the strong interaction yielding a linear potential was created. In parallel with this development, Kenneth Wilson (later a Nobelist) was developing lattice gauge theory. Because lattice lines focus the field of gauge boson, lattice gauge theory exhibited a linear potential as well between quarks. Thus it offered an alternative to our gauge theory. However, the linearity of the lattice potential was "built-in" by the lattice theory formulation and thus was an artifact of the lattice formulation. This approach, and other proposed approaches, all share the problem that the linear potential that they produce cannot be proven to truly be a consequence. Rather the linear potential is the "likely result."

On the other hand our higher dimensional theory produces the linear potential if the standard rules of quantum field theory are followed with the proviso that gauge field propagators are principal-value propagators.

This author had several discussions with Professor Wilson in late 1974 in the author's office and while walking to lunch at the Cornell Faculty Club. Professor Wilson proposed possible flaws in the author's theory on almost a daily basis. The author was able to show these suggested flaws were not flaws but physically acceptable.

---

[72] S. Blaha, Phys. Rev. D**10**, 4268 (July, 1974) and Phys. Rev. D**11**, 2921 (December, 1974). These papers appeared before the charmonium calculations of the Cornell and Harvard groups in 1975.

The final discussion with Wilson ended with Wilson stating words to the effect, "Your theory may be a correct phenomenological approximation to my theory of the strong interaction and quark confinement. But my theory is the correct one. Your theory is only a phenomenology." In the forty plus years since this concluding discussion no one has proved that the conventional strong interaction theory truly has a linear potential and quark confinement although some approximations suggest it does.

In the absence of a demonstration of a linear potential in the standard strong interaction model we suggest our theory is a viable alternative. Since the linear potential appears to fairly successfully describe much of the charmonium spectrum we feel our theory with its explicit derivation of a linear potential is worthy of interest – especially because it is in agreement with experiment as far as we know. *An experimentally completely correct phenomenology is a theory*

13.5 Numeric Constants of the Gravitational and Strong Sectors

The numeric constants appearing in

$$\mathcal{L}_G = \sqrt{g}[MD_\nu R'^1{}_{G\sigma\mu}D^\nu R'^2{}_G{}^{\sigma\mu} + aR'^1{}_{G\sigma\mu}R'^2{}_G{}^{\sigma\mu} + bH + cg^{\sigma\mu}g^2{}_{\sigma\mu} + eg^{2\sigma\mu}g^2{}_{\sigma\mu}]$$

and

$$\mathcal{L}_{SU(3)} = \text{Tr}\,\sqrt{g}\{M[\partial_\nu + ig_1(A_{SU(3)}{}^1{}_\nu + A_{SU(3)}{}^2{}_\nu)]F_{SU(3)}{}^1{}_{\sigma\mu}[\partial^\nu + i\,g_1(A_{SU(3)}{}^{1\nu} + A_{SU(3)}{}^{2\nu})]F_{SU(3)}{}^{2\sigma\mu} + aF_{SU(3)}{}^1{}_{\sigma\mu}F_{SU(3)}{}^{2\sigma\mu} - dA_{SU(3)}{}^2{}_\mu A_{SU(3)}{}^{2\mu}\}$$

are

$$M = 1.61 \times 10^{163}\ \text{eV}^{-2}$$
$$a = 7.47 \times 10^{181}$$
$$b = 2.364 \times 10^{55}\ \text{eV}^2$$
$$d = 1.99 \times 10^{383}\ \text{eV}^2$$
$$c = e = 0$$
$$g_1 = 0.552$$

They have remarkably large values. Yet they conspire to yield small constants in the gravity and Strong potentials.

Appendix 13-A. Comparison of S. Blaha, Phys. Rev. D11, 2921 (1974) and the Present Work

The eq. 13.1 lagrangian is approximately the same as eq. 17 of S. Blaha, Phys. Rev. D**11**, 2921 (1974). We now discuss the relation of the 1974 paper to the present work except for additional terms $MD_\nu F^1{}_{\sigma\mu}D^\nu F^{2\sigma\mu}$ and $[A^2{}_\varkappa, A^2{}_\mu]$; and the following changes in parameters:

$$a = -\tfrac{1}{2} \qquad\qquad d = \tfrac{1}{2}\lambda^2$$

Since that paper essentially contains a complete description of our Strong Interaction theory (modulo the additional terms) we refer the reader to it and to its predecessor paper referenced therein. There are a few additional changes required to bring the 1974-5 papers into agreement with our current theory:

1. Eq. 30 of the above referenced paper must be modified to

$$F^2{}_{\varkappa\mu} = \partial A^2{}_\mu/\partial x^\varkappa - \partial A^2{}_\varkappa/\partial x^\mu + \mathbf{ig_1[A^2{}_\varkappa, A^2{}_\mu]} + ig_1[A^1{}_\varkappa, A^2{}_\mu] + ig_1[A^2{}_\varkappa, A^1{}_\mu] \tag{13-A.1}$$

with the addition of the 'bolded' term $if[A^2{}_\varkappa, A^2{}_\mu]$. There is also a trivial change of notation of coupling constant from 'g_1' to 'f'.

2. Eqs. 6 and 18 should have the interaction term expanded to

$$g_1 A\!\!\!/^1 \rightarrow g_1(A\!\!\!/^1 + A\!\!\!/^2) \quad ``` \tag{13-A.2}$$

and similarly in eq. 20. Eqs. 38 – 41 directly show that the additional interaction term leads to a gluon propagator[73] $<A^1 + A^2, A^1 + A^2> = 2<A^1, A^2> + <A^1, A^1>$, and introduces a 1/r term in the potential part of the gluon propagator.

As a result the effective gluon propagator in the theory, **if the $Mf^2D_\nu F^1{}_{\sigma\mu}D^\nu F^{2\sigma\mu}$ term is neglected**, combines eqs. 38 and 39 to give the *short-distance*[74] gluon propagator between quarks:

$$g_{\mu\nu}\delta_{ab}P[\lambda^2/k^4 - 1/k^2] \tag{13-A.3}$$

up to a constant factor.[75]

These changes *explicitly* leads to a Strong Interaction potential of the form

$$V(r) = -2g^2/r + g^2\lambda^2 r \tag{13-A.4}$$

Naturally one can expect perturbative corrections to eq. 13.6 in higher order in f. However the apparent relative smallness of f suggests eq. 13-A.4 is a good approximation to the **short-distance,** ***inter-quark interaction.***

13-A.1 Canonical Equal Time Commutation Relations

The Euler-Lagrange equations of motion, eqs. 27 – 31 in the author's 1975 paper, are modified most significantly by the $Mf^2D_\nu F^1{}_{\sigma\mu}D^\nu F^{2\sigma\mu}$ lagrangian term in eq. 13.1. In order to use the canonical method to obtain the contributions to the equations of motion of this higher derivative term we use integration by parts and discard surface terms for eq. 13.1, as is usually done in quantum field theory.

---

[73] Eqs. 40-41 in the above referenced paper.

[74] We see that the $Mf^2D_\nu F^1{}_{\sigma\mu}D^\nu F^{2\sigma\mu}$ term will affect the short-distance behavior of the inter-quark interaction. The equivalent term in the gravitation sector influences the long-distance form of the gravity potential and leads to a MoND-like behavior. See chapter 12.

[75] This propagator is taken in Principal value to avoid potential unitarity problems. This topic is described in detail in earlier papers and books by the author.

14. Other Effects of New Interactions Between Bosons

14.1 Missing Nucleon Spin Puzzle

The estimates of nucleon spin that are obtained from parton analyses of deep inelastic electron – nucleon interactions are woefully short of the spin expected in quark models of nucleons. The missing spin has been attributed to a number of causes. However the Missing Spin Puzzle remains.

From eq. 11.28 - 11.30 it is clear that there are important new interaction terms between the Electromagnetic and Strong interaction fields. After taking traces we find

$$\mathcal{L}_{\text{intEM-S}} = -\text{Tr}\; iM\{(A_E{}^1{}_\nu + A_E{}^2{}_\nu)\, F_{SU(3)}{}^1{}_{\sigma\mu} D^\nu F_{SU(3)}{}^{2\sigma\mu} + iD_\nu F_{SU(3)}{}^1{}_{\sigma\mu}(A_E{}^{1\nu} + A_E{}^{2\nu}) F_{SU(3)}{}^{2\sigma\mu}\} \tag{14.1}$$

$\mathcal{L}_{\text{intEM-S}}$ generates a combined photon-gluon vertex insertion in gluon interactions between quarks within a hadron. Figs. 14.1 and 14.2 show two simple possible vertex insertions in a gluon line.

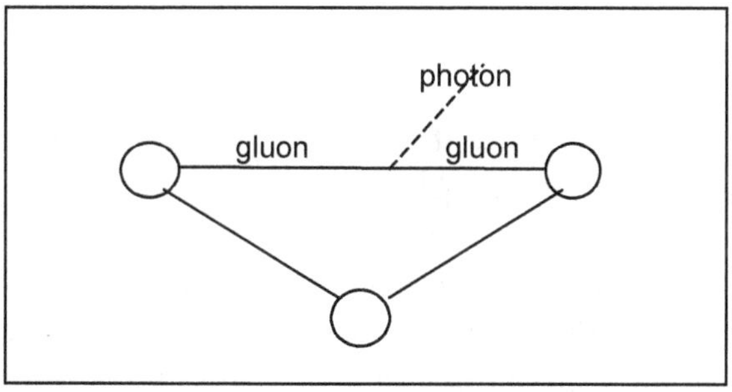

Figure 14.1 A single 'outgoing' photon vertex insertion in a gluon line. Only single gluon lines between the three quarks are displayed. This gluon-gluon-

photon interaction is possible because the photon field is an SU(1) field. Thus the diagram is one of the possibilities embedded in eq. 14.1.

The gluon line, by itself, has $1/k^4$ and $1/k^2$ momentum space propagator terms. *The insertion of the vertex in Fig.14.1 generated by* $\mathcal{L}_{intEM\text{-}S}$ *yields a combined momentum factor of* $k^3(k^4k^4)^{-1} = k^{-5}$ *which would make it (summed over all gluon lines) a significant contribution to the proton spin determination in deep inelastic electron-nucleon scattering.*[76] *The insertion of the vertex in Fig. 14.2 generated by* $\mathcal{L}_{intEM\text{-}S}$ *yields a combined momentum factor of* $k^2(k^4k^4)^{-1} = k^{-6}$ *which may have a less significant effect.*

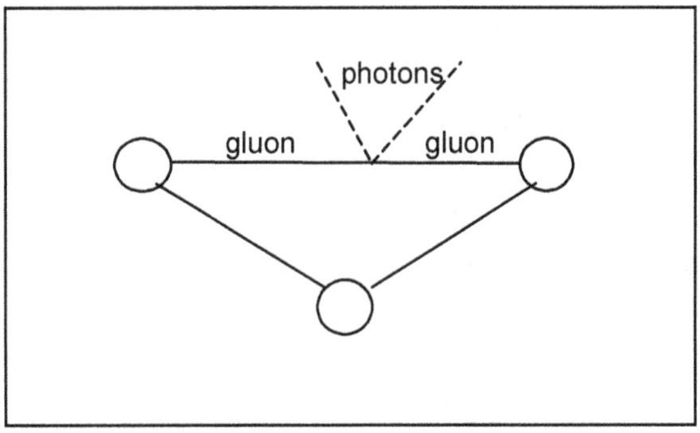

Figure 14.2 An 'outgoing' two photon vertex insertion in a gluon line. Only single gluon lines between the three quarks are displayed.

Thus our unified theory may help solve the nucleon spin puzzle. The interactions in Figs. 14.1 and 14.2 introduce a new direct connection between photons and spin one gluons. Thus their contributions to the summations of proton spin interactions in parton

[76] See C. A. Aidala, S. D. Bass, D. Hasch, and G. K. Mallot, arXiv: 1209.2803v2 (2013) and references therein for a review of the 'missing' nucleon spin puzzle.

models may account for the 'missing' two-thirds of proton spin. Our unified theory has a new gluon-photon interaction that is not found in the conventional Standard Model.

These considerations apply also to the Dark fermion sector, and also apply layer by layer to the four normal and Dark layers of fermions.

14.2 Vector Meson Dominance (VMD) Due to New Gluon – Electromagnetic Interactions

Vector Meson Dominance (VMD) is a model describing the hadronic part of the physical photon consisting of a purely electromagnetic part and a hadronic part.[77] The hadronic part contains of light vector mesons: ρ, ω, and φ. Consequently physical photon-hadron interactions have a hadonic part as well as a conventional electromagnetic photon part. Features of the interaction include a greater intensity than a purely photon part and a similarity of the physical photon interactions between protons and neutrons despite the difference in their charge structures. The physical photon thus appears to be a superposition of a purely electromagnetic photon and a vector meson.

The bosonic lagrangian term

$$\mathcal{L}_M = \mathrm{Tr}\ \sqrt{g} M D_\nu R''^{1}{}_{\sigma\mu} D^\nu R''^{2\sigma\mu} \qquad (14.2)$$

has a part with photon-gluon interactions. One of these terms has the form

$$\mathrm{AAGG} \qquad (14.3)$$

where A represents a photon and G represents a gluon, with indices and derivatives not shown. This lagrangian term (and other similar terms in eq. 14.3) has the Feynman diagram:

---

[77] J. J. Sakurai, Ann. Phs. 11 (1960).

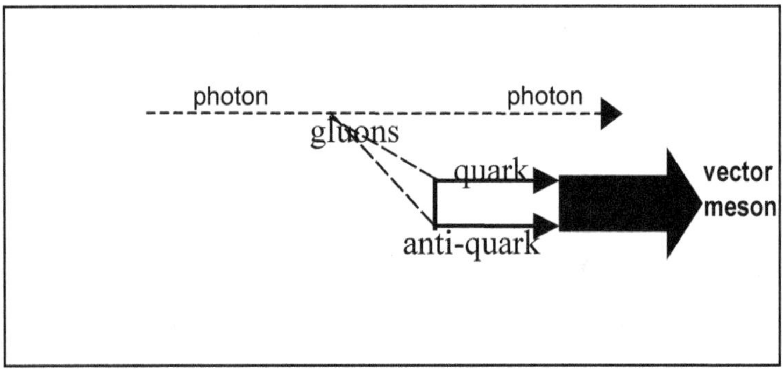

Figure 14.3 An energetic purely electromagnetic photon emitting a gluon pair that generate a quark – anti-quark pair that then combine to form a vector meson of the VMD type. The physical photon thus combines a pure electromagnetic part with a vector meson.

Thus VMD is part of the bosonic lagrangian terms in eq. 14.3.

These considerations apply also to the physical Dark photon consisting of a Dark electromagnetic photon part and a Dark hadronic part consisting of a Dark vector meson part. They also apply layer by layer to the four normal and Dark layer vector bosons.

14.2.1 Gluon – SU(2) Weak Vector Boson Interaction – A New Form of Vector Meson Dominance (VMD)

The preceding discussion of this section 14.2 also applies to SU(2) Weak Interaction vector bosons, denoted U, if one simply substitutes each of the three SU(2) Weak Interaction vector bosons above. The terms of the form

$$UUGG \tag{14.4}$$

where U represents one of the SU(2) Weak vector bosons and G represents a gluon, with indices and derivatives not shown. The types of Feynman diagrams that result are analogous to Fig. 14.4 with the photon lines replaced with U vector boson lines. Since gluons do not have electric charge the produced vector mesons are neutral.

One concludes that charged and uncharged SU(2) Weak vector bosons are each accompanied by a neutral vector meson yielding a form of VMD in the Weak Interaction sector. This applies to 'normal' and Dark Weak vector bosons in each of the four layers separately.

14.3 Layer Field – Gluon Interactions

The lagrangian in the four layer extension of eq. 11.23 indicates that there are direct interactions between gluons and the four[78] Layer group gauge fields, V_a, such as

$$V_a{}^{1i}{}_v V_a{}^{2iv} \partial_\alpha A_{bSU(3)\mu}{}^1 \partial^\alpha A_{bSU(3)}{}^{2\mu} \tag{14.5}$$

where a labels the Layer group for generation a, and b labels the specific layer[79] with other indices and derivatives not shown. These interactions result in Feynman diagrams that modify the Strong Interactions between quarks to include intermediate transitions between quarks in different layers. Fig. 14.5 shows the simplest forms of this interaction between two quarks.

---

[78] There is a separate Layer group field, labeled with subscript a, for each of the four generations comprising each layer.

[79] With implied summations over a nd b.

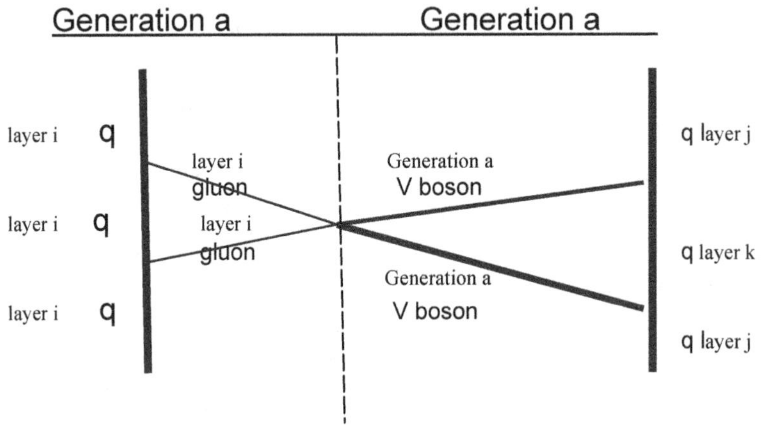

Figure 14.4 Gluon-Layer V gauge particle interaction between two quarks (one of layer i and one of layer j) with each of generation a (for one of the four generations appearing in each layer.) The layer k quark is generated by the V boson and then transformed back to layer j. The result is an interaction between quarks in different layers. Since only layer 1 is experimentally known this interaction must be quite weak and/or the masses of quarks in higher layers must be very large.

These considerations apply also to the Dark quarks sector, and also apply layer by layer to the four normal and Dark layers of quarks.

14.4 Summary of the Other New Gluon – Boson Interactions

The eq. 11.22 lagrangian term in eq. 14.3 above contains additional boson interactions. Since

$$R''^{1}{}_{\sigma\mu} = R'_{SU(3)}{}^{1}{}_{\sigma\mu} + R^{1}{}_{\sigma\mu} \qquad (11.19)$$
$$R''^{2}{}_{\sigma\mu} = R'_{SU(3)}{}^{2}{}_{\sigma\mu} + R^{2}{}_{\sigma\mu}$$

and since

$${}^a\mathbf{A}_I{}^\mu(x) = ({}^ag_1{}^a\mathbf{A}_{SU(3)}{}^\mu(x_C),\ {}^ag_2{}^a\mathbf{W}^\mu(x),\ {}^ag_3{}^a\mathbf{A}_E{}^\mu(x),\ {}^ag_4{}^a\mathbf{W}_D{}^\mu(x),\ {}^ag_5{}^a\mathbf{A}_{DE}{}^\mu(x),\ {}^ag_6{}^a\mathbf{U}^\mu(x),\ {}^ag_7{}^a\mathbf{V}^\mu(x)) \qquad (11.1)$$

$$\mathbf{C}_I^{\mu}(x) = \Sigma_{a,i}\, {}^a\mathbf{A}_{Ii}^{\mu}(x) + g_8\mathbf{A}_S^{\mu}(x) + g_\Theta\mathbf{A}_\Theta^{\mu}(x) \quad (11.2)$$

$$\begin{aligned} D_\nu V_\mu &= (\partial_\nu + iF_\nu)V_\mu - H^\sigma{}_{\nu\mu}V_\sigma \\ &= [g^\sigma{}_\mu\partial_\nu + ig^\sigma{}_\mu F_\nu - H^\sigma{}_{\nu\mu}]V_\sigma \\ &= [g^\sigma{}_\mu\partial_\nu + iD^\sigma{}_{\mu\nu}]V_\sigma \end{aligned} \quad (11.3)$$

$$F^\mu = C_I^{1\mu}(x) + \mathbf{C}_I^{2\mu}(x) + B^{1\mu} + B^{2\mu} \quad (11.4)$$

$$H^\sigma{}_{\nu\mu} = \Gamma_{GR}{}^\sigma{}_{\nu\mu} + \Gamma_{GR}{}^{2\sigma}{}_{\nu\mu} \quad (11.5)$$

$$D^\sigma{}_{\mu\nu} = g^\sigma{}_\mu F_\nu + iH^\sigma{}_{\nu\mu}$$

we see the remarkable 'new' 'normal' and Dark gluon – vector boson interactions:

Gluon – Gluon Interaction (chapter 11)
Gluon – Weak Vector Boson Interaction (described in section 14.3.1)
Gluon – Electromagnetic Interaction (Sections 14.1 and 14.3 above)
Gluon – Dark Weak Vector Boson Interaction (described in section 14.3.1)
Gluon – Dark Electromagnetic Interaction (Sections 14.1 and 14.3 above)
Gluon – Generation Group Vector Boson Interaction (Section 14.2 above)
Gluon – Layer Group Vector Boson Interaction (Section 14.4 above)

plus:

Gluon – Graviton Interaction (Possibly relevant for black holes, and quark and neutron stars)

14.5 New Graviton – Vector Boson Interactions

Eq. 14.2 also contains 'new' graviton – boson interactions:

Graviton – Weak Vector Boson Interaction
Graviton – Electromagnetic Interaction

Graviton – Dark Weak Vector Boson Interaction
Graviton – Dark Electromagnetic Interaction
Graviton – Generation Group Vector Boson Interaction
Graviton – Layer Group Vector Boson Interaction

plus an additional

Graviton – Graviton Interaction

The above new boson – graviton interactions appear to be of cosmological interest in some situations with high gravity such as black holes, and quark and neutron stars.

15. Reprise of the Derivation at this Point

15.1 Summary of Derivation Progress

We have seen that the Unified SuperStandard Theory, which contains the Standard Model, emerges in a natural way from the five axioms of chapter 1 and relevant appendices.We found its symmetry, gauge fields, fermion spectrum, and the fundamental interactions of elementary particles and gravitation.

We also were able to determine the deviations of gravity from Newtonian gravity (such as suggested by MoND theories) and to determine quark confinement features due to a confining "r" potential. The features of both these sectors are related in the Unified SuperStandard Theory through higher derivative lagrangian terms. Surprisingly the large lagrangian constants required by gravity "conspire" to give the small constants of phenomenological quark confinement models such as charmonium models.

We now turn to the remaining major issues: the value of coupling constants and the features of the Higgs particles and mechanism. Afterwards we will consider the evolution of the universe in detail (including the Hubble parameter) and suggest the universe has particle-like features. Then we will show that other universes—should they exist—will be similar to our universe.

15.2 Coupling Constants of Interactions

Our calculation of coupling constants will be based on the form of an apparently "universal" eigenvalue function[80] which, for each particular eigenvalue, has a zero at the value of the eigenvalue. We shall determine the coupling constant α_{QED} of QED exactly (as far as it is known – 13 decimal places) as well as the Weak Interaction coupling constant and the Strong Interaction coupling constant approximately.

---

[80] S. Blaha, Phys. Rev. **D9**, 2246 (1974) and Blaha (2019f) and references therein.

We begin by discussing the universal coupling constant eigenvalue function. Then we use it to calculate other coupling constants, generalize it further to another universal function, and finally calculate running coupling constants.

15.3 Higgs Mechanism

We shall show the origin of Higgs fields in in the phase of non-abelian fields. We take non-abelian fields to be complex in principle with the exception of Strong Interaction gauge fields (gluons) which are compelled to have complex values by their sources (quarks). Thus the origin of Higgs fields.

15.4 Yang-Mills Theories may Self-Determine their Coupling Constants and Higgs Couplings

The essence of our coupling constant and Higgs mechanism studies is:

Self-Determination of the Features of Non-Abelian Fields

16. Universal Coupling Constant Eigenvalue Condition

In a series of remarkable papers Johnson, Baker and Willey[81] developed a finite theory of massless QED (called JBW) without divergences if a certain function $F_1(\alpha)$ of the fine structure constant α called the eigenvalue function were zero. (A zero would imply Z_3, the divergent vacuum polarization constant of the electron, was zero.)

Adler[82] refined the discussion by pointing out that a zero of the eigenvalue function would be an essential singularity with:

$$F_1(\alpha) = 0 \qquad (16.1)$$
$$d^n F_1(\alpha)/d\alpha^n = 0$$

The calculation of the eigenvalue function was reduced by JBW to the sum of all single loop vacuum polarization diagrams with any number of free photon propagators having the general form of Fig. 16.1.

In 1973 the author[83] calculated $F_1(\alpha)$ approximately to all orders in α. *The calculation was exact to order α^2 (the first three terms).* A search for an essential singularity proved fruitless. Recently the author noticed that the vacuum polarization of the electron is manifest in experiment with the effective value of α increasing at higher energies. Thus Z_3 is not zero and has a divergent piece.

On this basis the author proposed, in a series of books in 2019, that the *appropriate* eigenvalue condition was

---

[81] Summarized in some detail in K. Johnson and M. Baker, Phys. Rev. **D8**, 1110 (1973). Also in Blaha (2019b) and (2019c).
[82] S. Adler, Phys. Rev. **D5**, 3021 (1972).
[83] S. Blaha, Phys. Rev. **D9**, 2246 (1974).

$$F_2(\alpha) = 0$$

where

$$F_2(\alpha) = F_1(\alpha) - [2/3 + \alpha/(2\pi) - (1/4)[\alpha/(2\pi)]^2] \qquad (16.2)$$

The additional terms[84] are those appearing in the *exact* low order calculation of $F_1(x)$:

$$F_{1\ \text{low order}}(\alpha) = 2/3 + \alpha\ /(2\pi) - (1/4)[\alpha/(2\pi)]^2 \qquad (16.3)$$

In terms of F_2 the renormalization constant Z_3 is

$$Z_3 = 1 + F_1(\alpha)\ln(p/\Lambda) = 1 + F_2(\alpha) + \text{divergent terms} = 1 + \text{divergent terms} \qquad (16.4)$$

The original goal of the JBW Model was to solve massless QED in a manner that made all renormalization constants either 1 or at least finite.

We modified this goal. We obtained a physically better eigenvalue function F_2 that has a zero at the known fine structure constant α. Until now we have not specified the value α that appears in the preceding equations. We now define α as a partially renormalized quantity that is related to the bare fine structure constant α_0 by

$$\alpha = \alpha_0[2/3 + \alpha_0/(2\pi) - (1/4)[\alpha_0/(2\pi)]^2] \qquad (16.5)$$

We will show in chapter 17 that the evaluation of the F_2 eigenvalue function gives the known approximate[85] physical value[86] of the fine structure constant:

$$\alpha = 0.007297352\ 5\ \ 693\ (11) \qquad (16.6)$$

The renormalized expressions appearing below are not fully finite. However the intermediate renormalized finite α is physically sensible—more so than the completely finite renormalization constants goal of the JBW Model.

---

[84] These terms do not diminish the essential singularity feature.

[85] The constant α is an irrational number.

[86] 2018 CODATA: P. J. Mohr *et al* CODATA group (2019). Thirteen places known.

The bare charge constant α_0 is known to approach ∞ at very short distances. The simplest examples of this phenomenon are the physical Coulomb scattering amplitudes and the first order change in hydrogen-like atomic energy levels.[87] Thus our modified JBW Model with a partial renormalization conforms to physical reality:

$$Z_3 = 1 + \{\alpha F_2(\alpha) + \alpha[2/3 + \alpha/(2\pi) - (1/4)[\alpha/(2\pi)]^2]\}\ln(p/\Lambda)$$
$$= 1 + \alpha\{2/3 + \alpha/(2\pi) - (1/4)[\alpha/(2\pi)]^2\}\ln(p/\Lambda) \qquad (16.7)$$

at α = the physical fine structure constant where $F_2(\alpha) = 0$.

Our approximate 1973 solution, which summed one loop pieces of the vacuum polarization yielded the algebraic equations, is:[88]

$$A_1 = (g + 1)(1 - 2g^2)/[(g + 2)(g - 1)] \qquad (16.8)$$

$$A_2 = [8g^2(2g + 1) - (2g^3 + 2g^2 + g - 2)(g^2 + 2g + 2)]/[2(g^2 - 1)(g^2 - 4)]$$

$$A_3 = -2(1 + 3g + 6g^2 + 2g^3)/[g(g + 1)]$$

$$A_4 = -(g + 2)(1 + 5g + 6g^2 + 2g^3)/[g(g^2 - 1)] - 1/(g + 1)$$

$$\psi = [gA_3 - (4 + 2g)A_1]/[(4 + 2g)A_2 - g A_4]$$

$$(\alpha/2\pi) = [gA_4 - (4 + 2g)A_2]/(A_4A_1 - A_2A_3)$$

$$F_1(g) = (2/3)(1 - 3g^2/2 - g^3) - (\alpha/4\pi)[(2 + 4g + 4g^2)(g - 2) + \alpha\psi g^3]/[(g^2 - 1)(g - 2) +$$
$$+ \alpha(2 + 4g + 4g^2)(g - 2) + \alpha\psi g^3]$$

expressed as a function[89] of g (the power of the divergent factor p/Λ) with ψ specifying the gauge, and with the renormalization definitions

---

[8787] See E. A. Ueling, Phys. Rev. **48**, 55 (1935) and R. Serber, Phys. Rev. **48**, 49 (1935).

[88] Blaha *op. cit.*

$$\Gamma_\mu(p) = f(\gamma_\mu + 2g\gamma\cdot pp_\mu/p^2)(p/\Lambda)^{2g} \quad (16.9)$$

$$S_F = [f\gamma\cdot p(p/\Lambda)^{2g}]^{-1} \quad (16.10)$$

$$\Gamma_{\mu\alpha}(p) = (f_3/p^2)(\gamma\cdot p\gamma_\mu\gamma_\alpha - \gamma_\alpha\gamma_\mu\gamma\cdot p)(p/\Lambda)^{2g} \quad (16.11)$$

and

$$F_1 = (2/3)(1 - 3g^2/2 - g^3) - f_3/f \quad (16.12)$$

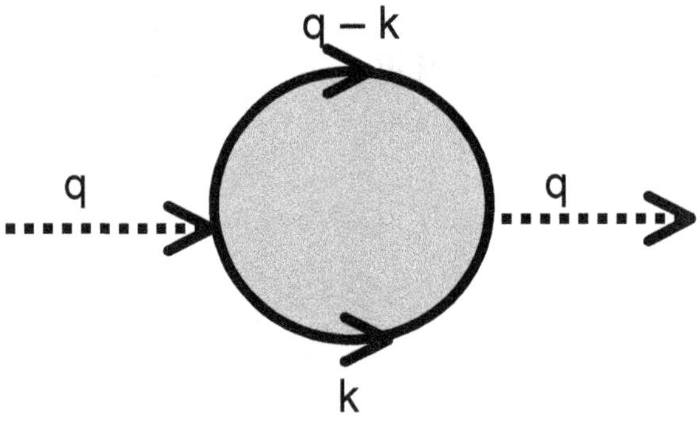

Figure 16.1 One loop vacuum polarization Feynman diagrams with internal free photon propagators.

We thus have an expression for the eigenvalue function F_2 within the framework of massless QED. We shall see that the eigenvalue function generalizes to all Standard Model gauge interactions. Later we shall also see that it applies to the expansion and contractions of universes upon the introduction of a Dark Energy gauge interaction for universes.

---

[89] The solution for the eigenvalue function is clearly best expressed in terms of the g factor in the exponents of the divergent renormalization factors. We use $F_1(g)$ and $F_1(\alpha(g))$ interchangeably.

16.1 A Universal Eigenvalue Function

In Blaha (2019b) we generalized eqs. 16.8 to a "universal" eigenvalue function F_2 to include the Weak interaction, the Strong interaction, and other interaction coupling constants by inserting an interaction specific factor in the α_G equation:

$$(\alpha_G/2\pi) = c_G^{-1}[gA_4 - (4 + 2g)A_2]/(A_4A_1 - A_2A_3) \qquad (16.13)$$

For a non-abelian group we set

$$c_G^{-1} = [(11/3)C_{ad} - 2C_f/3]/(16\pi)^3 \qquad (16.14)$$

where C_{ad} is the dimension of the fundamental representation of the group and C_f is the number of fermions (fermion flavor) of the interaction.

We also extend it to the universe Dark Energy interaction in Blaha (2019c).

In chapter 17 we will see that we obtain the exact QED Fine Structure Constant to the known 13 place accuracy with $F_2 \cong 0$. Thus our "approximate" eigenvalue condition $F_2(\alpha) = 0$ appears to be remarkably accurate. It generalizes directly in chapters 18 and 19 to the Standard Model interactions and the universe scale factor.

Blaha (2019b) describes the universal eigenvalue function F_2 in detail including numerous plots. It also proposes a generalization of F_2 to an "exact" tangent form.

16.2 Generalization to All Non-Abelian Interactions

Our QED calculation of α had no free (adjustable) parameters unlike other attempts in the past. It also is totally based on Quantum Field Theory. The calculations of the non-abelian coupling constants seen later also have ***no*** *free (adjustable) parameters.*

Thus, in our view, coupling constant eigenvalue functions depend only on inherent perturbation theory based on dynamics. Coupling constant values cannot be "tweaked" to their known values by adjusting input parameters.

The ability of our 1973 calculation of the JBW eigenvalue function to generalize to non-abelian gauge theories, together with the new insights into understanding of the precise method to obtain its "fine structure constant" eigenvalues, is also encouraging. It opens the possibility that the SuperStandard Theory (and the Standard Model within it) has within itself the mechanism for determining the constants appearing within it. It raises the hope that a similar self-determination mechanism may also exist within the theory to determine the masses appearing in the Higgs particles sector of the theory.

16.3 The Standard Model Exists in All Universes

Since we will see in the Chapters following chapter 23 that 4D universe(s) have the same evolutionary pattern of expansion and contraction, and since the Standard Model[90] *coupling constants (which govern all macroscopic and chemical interactions) are determined internally within quantum field theory we can assert that the Physics, Chemistry, and Biology of all 4D universes are the same.*

---

[90] One could suggest that the Standard Models of other universes are different. However the simplicity of the group structure would argue otherwise—as would consideration of the case of colliding universes.

17. QED α Eigenvalue Calculation

We have examined the values of the quantities in eq. 16.8 looking for an essential singularity (eq. 16.2) or its approximation. Fig. 17.1 below plots $F_2(\alpha)$ as a function of g. It displays a "flat region." While essential singularities usually are thought to imply a transcendental function such as exp(-1/α), a constant function with value zero fulfills the essential singularity conditions in eq. 16.1. Therefore we take the "flat region" to indicate an essential singularity.

Fig. 17.2 shows a "close up" of the flat region[91] where F_2 is approximately zero. Upon close numeric analysis we find the results in Tables 17.1 and 17.2.

g =	-0.0005805369 0000	-0.0005805369 1948	-0.0005805369 5000
α =	0.007297352	***0.0072973525693***	0.007297353
$F_2\times10^{10}$ =	3.26316 06817671	3.26316 025452474	3.26316 134861337

Table 17.1 Values of g, α and $F_2(\alpha) \times 10^{10}$. F_2 is very close to zero for the displayed range of values and throughout the flat region. F_2 has a local **minimum** at precisely the known value of α = 0.0072973525693 (11).

g =	-0.00058053700	-0.00058053705	-0.00058053710
α =	0.007297354	0.007297354	0.007297355
$F_2\times 10^{10}$ =	3.26316 299072544	3.26316 29663526	3.26316 408259273

Table 17.2 Other neighboring values of g, α and $F_2(\alpha) \times 10^{10}$ in the flat region *away* from g = 0 (where our approximate F_2 is exactly zero.) F_2 is very close to zero for the displayed range of values and throughout the flat region.

[91] These figures appeared in Blaha (2019a) and (2019b).

Thus we have a very good approximation with $F_2(\alpha) \cong 0$ at the experimentally known value of α that is exact to 13 places with a minimum in $F_2(\alpha)$ as anticipated.

F_2 is nearly zero, as are its derivatives, at the physical Fine Structure Constant. It closely approximates a trivial essential singularity of constant value zero in a neighborhood of the singularity.

Note $F_2(\alpha = 0) = 0$ as well. This zero can be viewed as a type of singularity. If QED could transition from positive α to negative α then it would lead to a catastrophe since like charges would then attract.[92,93]

It is extremely important to note the calculation is strictly QED. Thus α is space and time independent, and Anthropic due to QED primarily.

---

[92] Freeman Dyson has speculated on this possibility.

[93] $F_2(\alpha)$ may have more than one zero. One of the zeroes is at the value of the Fine Structure Constant as we show.

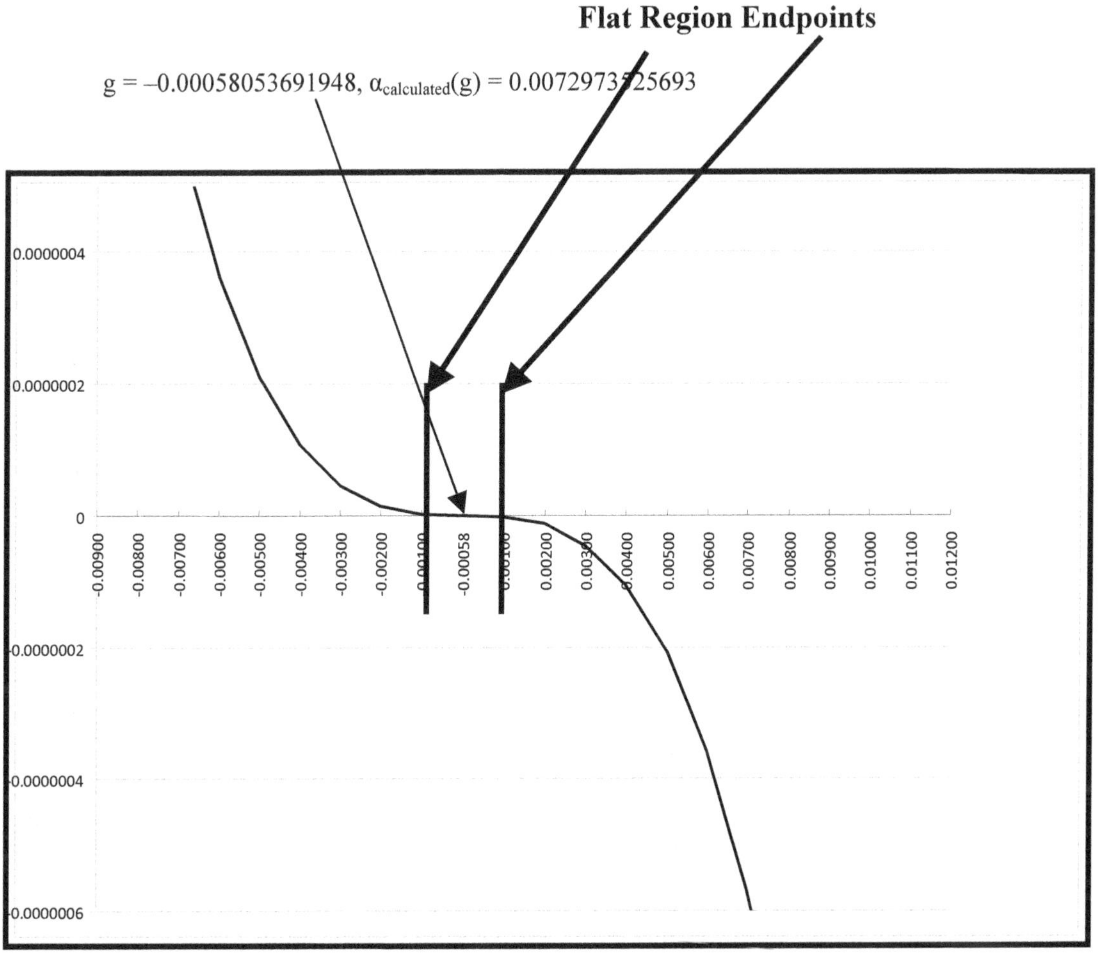

Figure 17.1 Close up plot of our eigenvalue function $F_2(g)$ (vertical axis) vs. g.

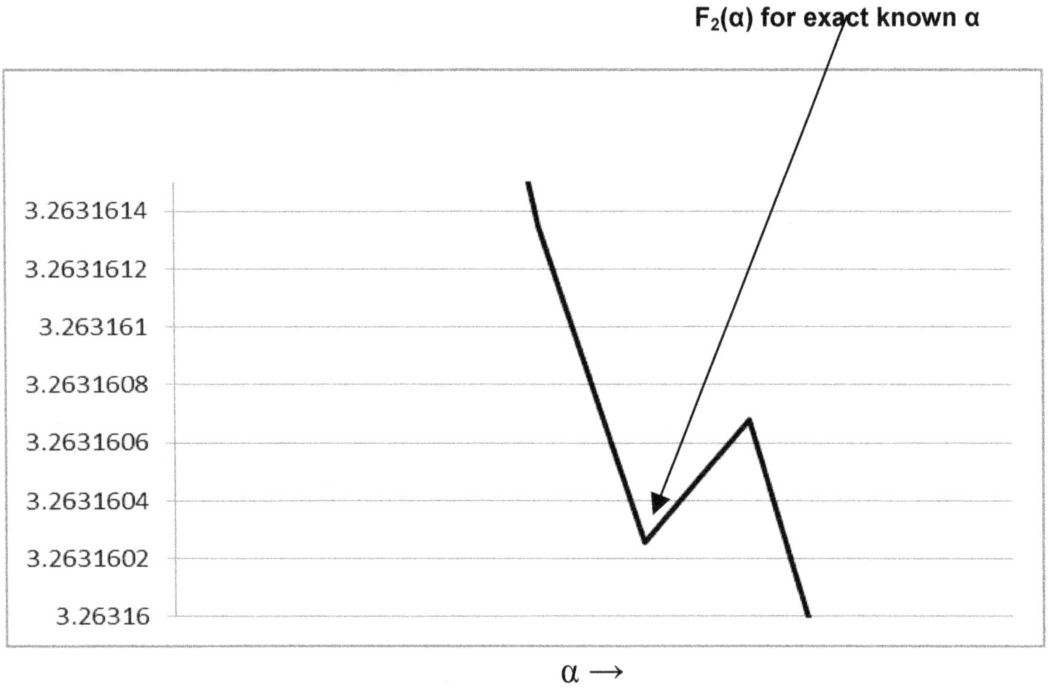

Figure 17.2 Detailed closeup plot of $F_2(\alpha) \times 10^{10}$ data in Tables 17.1 and 17.2. The local minimum of $F_2(\alpha) \times 10^{10}$ at g = -0.00058053691948 corresponds to the exact known value of α = 0.0072973525693.

17.1 Approximate Formulas for α in Terms of π and e

Our success in exactly determining α is very encouraging and suggests our approximate $F_2(\alpha)$ calculation in QED may have captured the essence of the full eigenvalue function. However it is of interest to consider α from a different perspective—approximate formulas based on fundamental mathematical constants such as π and e. Speculations of simple formulas for α have been considered for over 100 years.

We shall develop two expressions for α based on

$$e = 2.718281828$$
$$\pi = 3.141592654$$

A remarkably simple expression, yet quite accurate, is

$$\alpha = 1/(16\pi e) \quad (17.1)$$
$$= 0.007318729$$

It is accurate to three places.

A better formula, to which the above is an approximation, is

$$\alpha = 1/[16\pi e(1 + \alpha/e)] \quad (17.2)$$

Solving for α we obtain

$$\alpha = (-e + (e^2 + 1/4\pi)^{1/2})/2 \quad (17.3)$$
$$= 0.007299129$$

giving five place accuracy compared to the known value:

.

$$\alpha = 0.0072973525693\ (11)$$

18. Non-Abelian Interactions "Fine Structure Constants"

The Unified SuperStandard Theory has non-abelian interactions that go beyond QED by having group transformations that necessitate the introduction of cubic and quartic terms in the lagrangian of the theory. The calculation of non-abelian coupling constants have until now been limited to low order perturbative running coupling constant calculations.[94]

It would have been better to have calculations of coupling constants to all orders—even if only approximately—as we successfully saw in massless QED earlier. In this chapter and the following chapter we will attempt to calculate coupling constants in a manner analogous to that of earlier chapters for QED. Our attempt will be at very high energies where we can assume that all particle masses are negligible. We will make a number of assumptions which are not unreasonable:

1. We will assume that infrared divergences are not relevant.

2. We will assume that cubic and quartic Yang-Mills couplings can be neglected reducing Yang-Mills fields to multiplets of QED-like fields.

3. We will assume that each non-abelian interaction can be considered independent of all other interactions.

4. We will assume that each interacting field of a group is subject to the same vacuum polarization renormalization. This renormalization is assumed to be similar to that of JBW vacuum polarization.

---

[94] See, for example, H. Georgi, H. R. Quinn, S. Weinberg, Phys. Rev. Lett. **33**, 451 (1974).

5. Thus we assume that an eigenvalue function exists for each non-abelian interaction that results from a single logarithmic "divergence" in the vacuum polarization. This divergence stems from one fermion loop vacuum polarization diagrams summed over all fermion species of the fundamental representation of its group.

The justification for this series of assumptions is the similarity of the contributions of the vacuum polarization diagrams if cubic and quartic non-abelian interactions are neglected. The calculations of each Feynman diagram have the same characteristics of massless QED momentum space integrals up to an overall interaction constant factor in each order of perturbation theory.

Based on these assumptions we proceed to calculate non-abelian eigenvalue functions for SU(2), SU(3), and SU(4) in the following chapter.

We do make one further assumption we assume that the coupling constant for each interaction has the factor

$$c_G^{-1} = [(11/3)C_{ad} - 2C_f/3]/(16\pi)^3 \qquad (18.1)$$

to all orders where C_{ad} is the dimension of the fundamental representation of the non-abelian group and C_f is the number of fermions (fermion flavor) of the interaction.

The eigenvalue function F_1 for an interaction with coupling constant α can be expressed as a power series in α:

$$F_1(\alpha_G) = \sum_n a_n(\alpha_g)^n = \sum_n a_n(c_G\alpha)^n \qquad (18.2)$$

under the assumption that the interaction group constant is approximately c_G to all orders in n where the QED eigenvalue function has the form

$$F_1(\alpha) = \sum_n a_n\alpha^n \qquad (18.3)$$

Since non-abelian interactions are known to be asymptotically free—with coupling constants becoming finite at ultra-short distances we will not need to do intermediate renormalization such as we did for electric charge. Thus we will use F_1 *as the eigenfunction.*

Having established the framework for our approximate calculations of the eigenvalue functions and their eigenvalues we will proceed to calculate them in the next chapter for SU(2), SU(3) and SU(4).

19. Non-Abelian "Fine Structure Constant" Eigenvalue Conditions

19.1 Coupling Constants for Non-Abelian Interactions

The vector interactions and coupling constants of the Unified Superstandard Model are:[95]

- The strong interaction coupling constant[96] $g_S = 1.22$
- The Weak SU(2) coupling constant $g_W = 0.619$
- The Electromagnetic U(1) coupling constant $e = g_E = 0.303$
- The Dark Weak SU(2) coupling constant $g_{DW} = ?$
- The Dark Electromagnetic U(1) coupling constant $g_{DE} = ?$
- The U(4) Generation group coupling constant $g_G = ?$
- The U(4) Layer groups coupling constant $g_L = ?$
- The U(4) Species group coupling constant $g_{Sp} = ?$
- The U(192) Θ-interaction group coupling constant $g_\Theta = ?$

---

[95] All coupling constant values are based on data extracted from C. Patrignani *et al* (Particle Data Group), Chinese Physics **C40**, 100001 (2014).
[96] Based on the running coupling constant value $\alpha_s(M_Z^2) = 0.1193 \pm 0.0016$.

In this chapter we will calculate (approximately) g_S, g_W, and the "fine structure constant" for SU(4) g_4, which would seem to apply to the U(4) groups listed above, subject to the approximations listed in chapter 18.

We will use the below to calculate "fine structure constant"s α_G

$$A_1 = (g + 1)(1 - 2g^2)/[(g + 2)(g - 1)] \tag{19.1}$$

$$A_2 = [8g^2(2g + 1) - (2g^3 + 2g^2 + g - 2)(g^2 + 2g + 2)]/[2(g^2 - 1)(g^2 - 4)]$$

$$A_3 = -2(1 + 3g + 6g^2 + 2g^3)/[g(g + 1)]$$

$$A_4 = -(g + 2)(1 + 5g + 6g^2 + 2g^3)/[g(g^2 - 1)] - 1/(g + 1)$$

$$\psi = [gA_3 - (4 + 2g)A_1]/[(4 + 2g)A_2 - g A_4]$$

$$(\alpha_G/2\pi) = c_G^{-1}[gA_4 - (4 + 2g)A_2]/(A_4A_1 - A_2A_3) \tag{19.2}$$

$$F_1 = (2/3)(1 - 3g^2/2 - g^3) - (\alpha_G/4\pi)[(2 + 4g + 4g^2)(g - 2) + \alpha_G\psi g^3]/[(g^2 - 1)(g - 2) + \\ + \alpha_G(2 + 4g + 4g^2)(g - 2) + \alpha_G\psi g^3] \tag{19.3}$$

using eq. 18.1. Note the group factor c_G only appears explicitly in eq. 19.2. It does appear implicitly in eq. 19.3.

19.2 SU(2) ElectroWeak "Fine Structure Constant" Eigenvalue Function

The U(1) QED eigenvalue function and eigenvalue was discussed in chapter 16. In this section we discuss and show plots of the SU(2) ElectroWeak sector eigenvalue function subject to the discussion in chapters 16 and 18. In particular we calculate the SU(2) eigenvalue function and eigenvalues using eqs. 19.2 and 19.3.

Due to the asymptotic freedom of non-abelian interactions we anticipate that the value of the exponential factor g will be positive at the "fine structure constant" eigenvalue of the SU(2) eigenvalue function which we take to have the form shown above due to the assumptions of chapter 18.

We define $F_{1su(2)}(\alpha)$ using eq. 19.3 above with $c_{GSU(2)}^{-1} = -0.003023589$ from eq. 18.1 using $C_{ad} = 2$ and $C_f = 2$.

In our discussion of QED in chapter 17 we found a negative value for g and a consequent divergence in Z_3 as $\Lambda \to \infty$. We then found that $F_1(\alpha)$ did not yield the QED eigenvalue. We had to introduce $F_2(\alpha)$ by eliminating low order (in α) terms using an intermediate renormalization. Those terms have logarithmic divergences. Their elimination did not create a problem since the QED renormalizations also diverge due to $g < 0$ at the eigenvalue point. The total divergence is generated by combining the divergent factors due to the $(p/\Lambda)^{2g}$ factor in Z_3 with the omitted divergent terms of F_1. See eqs. 16.2 and 16.4.

In the case of non-abelian interactions we have $g > 0$, which implies renormalizations go to 1 as $\Lambda \to \infty$ (asymptotic freedom).Thus we will not truncate F_1 but will calculate using it (eq. 19.3) to find an approximation to the "fine structure constant" eigenvalues of non-abelian interactions. *As $\Lambda \to \infty$ non-abelian theories approach free field theories.*

The signature of the eigenvalue function essential singularity in the present case will be a divergence in F_1 for $g > 0$. It signals that we have an approximation to the expected essential singularity.

The below plots show the overall form of $F_{1SU(2)}(\alpha)$ plotted vs. g and the divergent region of $F_{1SU(2)}(g)$ specifying the "fine structure constant" eigenvalue. Our approximate "fine structure constant" values in this case and the SU(3) and SU(4) cases are quite reasonable. We find the "essential singularity" signal point at

$$g = 0.54$$

where $\alpha_{SU(2)} = g_W^2/(4\pi)$

$$\alpha_{calculatedSU(2)}(g) = 0.0425$$

compared to the actual measured "fine structure constant" value

$$\alpha_{SU(2)} = 0.0305$$

displaying a fairly close match given the approximate nature of the calculations.The positive value of the exponential factor g above indicates that $Z_3 \rightarrow 1$ as $\Lambda \rightarrow \infty$ showing the SU(2) interaction is asymptotically free.

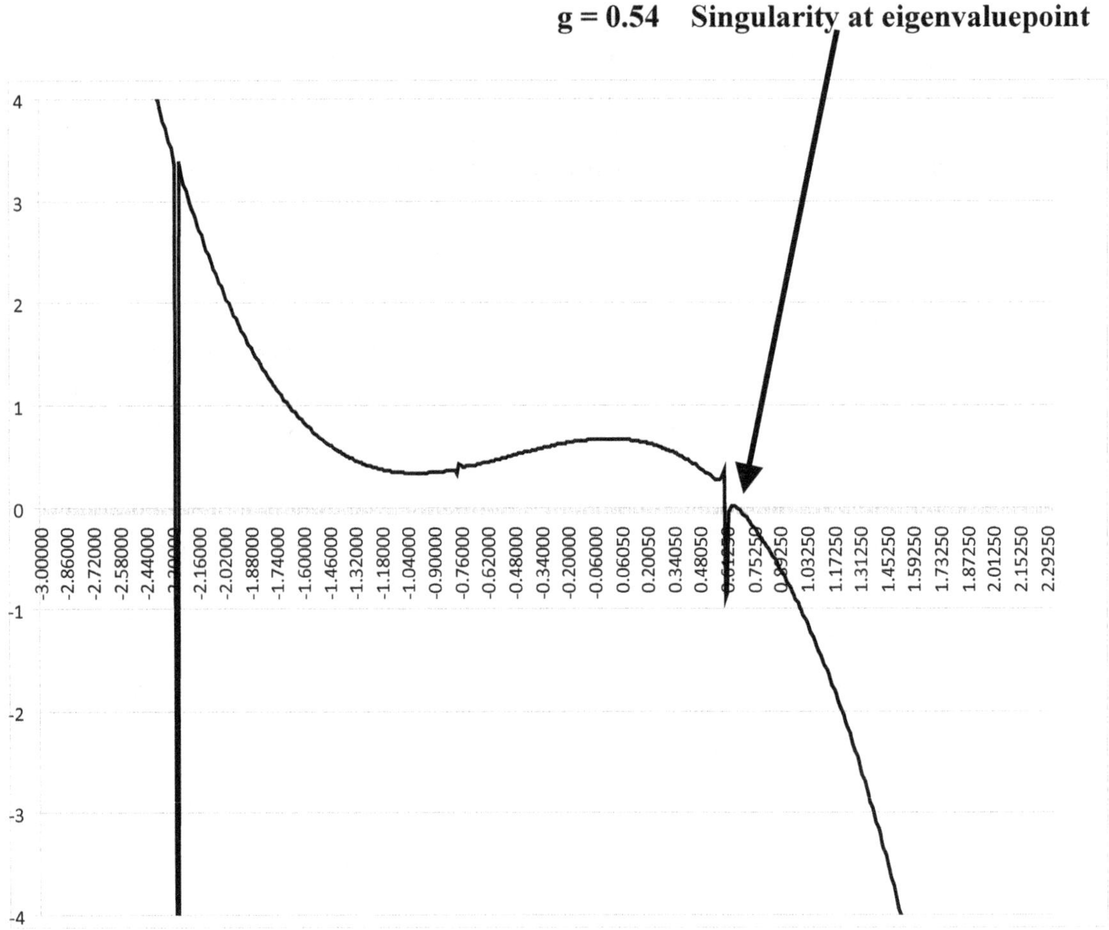

Figure 19.1 Plot of $F_{1su(2)}$ (vertical axis) as a function of g.

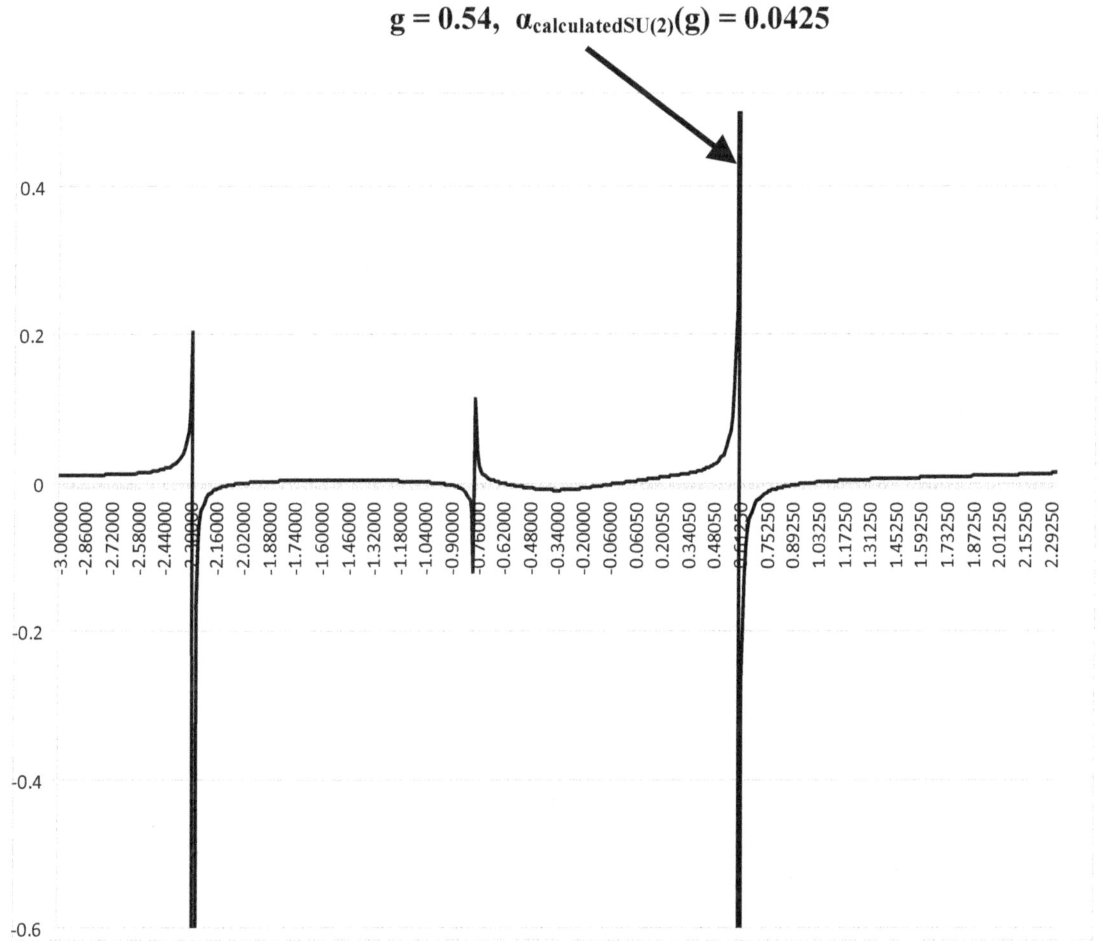

Figure 19.2 Plot of the "fine structure constant" $\alpha_{calculatedSU(2)}$ (vertical axis) as a function of g.

19.3 SU(3) Strong "Fine Structure Constant" Eigenvalue Function

The discussion of the SU(3) Strong interaction case is very much the same as the preceding SU(2) discussion. We define $F_{1su(3)}(\alpha)$ using eq. 19.3 above with

$$c_{GSU(3)}^{-1} = -0.004535384$$

from eq. 18.1 using $C_{ad} = 3$ and $C_f = 3$.

The signature of the eigenvalue function essential singularity in the present case will be a divergence in F_1 for $g > 0$. It signals that we have an approximation to the expected essential singularity.

The below plots show the overall form of $F_{1SU(3)}(\alpha)$ plotted vs. g and the divergent region of $F_{1SU(3)}(g)$ specifying the "fine structure constant" eigenvalue. We find the "essential singularity" point at

$$g = 0.5605$$

where

$$\alpha_{calculatedSU(3)}(g) = 0.086$$

compared to the actual "measured" "fine structure constant" value (at $Q^2 = 2$ GeV)

$$\alpha_{SU(3)} = 0.118$$

again displaying a fairly close match given the approximate nature of the calculations.

The positive value of the exponential factor g above indicates that $Z_3 \rightarrow 1$ as $\Lambda \rightarrow \infty$ showing the SU(3) interaction is asymptotically free.

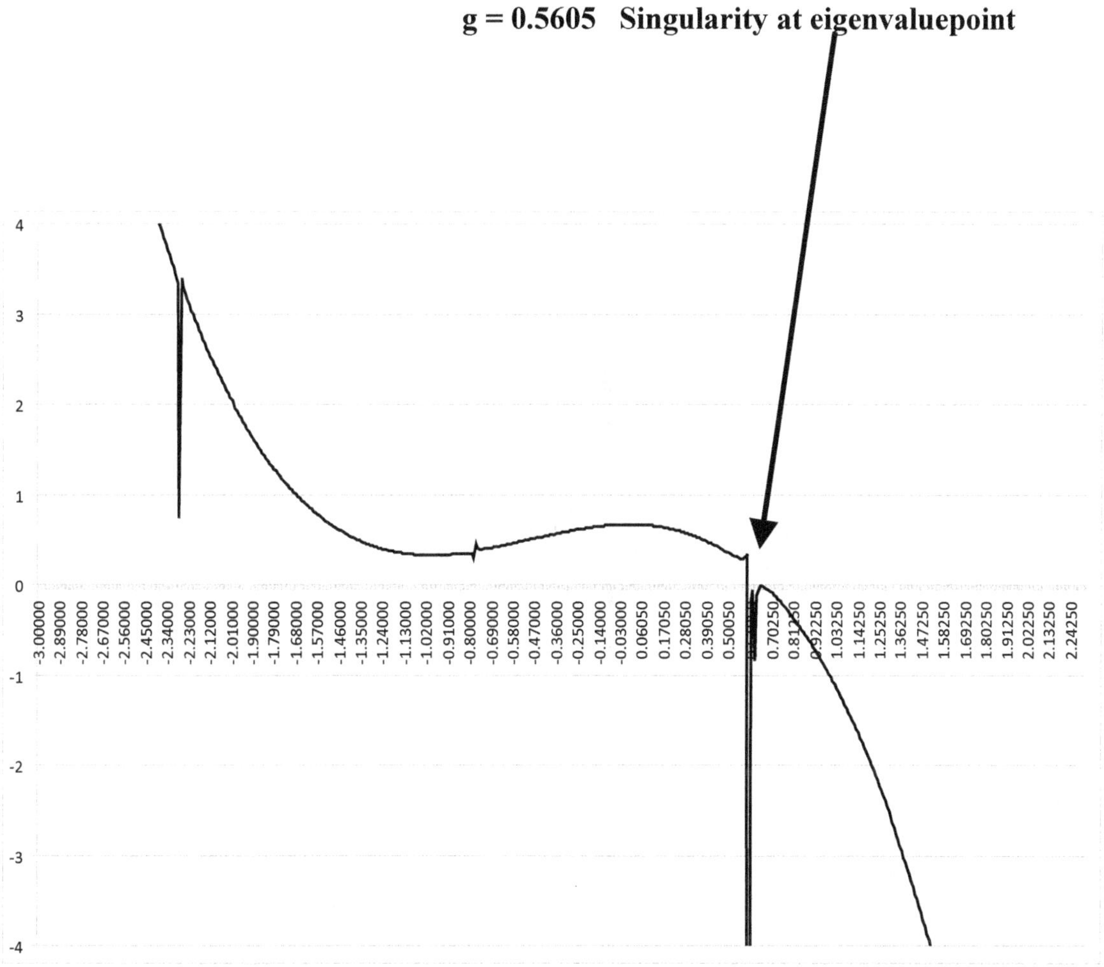

Figure 19.3 Plot of $F_{1su(3)}$ (vertical axis) as a function of g.

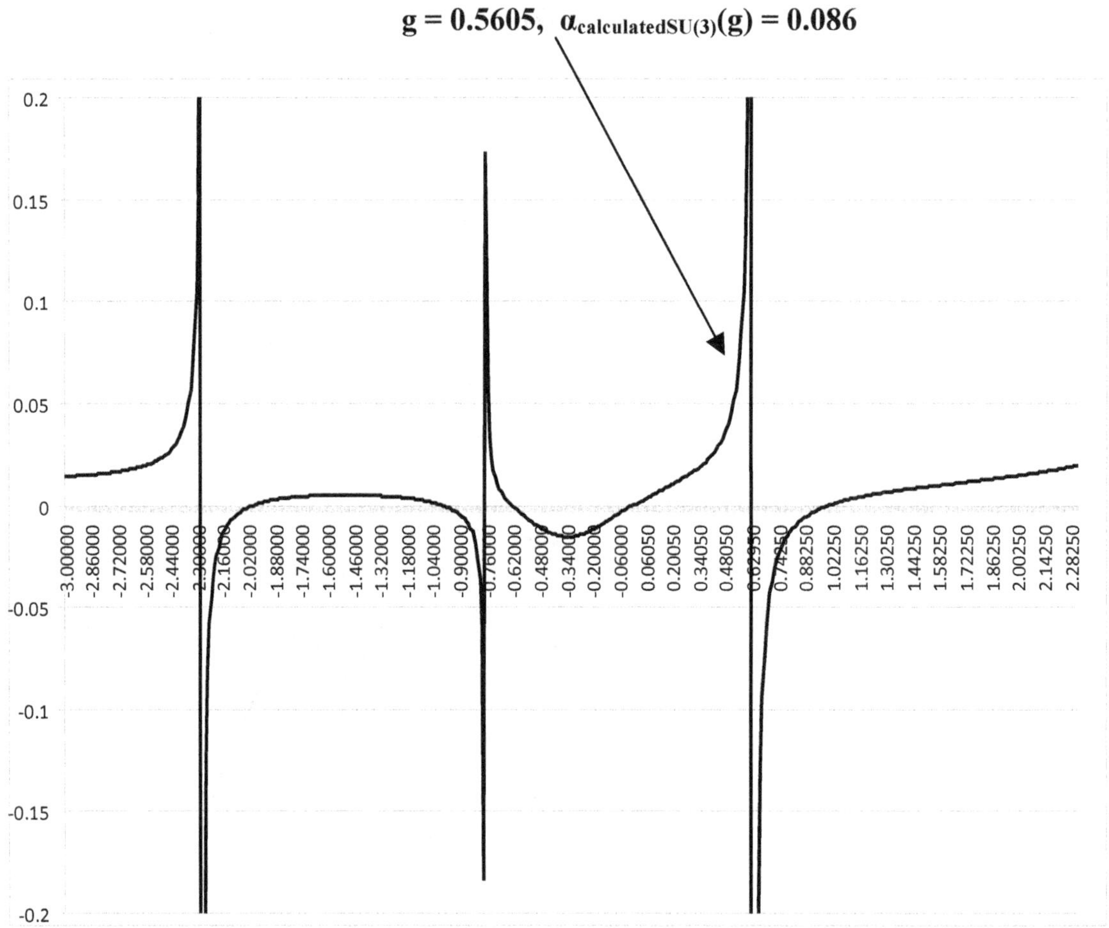

Figure 19.4 Plot of $\alpha_{calculatedSU(3)}$ (vertical axis) as a function of g.

19.4 SU(4) "Fine Structure Constant" Eigenvalue Function

The discussion of the SU(4) case of *other* interactions is very much the same as preceding discussions. We define $F_{1su(4)}(\alpha)$ using eq. 19.3 above with

$$c_{GSU(4)}^{-1} = -0.006047178$$

using $C_{ad} = 4$ and $C_f = 4$.

The signature of the eigenvalue function essential singularity in the present case will be a divergence in F_1 for $g > 0$. It signals that we have an approximation to the expected essential singularity.

The below plots show the overall form of $F_{1SU(4)}(\alpha)$ plotted vs. g, and the divergent region of $F_{1SU(4)}(g)$ specifying the "fine structure constant" eigenvalue. We find the "essential singularity" point at

$$g = 0.598$$

where

$$\alpha_{calculatedSU(4)}(g) = 0.384$$

compared to the conjectured[97] "fine structure constant" value

$$\alpha_{SU(4)} = 0.458$$

again displaying a fairly close match given the approximate nature of the calculations.

The positive value of the exponential factor g above indicates that $Z_3 \rightarrow 1$ as $\Lambda \rightarrow \infty$ showing the SU(4) interaction is asymptotically free.

[97] This value is based on the "doubling trend" that will be seen in the three known coupling constants in chapter 20.

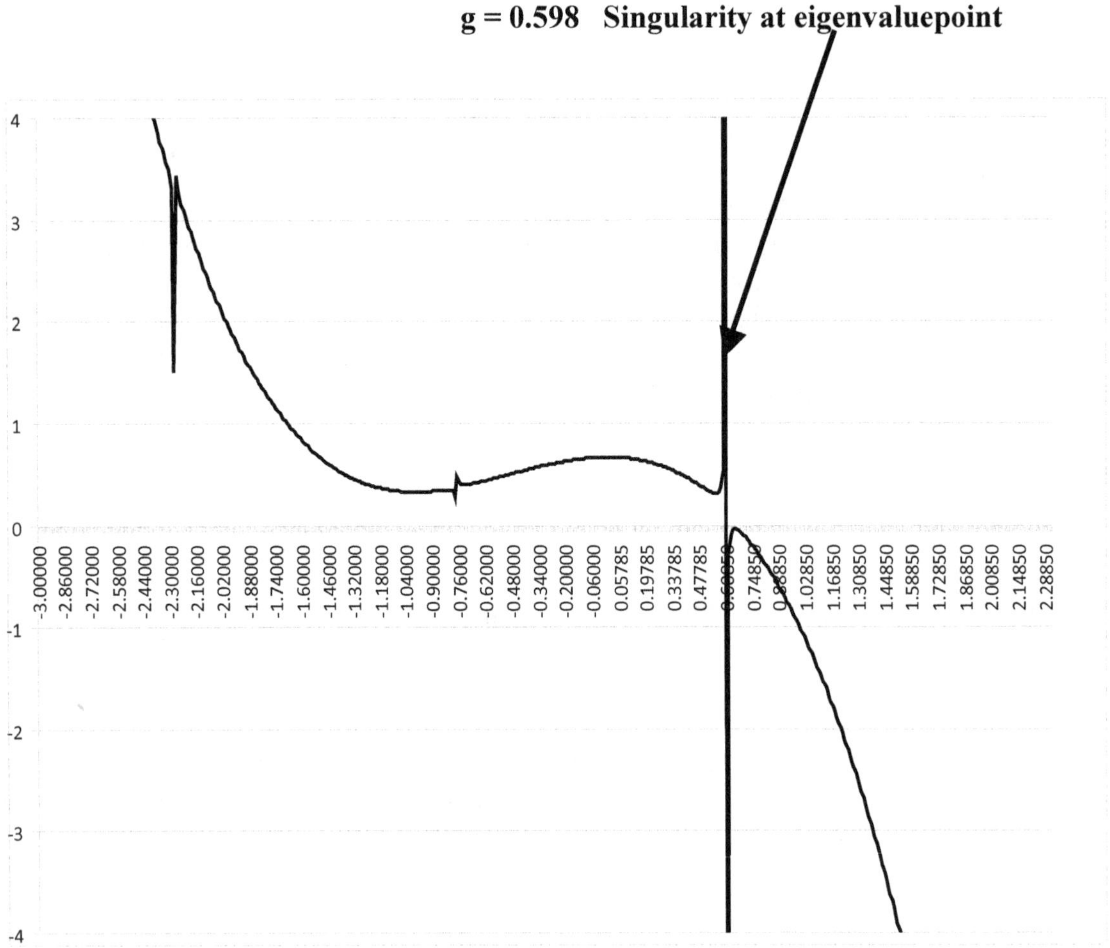

Figure 19.5 Plot of $F_{1su(4)}$ (vertical axis) as a function of g.

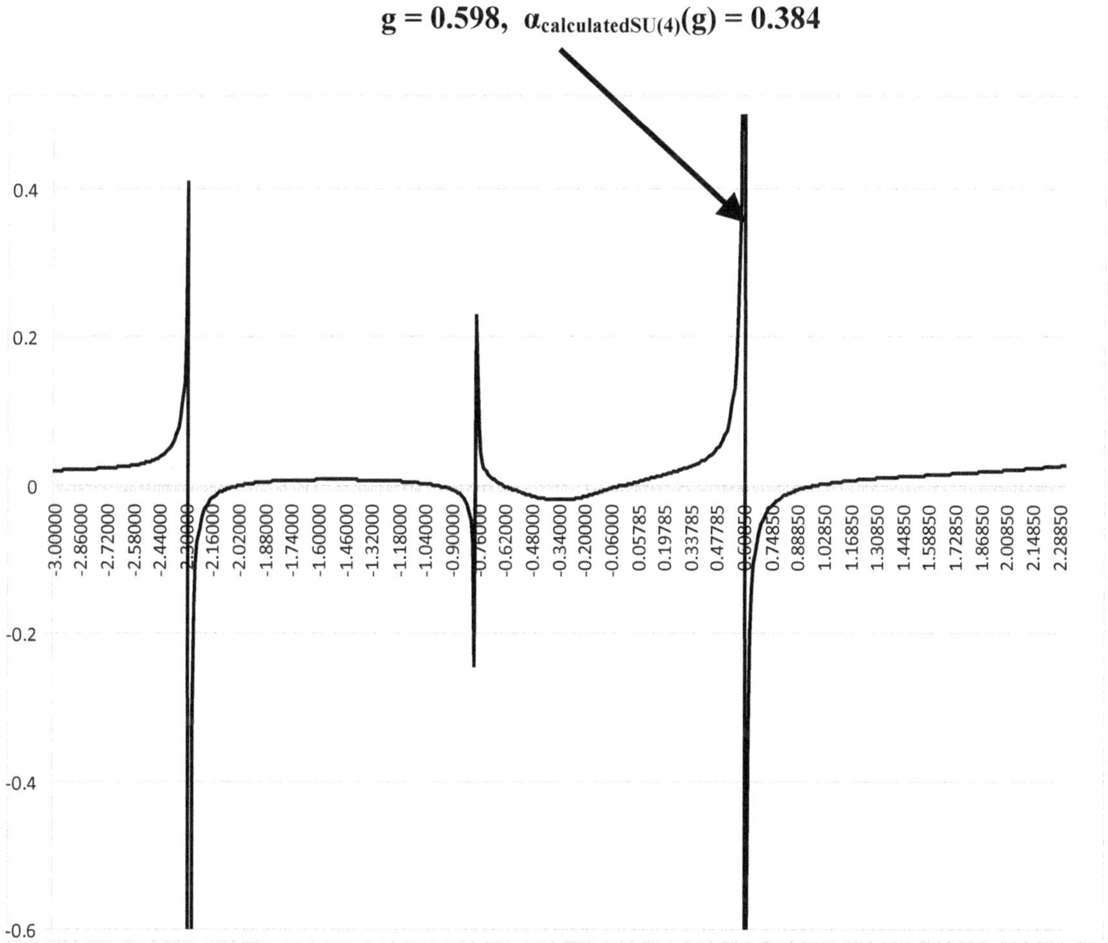

Figure 19.6 Plot of $\alpha_{calculatedSU(4)}$ (vertical axis) as a function of g.

20. "Fine Structure Constants" of the Unified SuperStandard and Standard Models

In this chapter we summarize the values of the approximate values of the α_G determined in the previous chapter.

The deeper significance of the observed doubling regularity seen below is not known.

Group	*Known Coupling Constant* e_G	*Known* $e_G^2/(4\pi)$	*Calculated* $\alpha_G = e_G^2/(4\pi)$	*Calculated*[98] *Exponent* g_G
QED, U(1)	0.30282212	α = 0.0072973525693	α = 0.0072973525693	-0.00058053691948
SU(2)	0.619	0.0305	0.0425	0.54
SU(3)	1.22	0.118	0.086	0.5605
SU(4)	2.4?[99]	0.458?	0.384	0.598

The relative closeness of the calculated values of "fine structure constants" to the experimentally known values is very encouraging—particularly in the case of the Electromagnetic fine structure constant α. It puts to rest other possible explanations for its value.

Our QED calculation of α has no free (adjustable) parameters unlike other attempts in the past. It also is soundly based on Quantum Field Theory. The calculation of the non-abelian coupling constants also has no free (adjustable) parameters.

---

[98] They appear in eqs. 5.5 – 5.8 in Blaha (2019b). See Blaha (2019b) for more details.

[99] This value is based on the "doubling trend" seen in the three known coupling constants above.

Thus the coupling constant eigenfunctions are self-determined and depend only on inherent perturbation theory based on dynamics. Coupling constant values cannot be "tweaked" to their known values by adjusting input parameters.

The ability of our 1973-4 calculation of the JBW eigenvalue function together with the new insights into understanding of the precise method to obtain its "fine structure constant" eigenvalues is also encouraging. It opens the possibility that the Unified SuperStandard Theory has within itself the mechanism for determining the constants appearing within it. It raises the hope that a similar self-determination mechanism may also exist within the theory to determine the masses appearing in the Higgs particles sector of the theory.

The result would be a self-contained all-encompassing fundamental theory.

21. Possible Exact Form of the Non-Abelian Eigenvalue Functions

Figs. 19.2-19.6 of chapter 19 display a repetitive pattern that is similar to those of the trigonometric functions. In particular they show a similarity to the tangent function. Our approximate F_1 eigenvalue functions for non-abelian interactions, plotted in these figures, is, in each case, an approximate sum of one loop vacuum polarization diagrams similar to that we calculated in our paper in 1974.

In this chapter we further approximate F_1. Our main purpose will be to study non-abelian interaction running coupling constants in chapter 22.

Since the F_1 eigenvalue functions that we calculated in chapter 19 yielded good approximate values for the Weak and Strong coupling constants, it seems reasonable to believe that these eigenvalue function approximations are close to the real F_1 eigenvalue functions that would have been obtained in precise calculations summing one fermion loop Feynman diagrams.

It is also reasonable to believe that the true F_1 functions are less complicated than the approximate ones. Some past perturbation theory summations to all orders have shown a remarkable simplicity in the resulting summation. A particular example is the author's leading logarithm summation for the deep inelastic e-p structure functions.[100] In this paper a remarkable cancellation of diagrams, due to a Stirling Numbers of the Third Kind identity,[101] led to a simple compact result.

In the present case we will take the periodic pattern in the approximate F_1 functions to be tangent functions indications in the exact F_1 eigenfunctions. We view our approximate F_1 eigenfunctions as generated by an approximation to the total one loop

---

[100] Stephen Blaha, Phys. Rev. D **3**, 510 (1971). This paper (the author's Ph.D. Thesis) showed that perturbation theory could not account for deep inelastic e-p scaling—an open question at the time.

[101] This type of Stirling number had not been encountered in perturbation theory before, or after, the author's paper.

contributions to the "vacuum polarization" of the non-abelian interaction coupling constants.

21.1 "Exact" Form of Non-Abelian Eigenvale Functions

We suggest the correct form of the F_1 vacuum polarization eigenfunctions for the group G is

$$F_{G1}(g) = \tan[\pi(g + d_{Gf})/d_{Gd}] \qquad (21.1)$$

and the coupling constant function is the absolute value[102]

$$\alpha_G(g) = |c_G \tan[\pi(g + d_{G\alpha})/d_{Gd}]| \qquad (21.2)$$

For some value of g, $F_{G1}(g_0) \rightarrow \infty$ and $\alpha_G(g_0)$ is the coupling constant for interaction group G. We found it reasonable to approximate the quantity d_{Gd} by d_{Gd} = 1.29911 - 0.08929g. Higher powers of g would be required to get a better fit—as we shall see in the following figures. We leave that issue to future work. The quantities d_{Gf} and $d_{G\alpha}$ are assumed to be constants. The graphs below show a good approximation of tangent fits to the approximate F_1 plots.

21.2 Similarity to the Madhava-Leibniz Representation of π

The form of the coupling constant eigenvalue function is comparatively simple compared to exact expressions found earlier. Remarkably it also suggests that α has a representation similar in character to the Madhava-Leibniz representation of π = 3.14159... :

$$\pi = 4 \arctan(x) \qquad (21.3)$$

for x = 1. Note the eigenvalue implied by eq. 21.2 is

---

102 The absolute value is physically required to maintain the reality of coupling constants. Eq. 21.2 is an approximation that does not exclude imaginary coupling constants.

$$\alpha_G = |c_G \tan(x)| \qquad (21.4)$$

(in absolute value) where

$$x = \pi(g_0 + d_{G\alpha})/d_{Gd} \qquad (21.5)$$

for some g_0.

The displayed range of values of g will be g ε [-3, 3].

21.3 Comparison of Approximate Eigenfunctions and Eigenvalues to the Tangent Representation

In this section we display the graphs of the approximate Eigenfunctions and Eigenvalue functions, and the proposed tangent graphs for SU(2), SU(3) and SU(4).

21.3.1 SU(2)

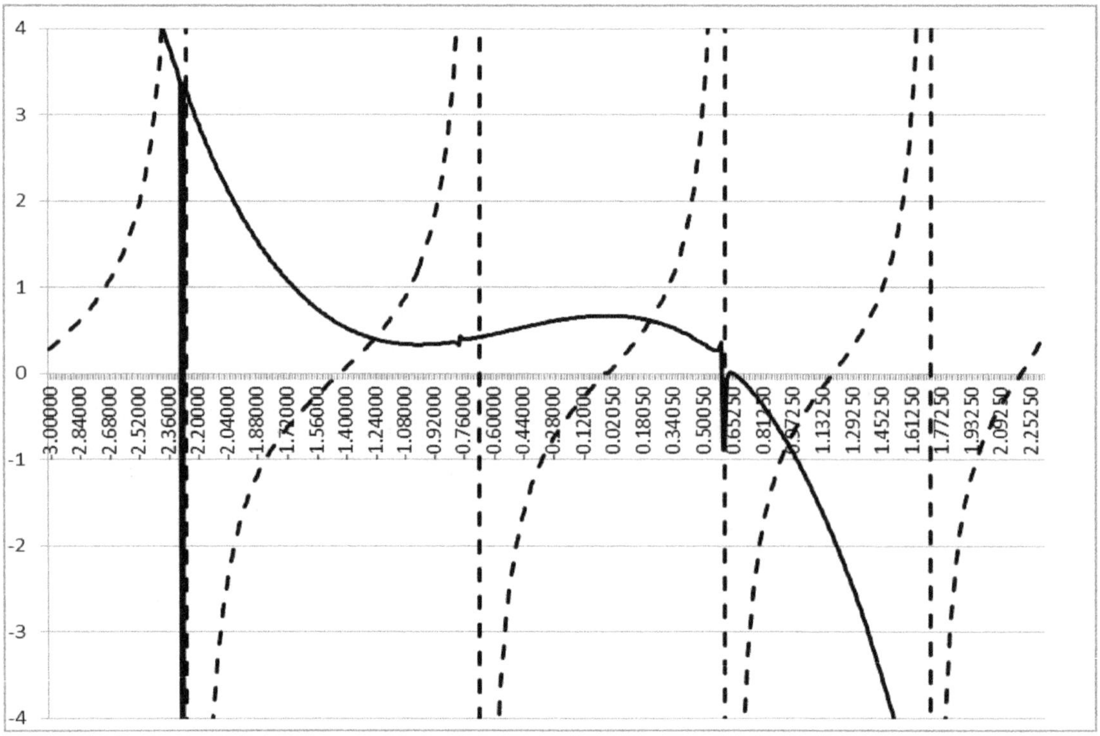

Figure 21.1 The SU(2) eigenvalue function F_1 is a solid line, while the tangent form of F_1 of eq. 21.1 is the broken line. The constants are d_{Gd} = 1.29911 - 0.08929g, and d_{Gf} = 0. F_1 is plotted vertically. The exponent g is plotted horizontally.

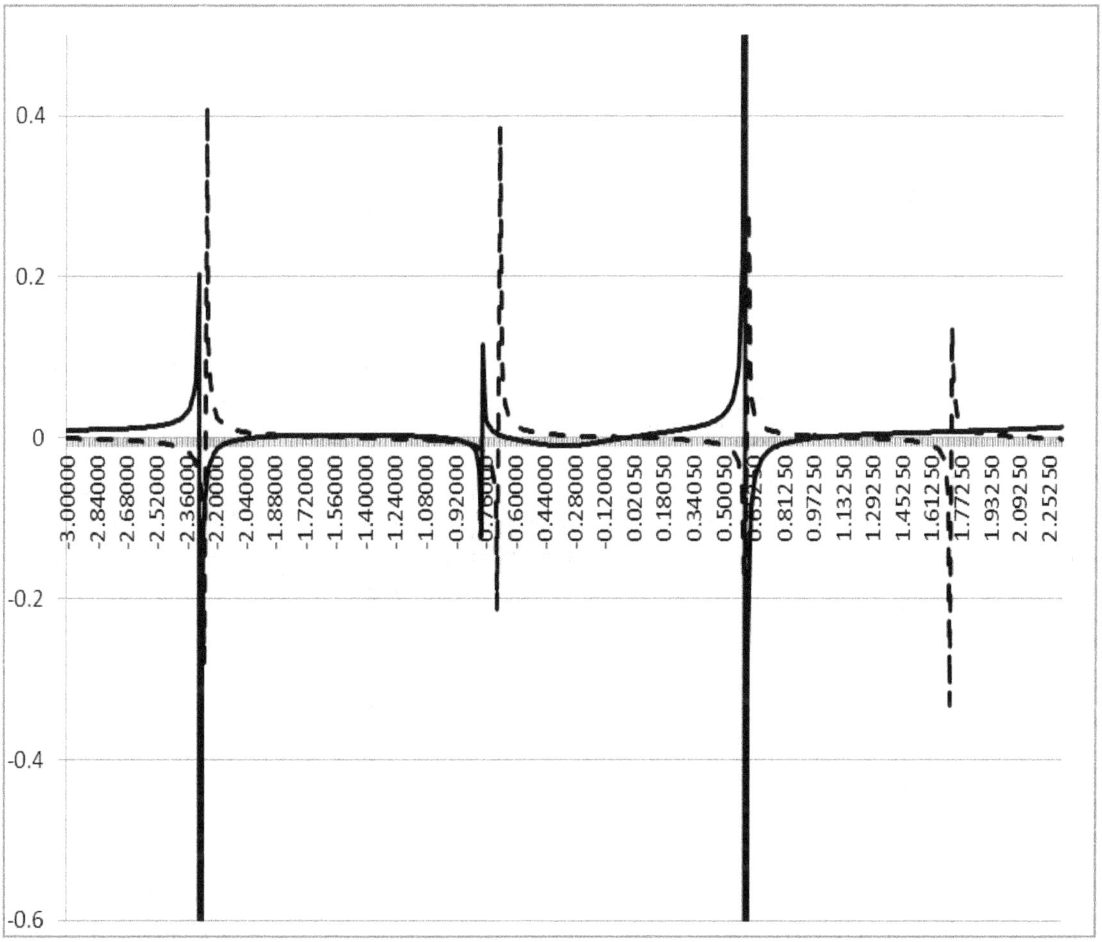

Figure 21.2 The plotted SU(2) eigenvalue $\alpha_G(g)$ is a solid line, while the tangent form of α_G of eq. 21.2 is the broken line. The constants are d_{Gd} = 1.29911 - 0.08929g, and the α_G constant is $d_{G\alpha}$ = 0.00348417. α_G is plotted vertically. The exponent g is plotted horizontally.

21.3.2 SU(3)

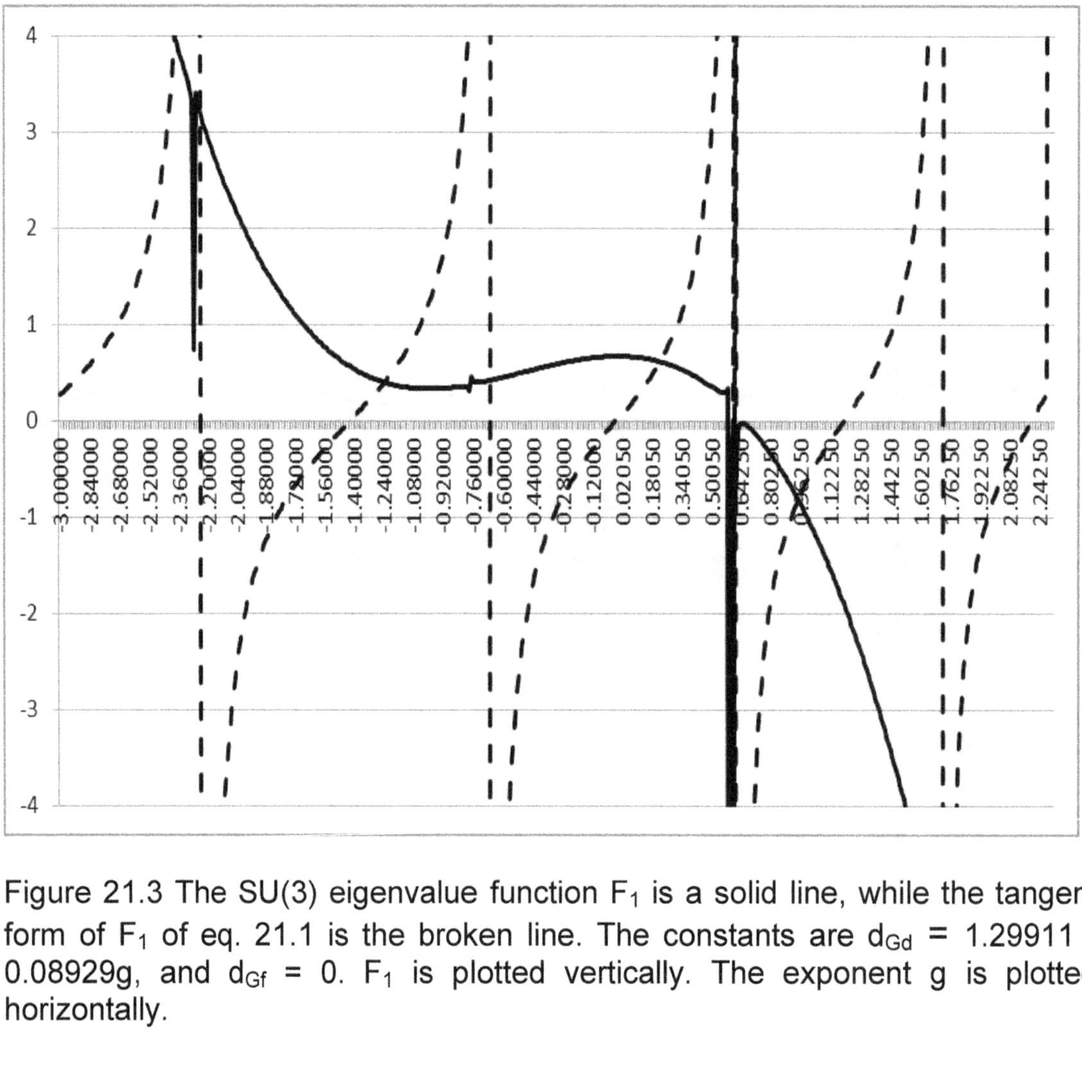

Figure 21.3 The SU(3) eigenvalue function F_1 is a solid line, while the tangent form of F_1 of eq. 21.1 is the broken line. The constants are d_{Gd} = 1.29911 - 0.08929g, and d_{Gf} = 0. F_1 is plotted vertically. The exponent g is plotted horizontally.

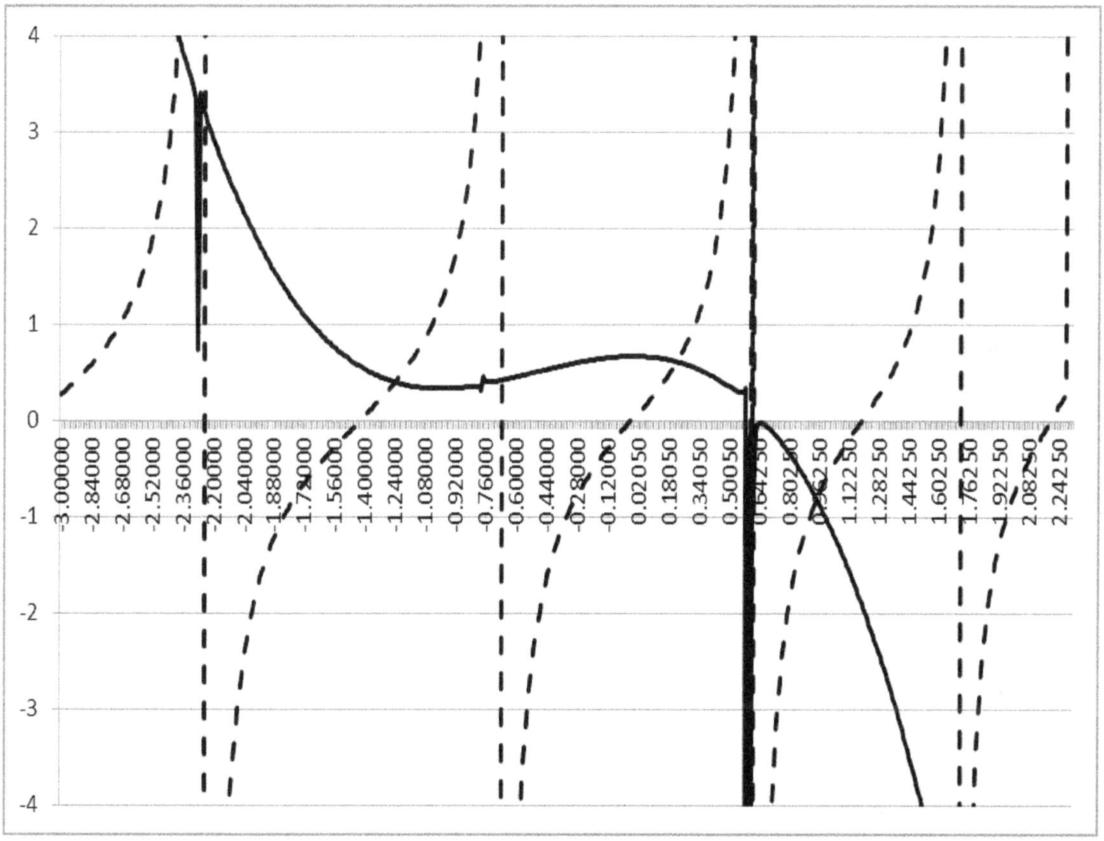

Figure 21.4 The plotted SU(3) eigenvalue α_G is a solid line, while the tangent form of α_G of eq. 21.2 is the broken line. The constants are d_{Gd} = 1.29911 - 0.08929g, and the α_G constant is $d_{G\alpha}$ = 0.00348417. α_G is plotted vertically. The exponent g is plotted horizontally.

21.3.3 SU(4)

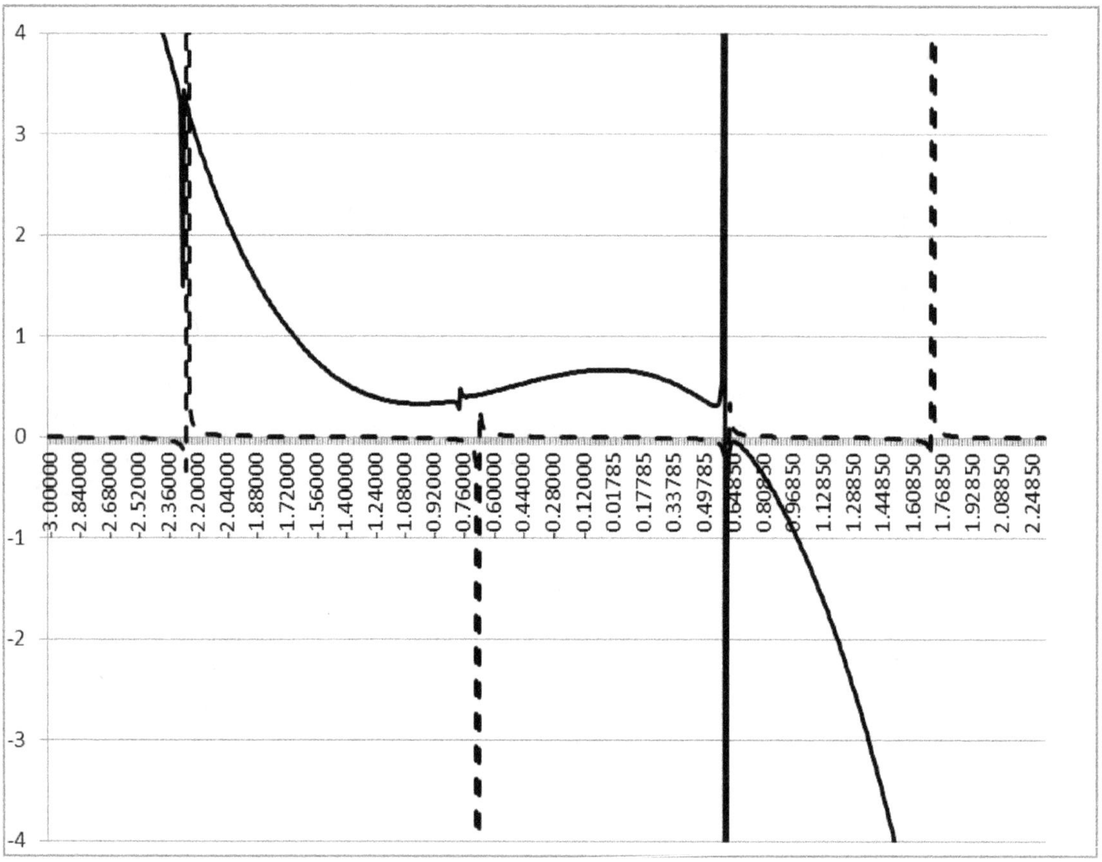

Figure 21.5 The SU(4) eigenvalue function F_1 is a solid line, while the tangent form of F_1 of eq. 21.1 is the broken line. The constants are d_{Gd} = 1.29911 - 0.08929g, and d_{Gf} = 0. F_1 is plotted vertically. The exponent g is plotted horizontally.

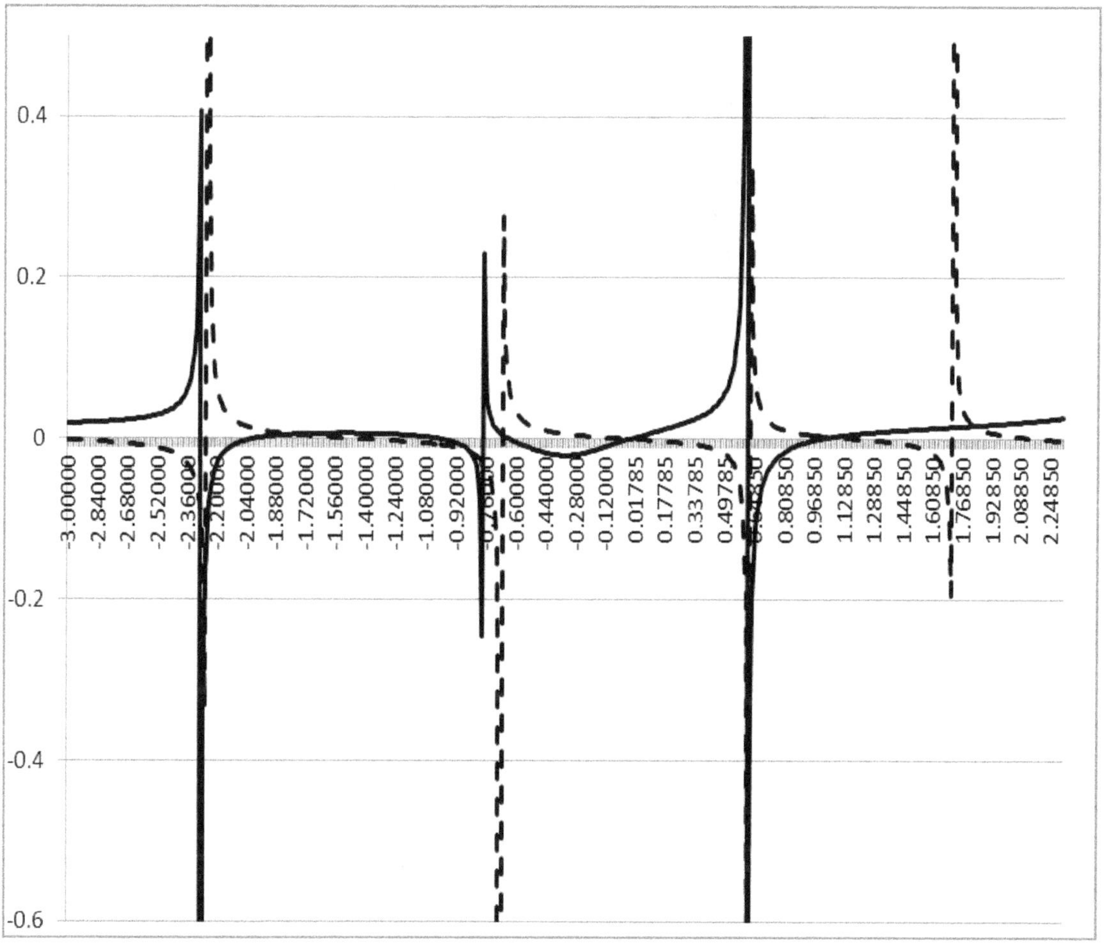

Figure 21.6 The SU(4) eigenvalue $\alpha_G(g)$ is a solid line, while the tangent form of α_G of eq. 21.2 is the broken line. The constants are d_{Gd} = 1.29911 - 0.08929g, and the α_G constant is $d_{G\alpha}$ = 0.00348417. α_G is plotted vertically. The exponent g is plotted horizontally.

21.4 Comments

The quantities d_{Gd} and $d_{G\alpha}$ have the same values for SU(2), SU(3) and SU(4). The group constants c_G differ from group to group. *The similarity in form for all non-abelian cases is due to our use of a "universal" form for eigenfunctions. Our justification is our success in the approximate calculation of all coupling constants.*

The similarity in the form, and the values of c_G and other constants in eq. 21.2, "explain" the different values in Standard Model coupling constants.

22. Approximate Non-Abelian Running Coupling Constants with Beta Functions to All Orders in Perturbation Theory

In this chapter we consider the one fermion loop approximation for the β function of the Callan-Symanzik equation:

$$-\beta_1(\alpha) + m\partial(\alpha\pi^{(1)}{}_c)/\partial m \approx 0 \tag{22.1}$$

where the vacuum polarization part is

$$\pi^{(1)}{}_c = F_1(\alpha)\ln(-q^2/m^2) + \ldots \tag{22.2}$$

resulting in the β function term

$$\beta_1(\alpha) = -2\alpha F_1(\alpha) \tag{22.3}$$

The integration of β_1 yields

$$\ln(m_2/m_1) = \int_{\alpha_1}^{\alpha_2} d\alpha/\beta(\alpha) \tag{22.4}$$

where

$$\alpha_i = \alpha(m_i) \tag{22.5}$$

22.1 Approximate Form of the Eigenvalue Function $F_1(\alpha)$ to all Perturbative Orders

We first calculate the eigenvalue function $F_1(\alpha)$ and then proceed to calculate $\beta_1(\alpha)$. Then eq. 22.4 becomes soluble and an expression for the running coupling constant can be calculated.

For the sake of a transparent physical result[103] we assume $d_{Gd} = 1.5$, which is a reasonable approximation for g in the range [-3, 3] (the domain of physical interest).

Eq. 21.2 can be inverted to yield $g(\alpha)$ for a given group G:

$$g(\alpha) = (d_{Gd}/\pi) \arctan(\alpha_G/c_G) - d_{G\alpha} \tag{22.6}$$

Substituting for g in eq. 21.1 using eq. 22.6 in gives

$$F_{G1}(\alpha) = [\alpha - c_G \tan(z_G)]/[1 + \alpha c_G \tan(z_G)] \tag{22.7}$$

where

$$z_G = \pi(d_{G\alpha} - d_{Gf})/d_{Gd} \tag{22.8}$$

Eq. 22.7 is a remarkably simple expression for the eigenvalue function in terms of α. It has the "eigenvalue zero" at[104]

$$\alpha_G = |c_G \tan(z_G)| \tag{22.9}$$

We will see in chapter 23 that the QED[105] Fine Structure Constant α has the same form making it analogous to the Madhava-Leibniz formula for π discussed in section 21.2.

For massless QED, given the known value of $\alpha = 0.0072973525693$ and using $\alpha = \tan(z)$ we find

$$z = 0.00729760570938695 \tag{22.10}$$

---

[103] The expression used in our plots $d_{Gd} = 1.29911 - 0.08929g$ could have been used and would lead to a tractable calculation. The essence of the physics is well brought out by approximating $d_{Gd} = 1.5$.

[104104] Eq. 10.9 uses an absolute value for α_G since coupling constants are real-valued.

[105] The use of intermediate renormalization and thus F_2 for QED does not change the character of the results.

and the QED quantities

$$z/\pi = (d_\alpha - d_f)/d_d = 0.00232290200484053 \quad (22.11)$$

22.2 Approximate Form of Running Coupling Constant to all Perturbative Orders

Eqs. 22.3, 22.4 and 22.7 determine the approximate running coupling constant

$$\ln(m_2/m_1) = \int_{\alpha_1}^{\alpha_2} d\alpha/(-2\alpha F_{G1}(\alpha)) \quad (22.12)$$

The running coupling constant integration yields

$$(m_1/m_2)^2 = [(\alpha_2 - c_G \tan(z_G))/(\alpha_1 - c_G \tan(z_G))]^y(\alpha_2/\alpha_1)^z \quad (22.13)$$

where

$$y = c_G \tan(z_G) + (c_G \tan(z_G))^{-1} \quad (22.14)$$

$$z = -(c_G \tan(z_G))^{-1} \quad (22.15)$$

If

$$\alpha_i >> |c_G \tan(z_G)|$$

for i = 1, 2, then eq. 22.13 becomes approximately

$$(m_1/m_2)^2 = [\alpha_2/\alpha_1]^{c_G \tan(z_G)} = [\alpha_2/\alpha_1]^{\alpha_G \mathrm{signum}(c_G)} \quad (22.16)$$

where the exponent α_G is the constant coupling constant value in eq. 22.9 and signum(c_G) is the sign of c_G giving the proper asymptotic freedom of non-abelian couplings and the infinite bare charge of the QED (abelian) case.

Eq. 22.16 can be reexpressed as

$$\alpha(m_2) = \alpha_1(m_1)\ (m_1/m_2)^{2\mathrm{signum}(c_G)/\alpha_G} \qquad (22.17)$$

22.3 Running Coupling Constant

The running coupling constant expression is applicable to QED, Weak SU(2), and Strong SU(3). Eqs. 22.13 and 22.17 show QED's running coupling constant gets stronger as the mass increases, while the non-abelian running coupling constants decrease as the mass increases yielding asymptotic freedom.

The growth/decline of the running coupling constant is a power law in this formulation while in other formulations, based on low order prturbation theory, it is logarithmic.[106] This difference impacts on Grand Unified Theories (GUT) theories.

---

[106] H. Georgi, H. R. Quinn, and S. Weinberg, Phys. Rev. Lett. **33**, 451 (1974); W. E. Caswell, Phys. Rev. Lett. **33**, 244 (1974).

23. Approximate Massless QED Eigenvalue Function to All Orders in Perturbation Theory

The approximate calculation of F_2 in perturbation theory to all orders in chapter 17 also resembles a tangent approximation with

$$d_d = d_{Gd} = 1.29911 - 0.08929*g$$

$$F_2(g) = \tan(\pi g/d_d)$$

and

$$\alpha(g) = \tan(\pi(g + 0.00348417)/d_d)$$

Figs. 23.1 and 23.2 show the close approximation given by the tangent curves.

The general form of the expression for the fine structure constant α is

$$\alpha_G = |c_G \tan(x)| \qquad (21.4)$$

where

$$x = \pi(g_0 + d_{G\alpha})/d_{Gd} \qquad 21.5)$$

for some g_0.

The QED α value is

$$\alpha = 0.00729773525693$$

We have shown the general form for α is

$$\alpha = \tan(z) \qquad (23.1)$$

in close analogy to the Madhava-Leibniz representation of $\pi = 3.14159\ldots$.

We use

$$z = 0.00729760570938695 \tag{22.10}$$

in eq. 23.1 based on the α calculation of chapters 16-17, and the QED quantity

$$z/\pi = (d_\alpha - d_f)/d_d = 0.00232290200484053 \tag{22.11}$$

Thus the fine structure constant α is a feature of the dynamics of QED and is not a cosmological result, or a result of any other theory.

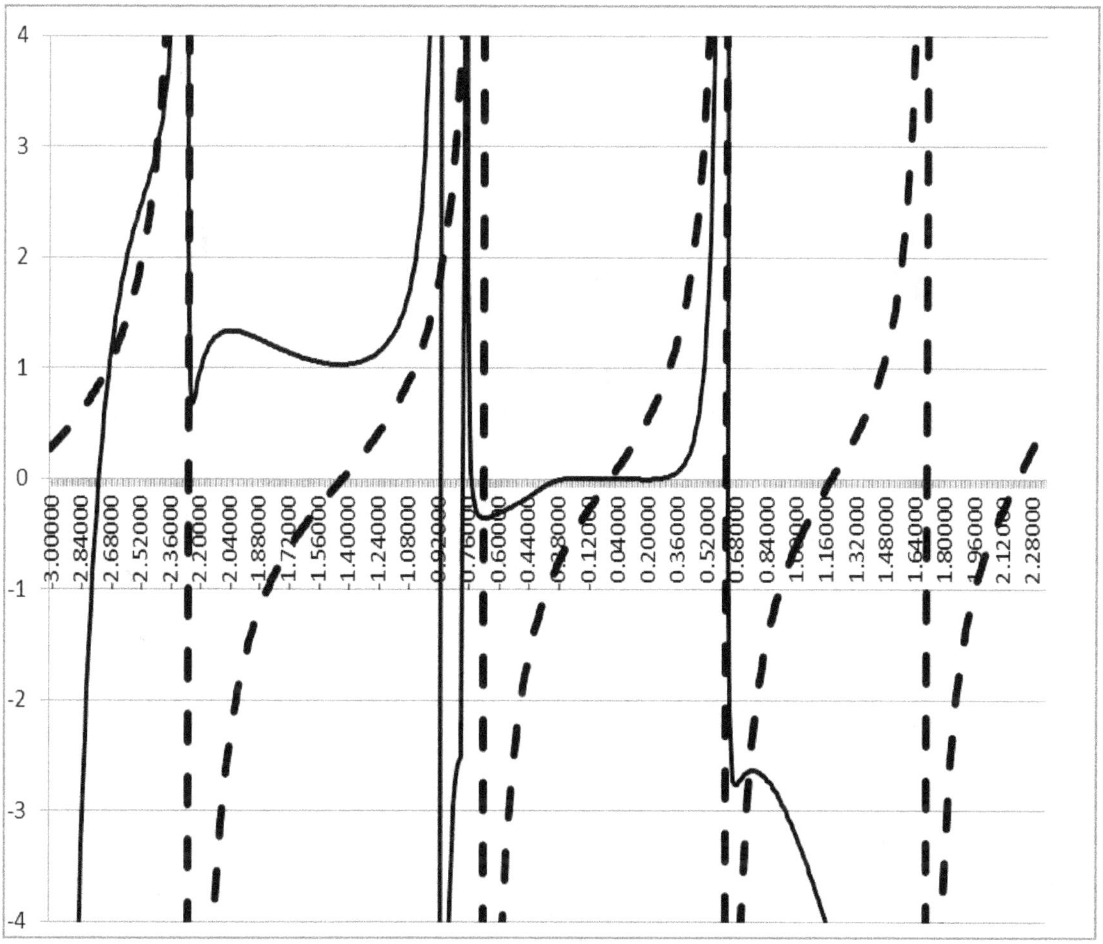

Figure 23.1 The massless QED eigenvalue function F_2 is the solid line, while the tangent form of F_2 is the broken line. The constants are d_{Gd} = 1.29911 - 0.08929g, and d_{Gf} = 0. F_2 is plotted vertically. The exponent g is plotted horizontally.

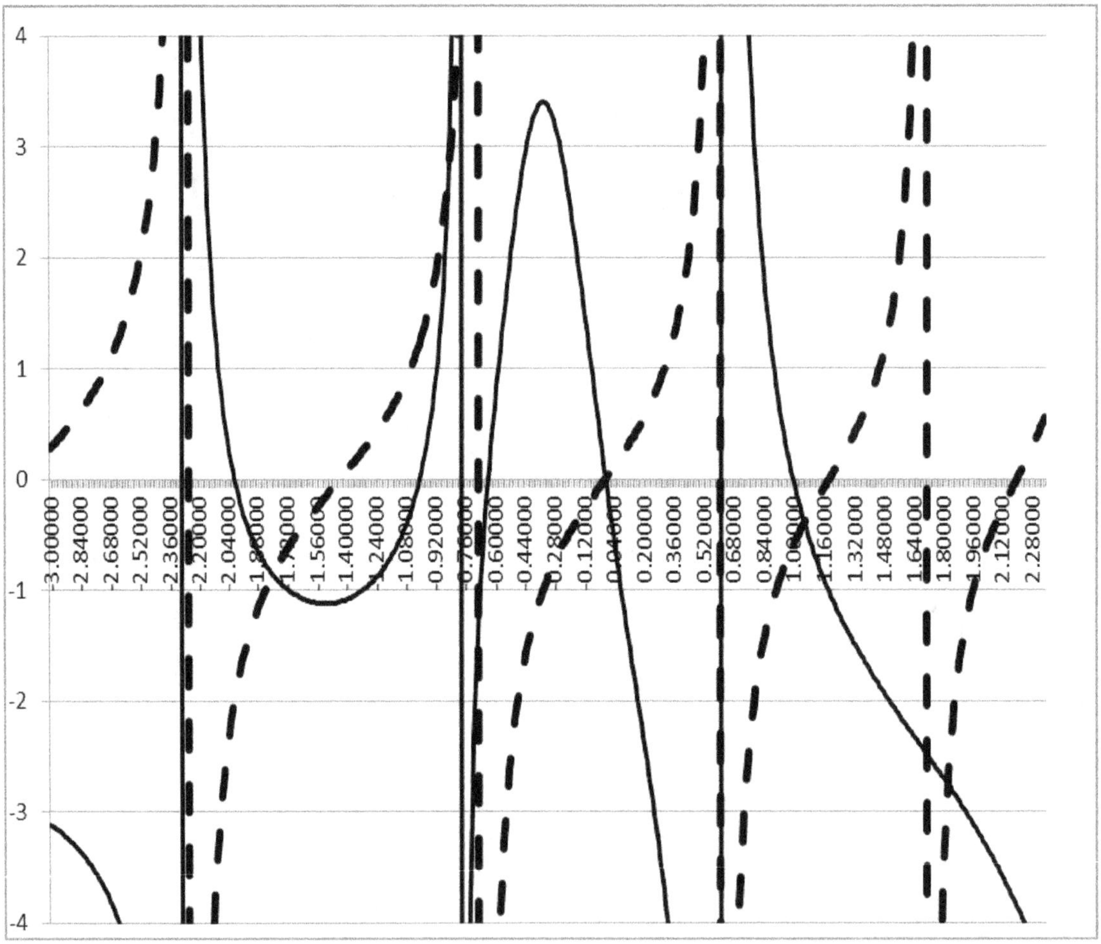

Figure 23.2 The QED fine structure eigenvalue α(g) is a solid line, while the tangent form of α(g) is the broken line. The constants are d_{Gd} = 1.29911 - 0.08929g, and the α_G constant is $d_{G\alpha}$ = 0.00348417 for G = QED. α(g) is plotted vertically. The exponent g is plotted horizontally.

24. Origin of Higgs Particles

Now we turn to the other major numerical issue of the Unified SuperStandard Theory (and the Standard Model): the origin Higgs fields and determination of Higgs Mechanism masses. As in the case of coupling constants, our goal is to answer these questions within the models without the introduction of new extraneous considerations.

Higgs particles appear in many contexts in the Unified SuperStandard Theory. In this chapter we consider a possible origin of Higgs particles in complex-valued gauge fields that explains why there are ElectroWeak sector Higgs fields but there no Strong Interaction sector Higgs particles.

We also show the Higgs Mechanism for fermion particle masses may explain the equality of inertial and gravitational mass – a topic of continuing interest for many years.

24.1 The Genesis of Scalar (Higgs?) Particle Fields from Complex Gauge Fields

In the past[107] we showed that scalar particles can be 'extracted' from all spin 1 gauge fields except color SU(3).

Since our Unified SuperStandard Theory is ultimately based on the Complex General Coordinate Transformations and the Complex Lorentz group (thus complex-valued coordinate systems), it appears reasonable to assume all spin 1 gauge fields to initially be similarly complex-valued. Most of these gauge fields can be rotated to real values. However we shall see that color SU(3) gauge fields are *necessarily* complex-valued. All other gauge fields can be rotated to real values. The price of rotation is the introduction of scalar fields. Some of these fields are Higgs particle fields and generate gauge boson masses (symmetry breaking) and fermion masses.

---

[107] Blaha (2015c) and (2016c).

Thus we can view scalar particles including Higgs particles as inherently associated with gauge fields. From the viewpoint of our derivation from basic axioms, the origin of Higgs bosons in complex-valued gauge fields gives a 'tighter' derivation of the overall theory. Higgs fields are necessarily part of our theory.

24.2 The Difference between the Strong Gauge Field and the Other Gauge Fields in the SuperStandard Theory

In our Unified SuperStandard Theory the only gauge field without an associated Higgs particle is the strong interaction gluon gauge field. *We view this exception as a particularly important clue as to the nature of the relation between gauge fields and Higgs particles.*

How does the strong interaction gauge field differ from all other gauge fields in the Unified SuperStandard Theory? An examination of the gauge fields dynamic equations (and other lagrangian terms) of our Unified SuperStandard Theory reveals that all gauge field dynamic equation kinetic terms *except those of the strong interaction gauge field* have the form:

$$\partial/\partial x_\mu F^a{}_{\mu\nu} + gf^{abc}A^{b\mu} F^c{}_{\mu\nu} = j^a{}_\nu \tag{24.1}$$

with

$$F^a{}_{\mu\nu} = \partial/\partial x^\nu A^a{}_\mu - \partial/\partial x^\mu A^a{}_\nu + gf^{abc}A^b{}_\mu A^c{}_\nu \tag{24.2}$$

where the coordinates x^ν *are real-valued,*[108] where a, b, c are structure constant indices, where g is a coupling constant, and where $j^a{}_\nu$ is the corresponding current.

The gauge field $A^a{}_\mu$ is real for all normal and Dark ElectroWeak gauge fields, Generation group gauge fields, and Layer group gauge fields. Thus the above equations are real-valued.

The strong interaction gauge field[109] in our Unified SuperStandard Theory differs from the other gauge fields by being *necessarily* complex[110] due to the complex 3-space complexon derivatives that appear in the corresponding dynamic equations:

---

[108] Before the introduction of the Two-Tier formalism.

$$D^{\mu}\, F_{C\ \mu\nu}^{\ a} + gf^{abc} A_C^{\ b\mu}\, F_{C\ \mu\nu}^{\ c} = j^{a}_{\ \nu} \tag{24.3}$$

with

$$F_{C\ \mu\nu}^{\ a} = D_\nu A_{C\ \mu}^{\ a} - D_\mu A_{C\ \nu}^{\ a} + gf^{abc} A_{C\ \mu}^{\ b} A_{C\ \nu}^{\ c} \tag{24.4}$$

where complexon derivatives have the form

$$D_k = \partial/\partial x_r^{\ k} + i\ \partial/\partial x_i^{\ k}$$

$$D_0 = \partial/\partial x^0$$

for k = 1, 2, 3 where $A_{C\ \mu}^{\ a}$ is the complexon color Strong interaction gauge field. Complexon spatial coordinates have the form $x_r^{\ k} + i\ x_i$. The time coordinate is real-valued. These equations are eqs. 10.16 and 5.162 of Blaha (2015a) for complexon gauge fields,[111] the carriers of the strong interaction in the Unified SuperStandard Theory.

This difference enables us to differentiate the strong gauge field from all other gauge fields in The Unified SuperStandard Theory. Thereby we can develop a unified formalism for the non-strong gauge fields and their corresponding Higgs particles.

The necessarily complex nature of the color SU(3) field is the reason that the Strong Interaction gauge fields do not acquire a mass via the Higgs Mechanism. As shown below, the necessary complexity of Strong Interaction gauge fields precludes the generation of Higgs fields from Strong Yang-Mills gauge fields.

24.3 Generation of Higgs Fields from Non-Abelian Gauge Fields

In the prior section we considered the difference between the strong gauge field and the other gauge fields of The Unified SuperStandard Theory. Unlike strong gauge fields the other gauge fields (ElectroWeak and so on) could be real or complex. In a

---

[109] This field is called a complexon gauge field in Blaha (2017b), (2015a) and earlier books.
[110] One cannot cleanly separate the real and imaginary parts of its dynamic equations.
[111] In The Unified SuperStandard Theory we also identify quark species particles as having complex 3-momentum. We call them complexon fermions.

manner similar to what we did in the preceding *Physics is Logic* books (and earlier books) we can assume gauge fields are initially complex, and then transform them to real-valued fields using a phase transformation that introduces scalar fields, some of which we will take to be Higgs fields.

We define a complex phase transformation for a gauge field $A^{b\mu}$ with

$$A'^{a\mu}(x) = \Phi(x)^a{}_b A^{b\mu}(x) \tag{24.5}$$

where $\Phi(x) = \text{diag}(\exp[i\varphi_1(x)], \exp[i\varphi_2(x)], \ldots , \exp[i\varphi_n(x)])$, and n is the number of symmetry components of $A^{b\mu}$. Inserting $A'^{a\mu}(x)$ above we find:

$$\partial/\partial x_\mu F'^a{}_{\mu\nu} + gf^{abc} A'^{b\mu} F'^c{}_{\mu\nu} = j^a{}_\nu \tag{24.6}$$

where

$$F'^a{}_{\mu\nu} = \partial/\partial x^\nu\{\exp[i\varphi_a(x)]A^a{}_\mu\} - \partial/\partial x^\mu\{\exp[i\varphi_a(x)]A^a{}_\nu\} + gf^{abc}\exp[i\varphi_b(x)]\, A^b{}_\mu \exp[i\varphi_c(x)]A^c{}_\nu \tag{24.7}$$

If we now assume that $\varphi_a(x)$ is small for all a then

$$\exp[i\varphi_a(x)] \simeq 1 + i\varphi_a(x)$$

to first order. Substituting above, and keeping terms to leading order yields the real part:

$$\partial/\partial x_\mu F^a{}_{\mu\nu} + gf^{abc} A^{b\mu} F^c{}_{\mu\nu} = j^a{}_\nu \tag{24.8}$$

where $F^a{}_{\mu\nu}$ is given above, and the imaginary part is:

$$\partial/\partial x_\mu F_i{}^a{}_{\mu\nu} + gf^{abc} A^{b\mu} F_i{}^c{}_{\mu\nu} = 0 \tag{24.9}$$

to leading order where

$$F_i{}^a{}_{\mu\nu} = \partial/\partial x^\nu\, \varphi_a(x)A^a{}_\mu - \partial/\partial x^\mu\, \varphi_a(x)A^a{}_\nu$$

Then we find

$$A^a{}_\nu\Box\varphi_a(x) - A^a{}_\mu\, \partial/\partial x_\mu \partial/\partial x^\nu\, \varphi_a(x) - gf^{abc}A^{b\mu}\, [A^c{}_\mu\, \partial/\partial x^\nu\, \varphi_a(x) - A^c{}_\nu\, \partial/\partial x^\mu\, \varphi_a(x)] = 0 \tag{24.10}$$

in the Lorentz gauge, with no sum over a. This equation is a form of Klein-Gordon equation having interaction terms with the gauge field. If the gauge field is weak then only the first two terms are important.

Note that only derivatives of $\varphi_a(x)$ appear above. Consequently shifts of the $\varphi_a(x)$ field by a constant still yield solutions. This feature makes $\varphi_a(x)$ a candidate to be a Higgs particle field.

Note also that complexon gauge fields such as the Strong Interaction fields cannot have such a phase change, with a subdivision into real and imaginary dynamic equations, due to the complexity of the spatial coordinates. This difference appears to be the reason why the strong interaction gauge fields do not have an associated Higgs particle.

The $\varphi_a(x)$ particles can be made into Higgs particles by adding an appropriate potential:

$$V = A\, \varphi_a{}^2(x) + B\, \varphi_a{}^4(x) \tag{24.11}$$

where A and B are constants. Approximating with its first two terms and inserting the potential term (with an $A^a{}_\nu$ factor) we find the Higgs-like equation:

$$A^a{}_\nu\Box\varphi_a(x) - A^a{}_\mu\, \partial/\partial x_\mu \partial/\partial x^\nu\, \varphi_a(x) + A^a{}_\nu \partial V/\partial\varphi_a = 0 \tag{24.12}$$

$\varphi_a(x)$ has a minimum at the minimum of the potential in the corresponding lagrangian. The second and third terms constitute the interaction. Neglecting these terms we see that we obtain the free, massless, field Klein-Gordon equation

$$\Box\varphi_a(x) = 0 \tag{24.13}$$

The pairing of Higgs particles with real-valued gauge fields is thus established.[112] The non-existence of a matching Higgs field for the Strong Interaction is due to the inherently complex nature of the strong interaction (complexon) gauge field in the Unified SuperStandard Theory also follows.

The derivation presented here is analogous to the derivation of Higgs fields in Complex General Relativity – also a gauge theory – in our *Physics is Logic Part II.*

One of the remarkable aspects of The Unified SuperStandard Theory is its ability to directly prove qualitative properties of elementary particles: four fermion species, Parity violation, the distinction between leptons and quarks, the match of the SuperStandard Theory's (broken) symmetries with the internal symmetry Reality group, and now the existence of Higgs gauge fields in all interaction sectors except for the strong interactions. We take these successes to be indicators of the correctness of The Unified SuperStandard Theory.

24.4 General Higgs Formulation of Gauge and Fermion Particle Masses

In Blaha (2018e) we present a formulation of the theory that applies the Higgs Mechanism to determine fermion and boson masses.

---

112 Some of the Higgs fields so generated may not have vacuum expectation values and so may only play a role in interactions.

25. Unified SuperStandard Theory Successes

In Blaha (2019e) and this companion volume we have developed the Unified SuperStandard Theory from Complex General Relativity and Quantum Field Theory (extended by the Two Tier and Pseudoquantum formalisms). From these beginnings we have postulated five axioms (chapter 1) that specify the generation of the Unified SuperStandard Theory.

The directly generated theory contains:

1. A direct product symmetry $[SU(2)\otimes U(1)\otimes SU(3)\otimes SU(2)\otimes U(1)]^4\otimes U(4)]^9$

2. A 192 fermion and a 192 vector boson spectrum of fundamental particles. The spectrum conforms to astrophysics data on the proportions of Dark and normal matter and energy.

3. A set of interactions based on the Riemann-Christoffel tensor that yield the known interactions including a quark confining Strong Interaction and a gravity with three regions of potential that agrees with MoND.

4. A set of coupling constants calculated from vacuum polarizations that agrees with known coupling constants.

5. A source for Higgs particles in complex-valued non-abelian gauge fields.

6. An elimination of all divergences in perturbation theory using our Two-Tier formalism

7. Support for second quantization in all reference frames, and for canonically implemented higher order interactions using our Pseudoquantum formalism.

8. Elimination of "spookiness" of quantum entanglement by dividing each quantum field into the product of a functional (qubes and qubas) and a fourier expansion. "Immediate action without a distance." See Blaha (2018e).

9. Support for a Physics for the universe based on internal characteristics of the theory.

Thus we have a theory of Physics (and Chemistry and Biology) that has the experimentally known features of the Standard Model but goes beyond it with new features that will eventually be found in experiment.

26. Major Implications For Universes: SuperUniverses

Our ability to calculate the SuperStandard Theory coupling constants and, as we shall see, the universe vector interaction coupling constant (and universe scale factor) raises important issues both within our universe and in possible other universes. (The experimental and theoretical basis of the possibility of other universes is described in earlier books such as Blaha (2018).)

In our universe we see that the Anthropic hypothesis, based on the precise value of the QED Fine Structure Constant, is not compelling. The value of α is set within QED not otherwise. In addition the other known Standard Model coupling constants, which were found to good approximation in our approach, are also *self-determined* in Quantum Field Theory. Since the Standard Model interactions are responsible for the vast majority of the features of Physics, Chemistry and Life we find the universe's detailed structure is set by the Unified SuperStandard Theory.

If, as we have proposed, there are other four-dimensional (4D) universes within a higher dimension space called the Megaverse, then, assuming that the overall physical theory of a four-dimensional (4D) universe[113] satisfies

1. The universe is described by Complex General Relativity.

2. The particles, and particle interactions, in the universe are those of the SuperStandard Theory.

we find 4D universes are based on the same Unified SuperStandard Theory. We call such universes SuperUniverses and their common theoretic underpinning the SuperUniverse Model.

---

[113] And possibly of other universes of different dimension.

The conclusions that follow from this line of reasoning are:

1. Standard Model features are the same in all 4D universes.

2. A SuperUniverse has the same form for the scale factor, and a similar evolution as other universes and parallels our universe. We therefore expect a SuperUniverse to have Dark Energy dominance, varying Hubble Constants; Big Dips, Superclusters, and voids in other universes. (The total energy of universes may differ.)

3. SuperUniverses have the same Physics, Chemistry and Biology as our universe. Life in other universes is possible but can be expected to take different forms.

4. SuperUniverses can differ in some ways that do not change our overall conclusions: in some SuperUniverses a) anti-particles may dominate; b) left-right symmetry may differ; c) Life may favor either dextrorotary or levorotary molecules;

5. Communication with creatures in other universes may be possible using more advanced quantum entanglement. Quantum entanglement of photons from earth and the sun has been found experimentally.

6. Universes, as a whole, have a composite particle nature and can be described in a second quantized framework. See Blaha (2018e), and comments there by DeWitt in particular on quantum universe features.

The following chapters begin by deriving and describing the expansion of our universe from the Big Bang. Then we provide experimental evidence for other universes and touch on their features.[114]

[114] These features are described in Blaha (2018e) and earlier books by the author together with suggestions for inter-universe travel.

27. The Expansion of Our Universe

Our universe is clearly expanding from an initial state called the Big Bang to its current state. The Hubble "Constant" measures the rate of growth. Most of the expansion data on the Hubble Constant is presented in section 22.1. After reviewing the data we propose a fit to the data in this chapter that explains the apparent growing rate of expansion. In chapter 28 we derive the form of the fit.

22.1 Hubble Constant Experimental Data

There are a number of astrophysical studies of the universe that suggest that the Hubble Constant is *not* constant. Although there are significant margins of error it appears that the early universe "beginning" epoch around 380,000 years had a Hubble Constant of 67.8 km s^{-1} Mpc^{-1}.[115] More recently, red shift studies of quasars have given a Hubble Constant of 73.2 km s^{-1} Mpc^{-1}.[116] And studies of binary black hole merger gravity waves[117] have given a Hubble Constant of 75.2 km s^{-1} Mpc^{-1} (and earlier of 78 km s^{-1} Mpc^{-1}). Another study of events at 1.8 billion ly yielded a Hubble Constant of 70.0 km s^{-1} Mpc^{-1}.[118] Further studies have given the Hubble Constants: 1) Of variable stars 73.2 km s^{-1} Mpc^{-1}, 2) Of light bent by distant galaxies 72.5 km s^{-1} Mpc^{-1}, 3) Of Magellan Cepheids 74.03 ± 1.42 km s^{-1} Mpc^{-1}, [119] 4) Of distant red giant[120] brightness 69.8 km s^{-1} Mpc^{-1},

[115] See, for example, K. Aylor *et al*, arXiv:1811.00537v1 (2018) based on studies of the cosmological sound horizon.
[116] M. Soares-Santos *et al* , arXiv:1901.01540 (2019).
[117] DES and LIGO collaborations *et al*, arXiv:1901.01540 (2019).
[118] B.P. Abbott *et al*, arXiv:1710.05835 (2017).
[119] J. T. Nielsen *et al*, Marginal evidence for cosmic acceleration from Type Ia supernovae, Nature Scientific Reports (2016); arXiv:1506.01354 (2015). A. Riess *et al*, The Astrophysical Journal **875**, 145 (2019) and references therein. A. Riess *et al*, arXiv:1903.07603 (2019).
[120] W. Freedman *et al*, The Astrophysical Journal **880** (July, 2019).

The only apparent conclusion at this time is that there was a Hubble Constant (Constant) H of approximately 67.8 km s^{-1} Mpc^{-1} early in the universe, and ranging up to 75.2 km s^{-1} Mpc^{-1} at the current time. Thus an increasing Hubble Constant.

For the purpose of discussing the apparent increase in H with time, we average the above eight "recent" values of H in the spirit of Bayesian equal probability to obtain a **recent time Hubble average of 73.24** km s^{-1} Mpc^{-1}.[121] Thus there appears to be a 7% - 9% increase in the Hubble Constant over time.

27.2 Fit to the Hubble Constant Data and Scale Factor

It is generally expected that the Hubble Constant will decline with time from the time of the Big Bang. It is generally believed that the Hubble Constant has recently been increasing with time. **The declining value in the past and the current growth of the Hubble Constant imply that it reached a minimum at some time in the past.**

Our fit to the data from Blaha (2019c) and (2019e) was

$$\begin{aligned} a(t) &= (t/t_{now})^{g + ht} \\ &= \exp[(g + ht)\ln(t/t_{now})] \end{aligned} \tag{27.1}$$

where g and h are constants. (The constant h is *not* the Hubble parameter.) There is an "ht" term in the exponent based on the rise in H(t) suggested by experimental data.

The basis of this choice was:

1. Power law behavior (in part) as in the radiation and matter dominated approximations.

2. The known shape of H(t) at early times, and at present, as described above

3. The simplicity of the fit. Two values of H(t) set the constants g and h.

4. Faster than exponential future growth with no Big Rip.

---

[121] In Blaha (2019c) and (2019e) we used an average estimate of 73.7 km s^{-1} Mpc^{-1}.

The Hubble Constant implied by eq. 27.1 is

$$H(t) = (da/dt)/a = g/t + h(1 + \ln(t/t_{now}) \qquad (27.2)$$

We set the value of H(t) by using its value at two values of time determining g and h. Based on experimental data:

$$H(t_c) \equiv H(380{,}000\text{ yr}) = 67.8 \text{ km s}^{-1} \text{ Mpc}^{-1} \qquad (27.3)$$
$$H(t_{now}) = 73.24 \text{ km s}^{-1} \text{ Mpc}^{-1}$$

and

$$h = (t_c\, H(t_c) - t_{now}H(t_{now}))[\, t_c - t_{now} + t_c \ln((t_c/t_{now})]^{-1} \qquad (27.4)$$
$$g = (H(t_{now}) - h)\, t_{now}$$

where t_c = 380,000 years after the Big Bang.[122] We obtained

$$h = 2.25983 \times 10^{-18} \text{ s}^{-1} = 1.49 \times 10^{-33} \text{ eV} \qquad (27.5)$$
$$g = 0.000282377 = 2.82377 \times 10^{-4}$$

27.3 Fluctuating Behavior of a(t) and H(t) – The Big Dip

An examination of the following Figs. 27.1 – 27.4 reveal a Big Dip in H(t) (and also in a(t)) which seems to have been unforeseen in astrophysical investigations.

The cause of the Big Dip is the form of the universal scale factor as described in section 27.2. It would be present (although slightly modified) even if the Hubble Constant were truly constant from the 380,000 year point to the present.

Figs. 27.6 through 27.9 plot H(t) (and its logarithm) from the Big Bang to the present and beyond. H(t) has a wide range of values from very large near the Big Bang through the Big Dip to the present and beyond. The recent growth in H is displayed.

27.3.1 Location of the Big Dip in H(t)

The locations in time of the Big Dip events are:

[122] Based on the data value of 67.8 km s^{-1} Mpc^{-1} at t = 380,000 years.

Big Dip low point (where H = -445 km s^{-1} Mpc^{-1}) at t = 8.71×10^{13} sec.

Big Dip low in a(t) has the value 0.69628 at t = 1.56×10^{17} sec.

The decrease in a(t) is to 69.6% of the initial value of 1. The delay from the H(t) Big Dip value to the low point of a(t) may be attributable to the time required to propagate the diminished H value throughout the universe.

27.3.2 Possible Reason for the Big Dip in H(t)

Since the H and Ω_T times match it appears that the Big Dip is a result of massive growth in the mass-energy (and pressure). See chapter 29.

Since the changeover from a radiation-dominated phase to a matter-dominated phase occurs at a slightly earlier time:

Radiation – Matter Domination Transition:[123] t = 1.48×10^{12} sec.

it seems reasonable to conclude the transition from radiation-dominated to matter-dominated causes the Big Dip to occur. The matter-dominated phase transition causes shrinkage as shown in a(t) in Fig. 27.4. *The universe contracts by one-third!* [124] We attribute the time delay between the transition and the Big Dip in a(t) to the time required for the transition to occur. (The universe is large at this time after all)

---

[123] In view of our universal scale factor formulation the time of the radiation-matter transition becomes questionable.
[124] Rather like the condensation of water vapor to liquid.

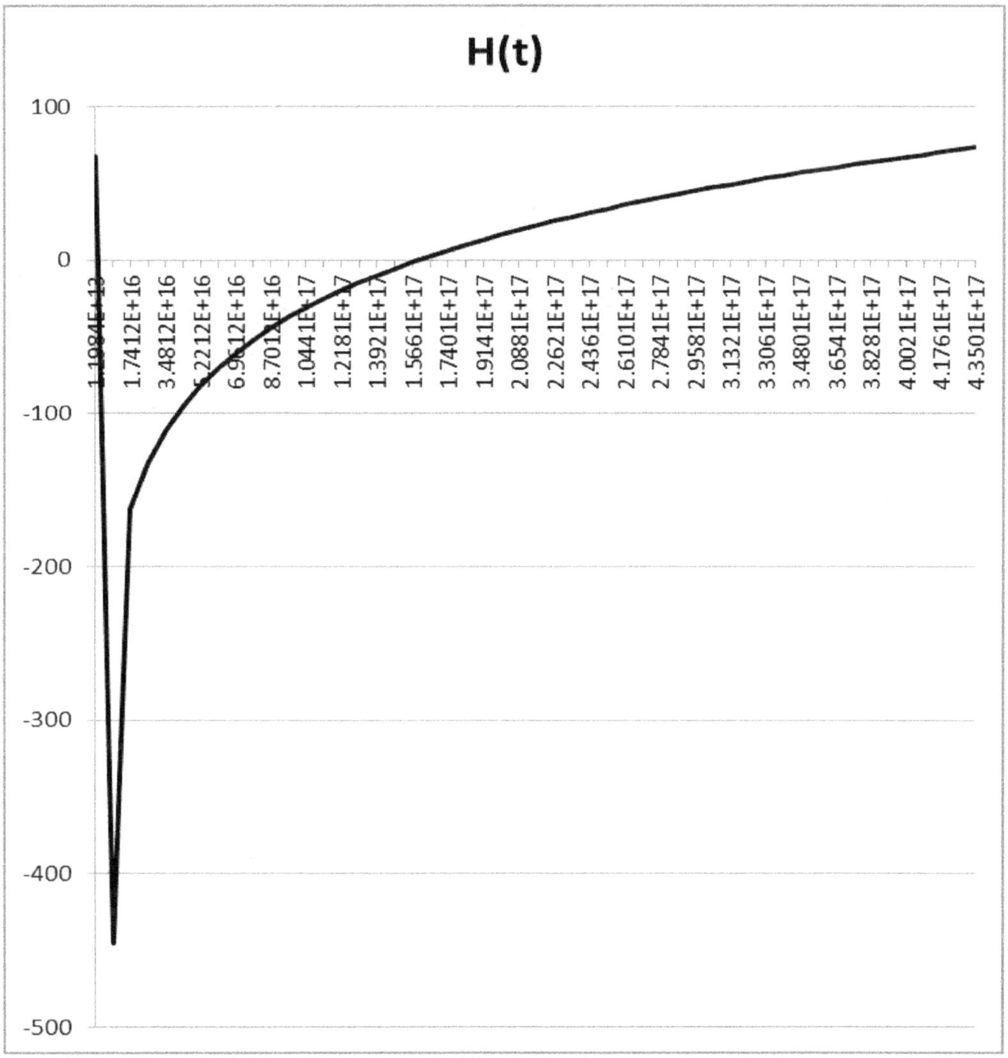

Figure 27.1 The Hubble Constant H(t) plotted vs. seconds from $t = 1.19 \times 10^{13}$ sec. to the present 4.35×10^{17} sec. The minimum is H = -445 km s^{-1} Mpc^{-1} at $t = 8.71 \times 10^{13}$ sec.

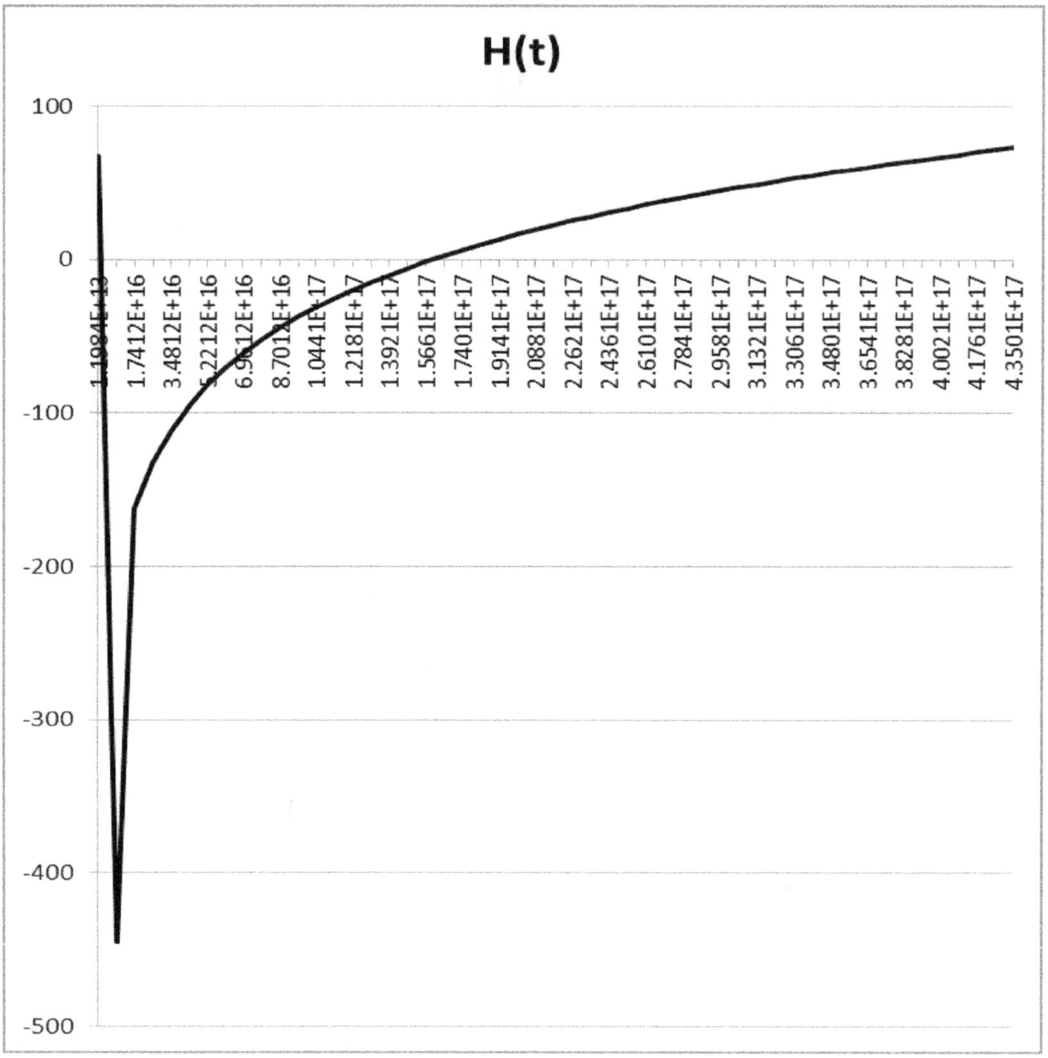

Figure 27.2 A closer view of H(t) plotted vs. seconds from t = 1.19 × 10^{13} sec. to the present 4.35 × 10^{17} sec. The minimum is H = -445 km s^{-1} Mpc^{-1} at t = 8.71 × 10^{13} sec.

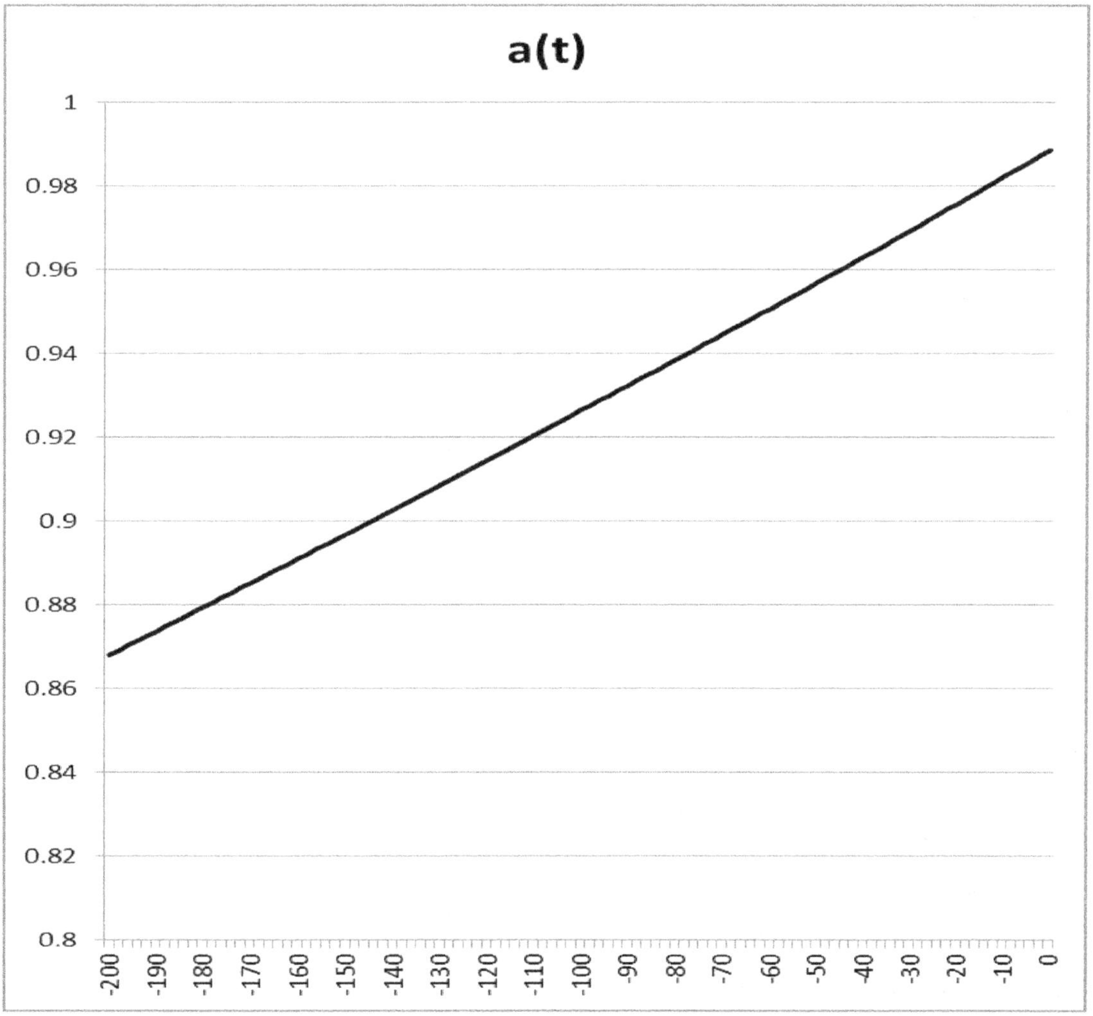

Figure 27.3 Universal Scale Factor a(t) plotted in $\log_{10}$ seconds from $t = 10^{-200}$ sec. to t = 1 sec.

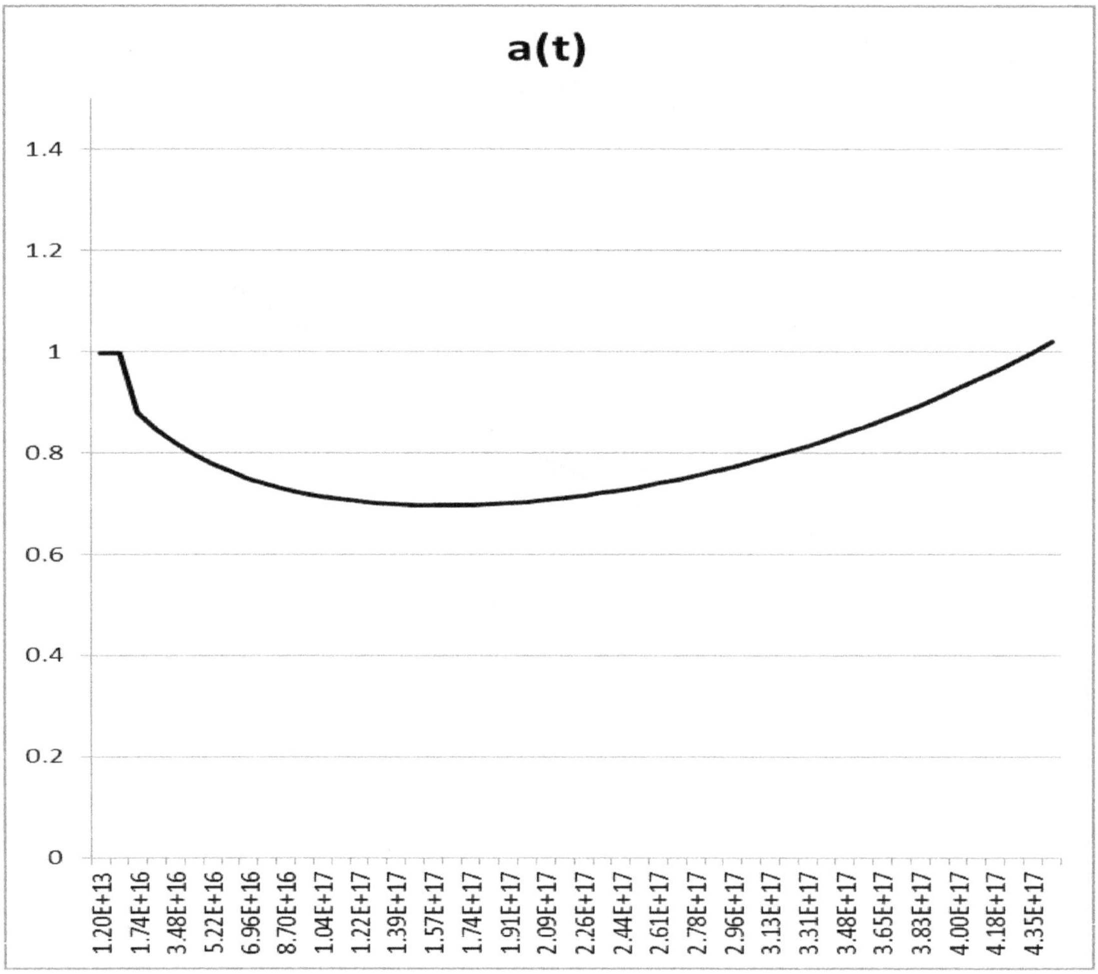

Figure 27.4 Universal Scale Factor a(t) plotted in seconds from $t = 1.19 \times 10^{13}$ sec. (380,000 years) to $t = 4.35 \times 10^{17}$ sec. (the present). Note the Big Dip to a = 0.69628 at $t = 1.56 \times 10^{17}$ sec. The decrease in a(t) is to 69.6% of the initial value of 1. (The almost flat part before 1.20×10^{13} sec is a roundoff effect. See Fig. 22.3.)

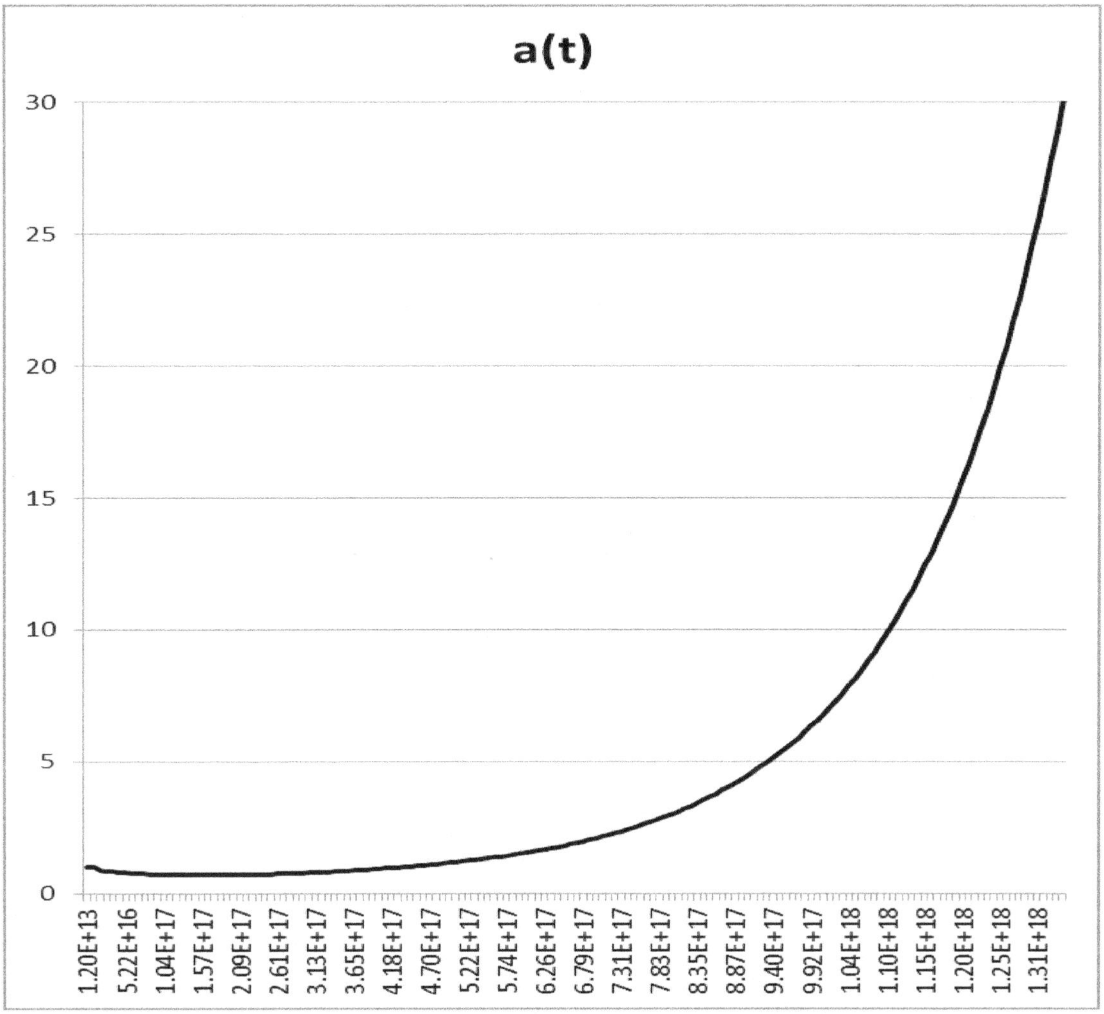

Figure 27.5 Universal Scale Factor a(t) plotted in seconds from $t = 1.19 \times 10^{13}$ sec. (380,000 years) to $t = 1.31 \times 10^{18}$ sec. (the distant future). Note the Big Dip appears diminished due to the scale of the plot.

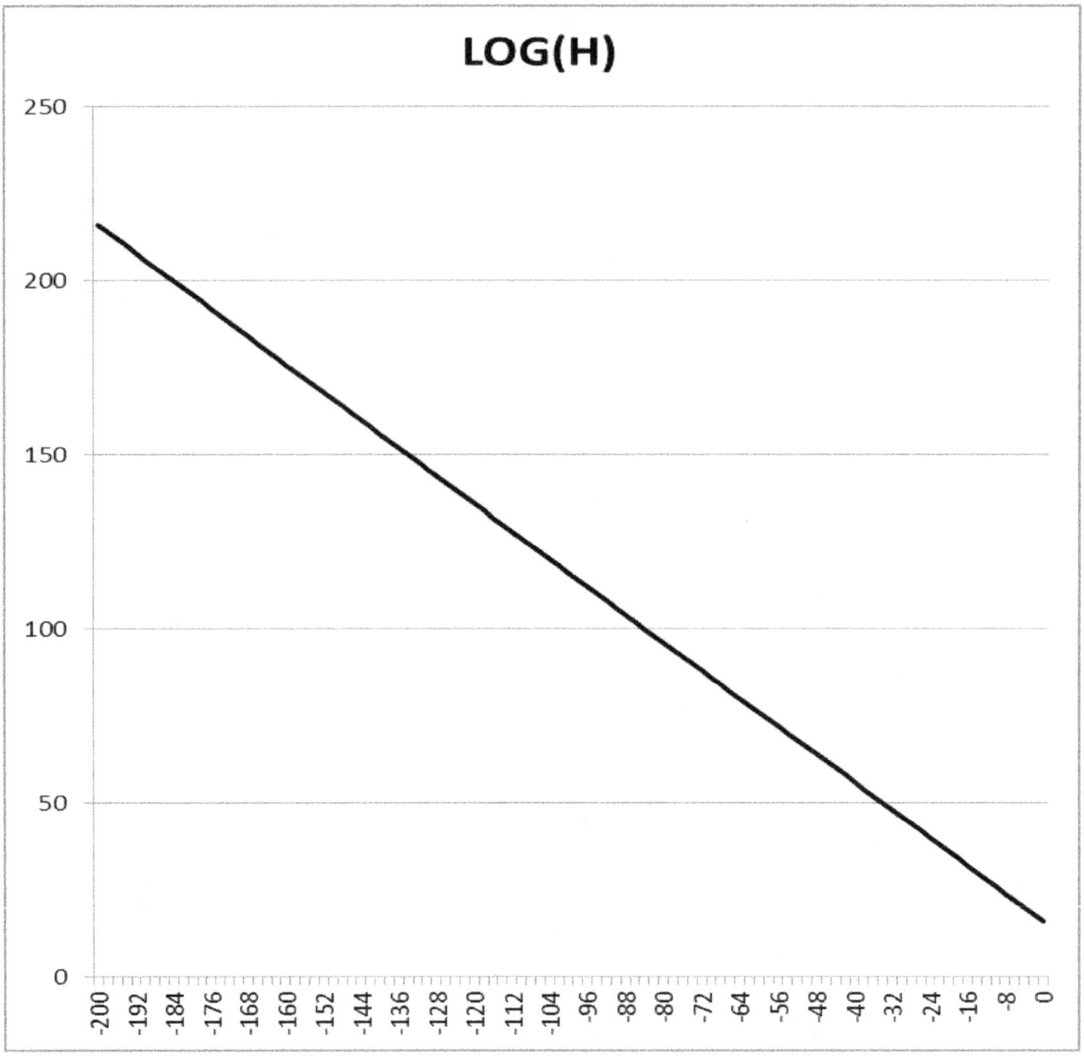

Figure 27.6 Log_{10} H(t) plotted in $\log_{10}$ seconds from t = 10^{-200} sec. to 1 sec.

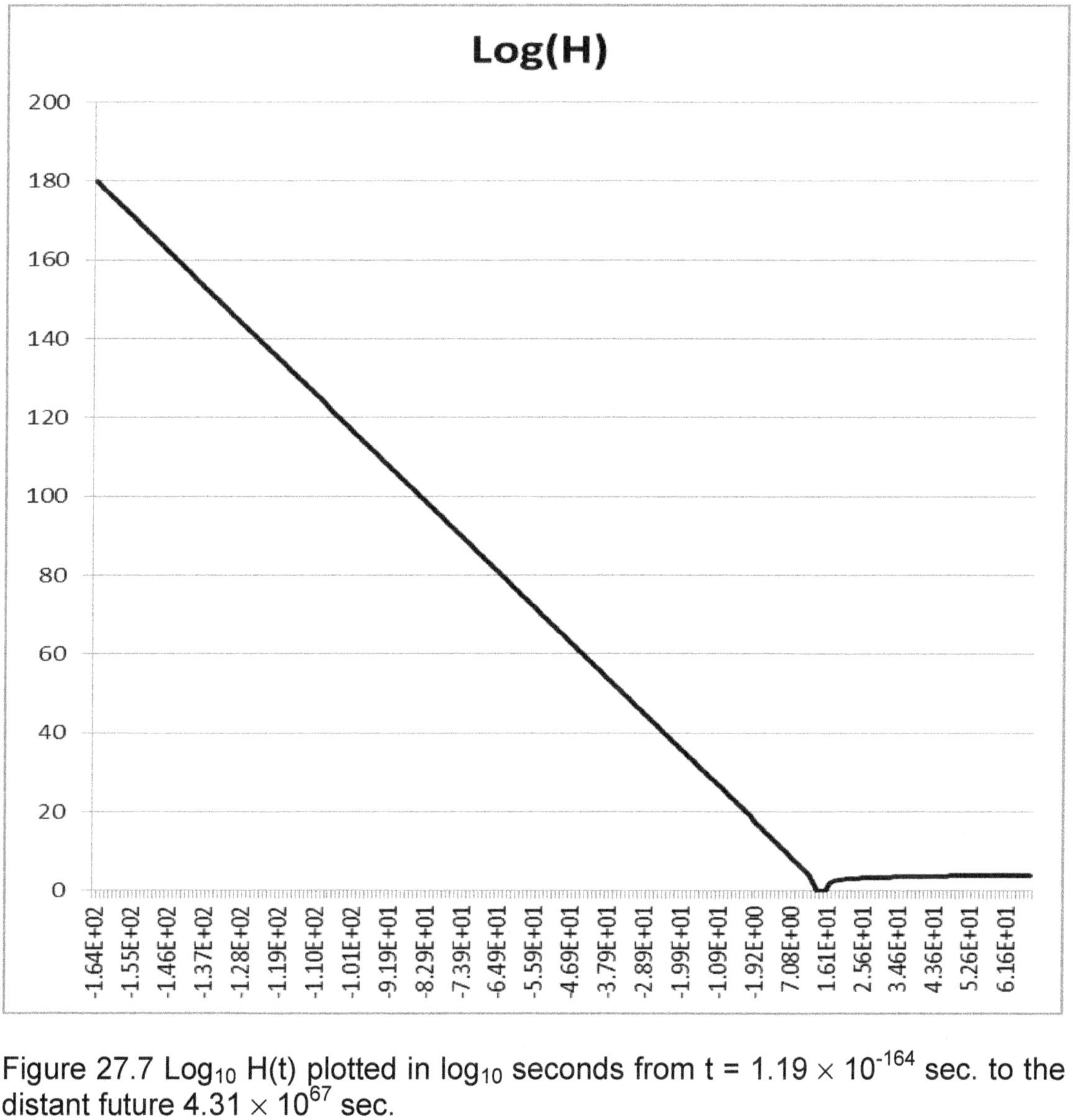

Figure 27.7 $\log_{10}$ H(t) plotted in $\log_{10}$ seconds from $t = 1.19 \times 10^{-164}$ sec. to the distant future 4.31×10^{67} sec.

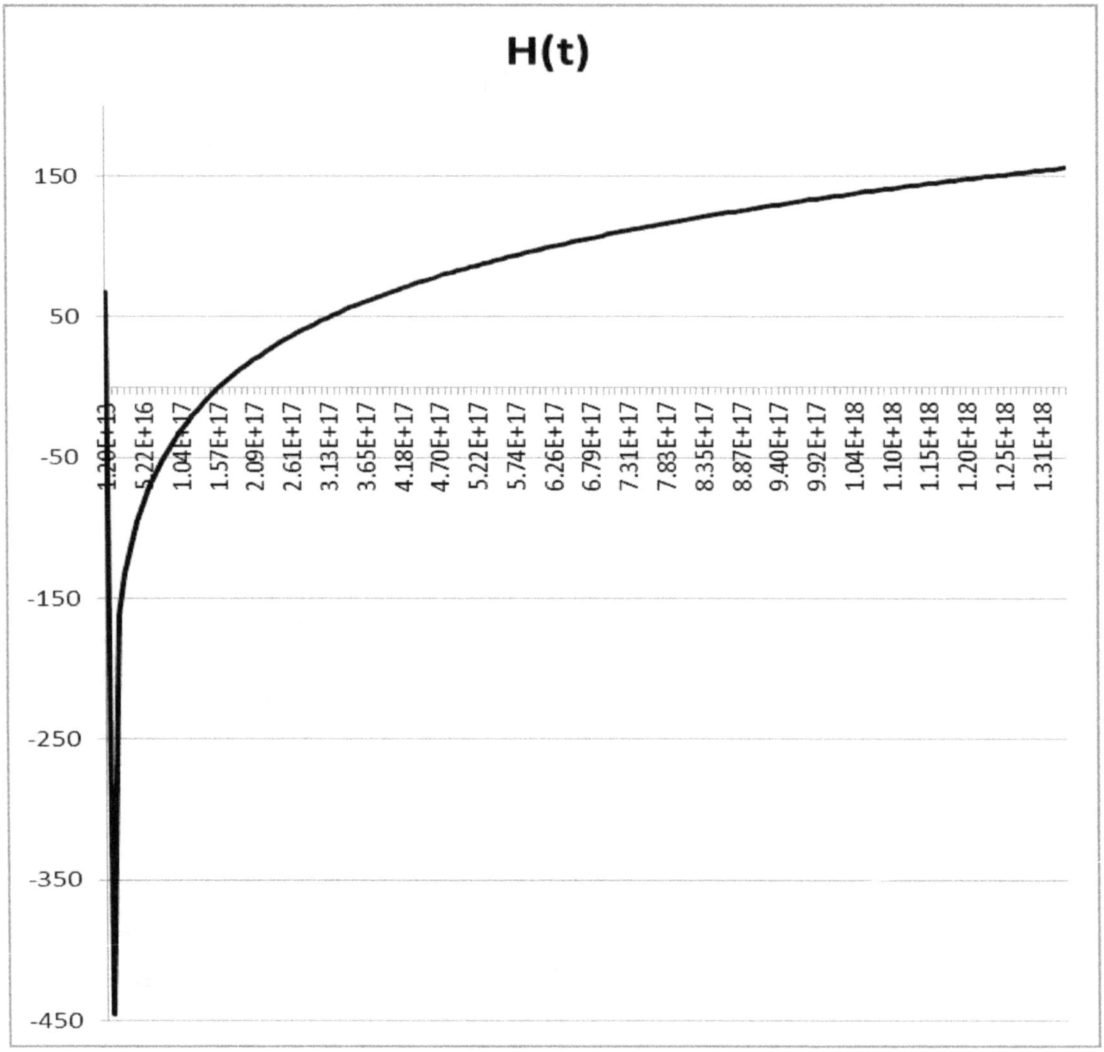

Figure 27.8 H(t) plotted vs. seconds from t = 1.19×10^{13} sec. (380,000 years) to 1.31×10^{18} sec. (three times the present time) showing the Big Dip in H(t).

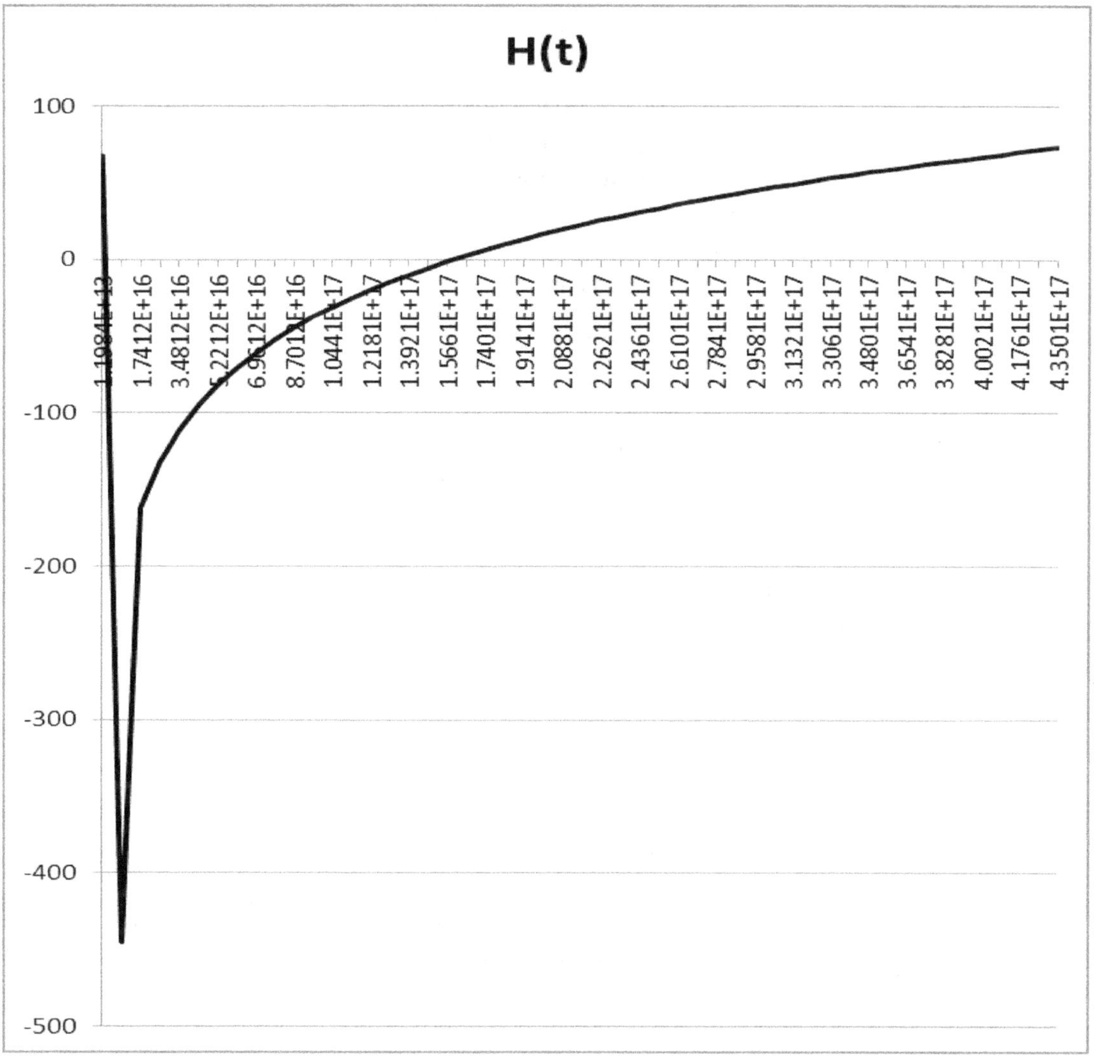

Figure 27.9 A closer view of H(t) plotted vs. seconds from t = 1.19×10^{13} sec. to the present 4.35×10^{17} sec. The minimum is H = -445 km s^{-1} Mpc^{-1} at t = 8.71×10^{13} sec.

27.4 Big Bang Metastate Model Scale Factor and Hubble Constant

In Blaha (2019c), (2019d) and (2019e) and earlier books the author developed a model of a Quantum Big Bang which avoided singularities at t = 0 using quantum effects. The results of this model appear below and they can be seen to approximately match at very early times with the scale factor and Hubble Constant values implied by eq. 27.1.

27.4.1 Big Bang Scale Factor a(t) and H(t)

The real part of the complex-valued Big Bang scale factor is

$$\text{Re } a_{BBRW}(t_c) \cong \gamma = a(t_c) \qquad (27.6)$$

from Blaha (2019c) where

$$\gamma = 1.632 \times 10^{-92}$$
$$t_c = 1.1984 \times 10^{-165} \text{ s}$$

We can see the removal of the singularity at t = 0 by combining the Big Bang metastate scale factor $a_{BBRW}(t)$ with a(t):

For $0 \leq t \leq t_c$

$$a_{combined}(t) = \gamma/2 + a(t)/2 = \text{Re } a_{BBRW}(t) \qquad (27.7)$$

For $t_c \leq t \leq t_{now}$

$$a_{combined}(t) = a(t) \qquad (27.8)$$

Note a(0) = 0 while $a_{combined}(0) = \gamma/2 = 8.16 \times 10^{-93}$. Thus a Big Bang catastrophe is averted. Fig. 27.10 contains relevant values of the Big Bang Model.

Time	0	t_c
Phase	Big Bang Metastate Beginning	Big Bang Metastate End
Time	0	1.26×10^{-165}
Re a(t)	8.16×10^{-93}	1.632×10^{-92}
Im a(t)	-3.16×10^{-93}	-5.24×10^{-93}
Re radius	4.278×10^{-65} cm	8.5×10^{-65} cm
Im radius	4.278×10^{-65} cm	8.5×10^{-65} cm
Volume	4.37×10^{-192} cm^3	2.6398×10^{-191} cm^3
Central Expansion Energy Density	1.63×10^{218} GeV/cm^3	4.08×10^{217} GeV/cm^3
Edge Expansion Energy Density	8.16×10^{217} GeV/cm^3	2.04×10^{217} GeV/cm^3
Total Expansion Energy[125]	5.34×10^{35} eV	8.07×10^{35} eV
Hubble Constant[126] (km s^{-1} Mpc^{-1})	1.79×10^{218}	1.14×10^{126}

Figure 27.10 Big Bang Model Metastate detailed data.

[125] The expansion energy does not include the mass-energy of particles in the Big Bang universe. It only includes Y field black body energy which drives the initial expansion. All particles are massless initially and all interparticle forces are zero.

[126] In the Big Bang metastable state we display the Hubble constant at the "expanding' edge.

27.5 Smooth Fit Connection to Big Bang Metastate

The parameters in eqs. 27.1 and 27.5 were set by the H(t) data at 380,000 years and the present. If we extrapolate back to the end of the Big Bang metastate then we find a good match between the values of a(t) and H(t) of the Big Bang metastate and the extrapolation:

		H(t)
Big Bang Metastate	Big Bang Center	8.95×10^{217} km s^{-1} Mpc^{-1}
	Big Bang Edge	1.14×10^{126} km s^{-1} Mpc^{-1}
H(t) for the universal scale factor (and much larger as t → 0)	9.149×10^{215} km s^{-1} Mpc^{-1} at $t = 10^{-200}$ sec.	

where the Big Bang values appear in section 13.6.2 of Blaha (2019c).. Note that the extrapolated value is within the range of values in the Big Bang Metastate and is close (within a factor of 100) to the Big Bang Center value. Thus our H(t) fit (eqs. 27.1 and 27.5) extends smoothly back to the Big Bang. H(t) is in a rather remarkable approximate agreement for the Big Bang Metastate and our fit.

27.6 Universe Contraction – Early Massive Galaxies

The contraction of the universe at the Big Dip has important consequences for galaxy formation and their distribution. It may be relevant to recent studies that suggest synchronous behavior of galaxies and quasars separated by great distances.

The Big Bang contraction would appear to "squeeze" the mass-energy in the universe giving it a "belly" (the Big Belly??? of "squeezed" mass-energy).This mass-energy contraction leads to the early formation of galaxies, and their correlated behavior at large distances. The subsequent expansion creates a "type" of wave that generates massive galaxies (bubbles of mass-energy), and also voids – bubbles of space devoid of galaxies .The galaxies have dispersed in the 13.5 Gyrs that followed.

Evidence[127] has been found for the existence of a huge population of very massive galaxies (39+ have been found so far) that were created within one billion years after the Big Bang. This population of early galaxies is inconsistent with the standard present-day models of galaxy formation. The Big Dip occurs at 2.76 million years – well before one billion years – consistent with the formation of early massive galaxies.

A concentration of mass-energy due to the contraction of the universe appears to present a possible solution. Universe contraction was not considered in the creation of these models of galaxy formation.

Another possible source of universe concentrations (and voids) of energy appears in our Quantum Big Bang Model seen earlier. The cause there is a large difference in expansion rates (Hubble Constant variations) at the center of the Big Bang compared to the outer edge of the Big Bang as shown earlier in the section 2 Big Bang Model.

27.7 An Interlude in the Aeons

Based on the above analysis it appears that the universe is currently in an *interlude* following a decline in growth rate after the Big Bang, and a new beginning of major growth.

[127] T. Wang *et al,* Nature **572**, 211 (2019).

28. Proof of the Universal Scale Factor for the Expansion of the Universe

There are two approaches to the universal scale factor fit of eq. 27.1. One approach is based on a remarkable coincidence between the power g in the fit and the QED power g seen earlier. It leads to a theory in which the expansion of the universe taken over all time is a vacuum polarization phenomenon. The other approach is based on the Einstein equation for the scale factor. We show that the Universal Scale Factor is consistent with the Einstein equation if additional (dark) energy is properly taken into account.

28.1 Vacuum Polarization Generation of the Early Times Part of the Universal Scale Factor

Perhaps the crowning achievement of our universal scale factor eigenvalue formulation for coupling constants is the successful relation of universe evolution to vacuum polarization due to a vector QED-like interaction between universes.

28.1.1 Recap of Massless QED Vacuum Polarization

In massless QED we found that the vacuum polarization had the form:[128]

$$F_1(\alpha)(p/\Lambda)^{2g_{QED}} \tag{28.1}$$

where $F_1(\alpha)$ is the "eigenvalue function" for the Fine Structure Constant[129] of the Johnson-Baker-Willey model of massless QED, p is the momentum, and Λ is the

[128] Eq. 12 in S. Blaha, Phys Rev **D9**, 2246 (1973).

[129] The author calculated $\alpha = 1/137\ldots$ exactly (within experimental limits) in Blaha (2019a) and (2019b).

ultraviolet cutoff. The value of g_{QED} that corresponded to the Fine Structure Constant is[130]

$$g_{QED} = -\, 0.00058053691948 \tag{28.2}$$

and the Fine Structure Constant was correctly found (well within experimental limits) to be

$$\alpha_{calculated}(g_{QED}) = \ 0.0072973525693 \tag{28.3}$$

to 13 digit accuracy according to the Particle Data Table of 2019.

Comparing our Universal Scale Factor g value (eq. 27.5) with g_{QED} we find

$$-g \ = 0.000282377 \cong -\tfrac{1}{2} g_{QED} = -0.000290268 \tag{28.4}$$

28.1.2 Comparison of QED Vacuum Polarization Exponent with Universe Vacuum Polarization Exponent

Eq. 28.4 shows the numeric values of the g powers are approximately equal up to a factor of -2. The QED exponent describes high energy vacuum polarization behavior. The universe power g describes the small time universe expansion (near the Big Bang). The relation between the values of g and g_{QED} clearly suggests a close analogy.

Further the low energy (infrared) behavior of the QED vacuum polarization which is mass dependent is analogous to the large time (recent time) behavior of a(t) which is governed by the h term in the exponent of a(t).

The problem now before us is to find the universe vacuum polarization due to a new vector interaction between universes, and show that it is related to the QED vacuum polarization by eq. 28.4.[131]

---

[130] Chapter 27 of this book.

[131] The following subsections appeared in Blaha (2019c).

28.1.3 A New Vector Interaction for Universe Particles

We assume universes can be treated as particles in 4-dimensional space-time.[132] Since experiments appear to have shown that our universe does not rotate (does not have spin)[133] we will assume the universe is a spin 0 boson. We assume that universes have a vector field interaction similar to QED.

Given this QED-like framework, then universe-antiuniverse pair production and vacuum polarization becomes possible. We assume the QED-like boson lagrangian

$$\mathcal{L} = \tfrac{1}{2}\,(\partial_\mu\varphi^\dagger\partial^\mu\varphi - m^2\varphi^\dagger\varphi) - ie_0\!: \varphi^\dagger(\overrightarrow{\partial_\mu} - \overleftarrow{\partial_\mu})\,\varphi\!: A^\mu + e_0{}^2\!:\!A^2\!:\, :\!\varphi^\dagger\varphi\!: + \delta m^2\!:\!\varphi^\dagger\varphi\!: \tag{28.5}$$

where $\varphi(x)$ is a "charged" quantum universe scalar particle field[134] and A^μ is a QED-like field. We now proceed to calculate the second order vacuum polarization of a universe particle. We will assume the term in $\mathcal{L}$ linear in A^μ is the relevant term since the quadratic term always is negligible compared to the linear term in each order α^n of perturbation theory by a factor of α. The neglected terms will be assumed to not affect the calculated eigenvalue function.

---

[132] Universes are composite entities but we can treat them as quantum particles in the same manner as physicists treated protons and neutrons etc. as quantum particles before quark theory was accepted. See Blaha (2018e) for a detailed discussion of universe particles.

[133] The lack of universe rotation (spin) is indicated by a study of Cosmic Microwave Background (CMB) by D. Saadeh *et al*, Phys. Rev. Lett. **117**, 313302 (2016).

[134] The charge is not electromagnetic charge.

28.1.4 Second Order Vacuum Polarization of a Scalar Universe Particle

The one loop vacuum polarization Feynman diagram is

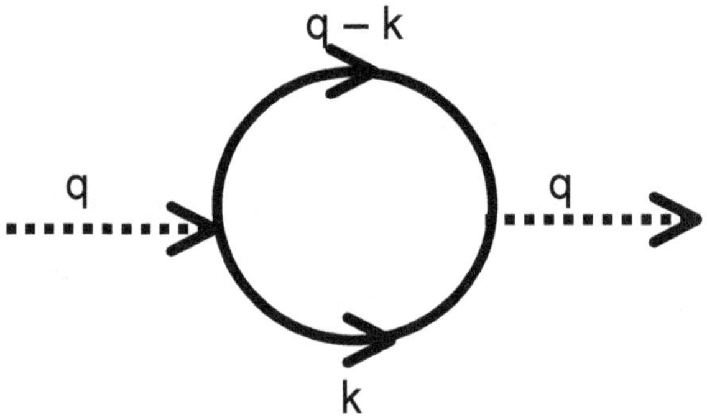

Figure 28.1 One loop vacuum polarization boson Feynman diagram.

Its evaluation is

$$I_{\mu\nu} = (-ie_0)^2 \int \frac{d^4k}{(2\pi)^4} \frac{i}{(k^2 - m^2 + i\varepsilon)} \frac{i}{(k^2 - m^2 + i\varepsilon)} (q - 2k)_\mu (q - 2k)_\nu \tag{28.6}$$

$$= \frac{\alpha}{2\pi} \int_0^\infty dz_1 \int_0^\infty dz_2 \frac{g_{\mu\nu}}{(z_1 + z_2)^3} \exp[i(q^2 z_1 z_2/(z_1 + z_2) - (m^2 + i\varepsilon)(z_1 + z_2))] + \text{gauge terms}$$

upon introducing parameters z_1 and z_2 to enable exponentiation and integration over k, where

$$\alpha = e_0^2/4\pi \tag{28.7}$$

Applying $q^2\partial/\partial q^2$ to $I_{\mu\nu}$ to eliminate the quadratic divergent part, and then using the identity

$$1 = \int_0^\infty d\lambda/\lambda \; \delta(1 - (z_1 + z_2)/\lambda)$$

and letting $z_i = \lambda x_i$ we obtain

$$I_{\mu\nu} = \frac{i\,\alpha}{2\pi}\, q^2 g_{\mu\nu} \int dx_1 \int dx_2 \int d\lambda/\lambda\; x_1 x_2 \exp[i\lambda(q^2 x_1 x_2 - (m^2 + i\varepsilon))]\; \delta(1 - x_1 - x_2) \tag{28.8}$$

up to gauge terms. The λ integration yields a logarithmic divergence which we cut off.

Then

$$I_{\mu\nu} = \frac{i\,\alpha}{2\pi}\, q^2 g_{\mu\nu} \int_0^1 dx\; x(1\text{-}x)\, \ln(q^2 x(1 - x) - m^2) + \ldots \tag{28.9}$$

which becomes

$$I_{\mu\nu} = \frac{i\,\alpha}{12\pi}\, q^2 g_{\mu\nu} \ln(\Lambda^2/m^2) + \ldots \tag{28.10}$$

with finite and other gauge terms not shown.

Thus we find the renormalization constant Z_{3U} for a scalar universe particle is

$$Z_{3U} = 1 - \alpha/12\pi \ln(\Lambda^2/m^2) \tag{28.11}$$

If we let

$$\alpha_U = \alpha/4 \tag{28.12}$$

then we obtain the form similar to the one loop value of Z_3 for spin ½ electron QED:

$$Z_{3U} \cong 1 - \alpha_U/3\pi \ln(\Lambda^2/m^2) \tag{28.13}$$

We now *provisionally assume* that α is the QED fine structure constant. We denote it as α_{QED}. We verify this choice later.

Thus the "fine structure constant" α_U for our vector interaction is

$$\alpha_U \equiv \alpha_{QED}/4 = 0.001824338 \tag{28.14}$$

We now turn to the Johnson-Baker-Willey (JBW) model of massless QED since at ultra-high energy our vector interaction theory with lagrangian eq. 28.5 becomes the JBW model for a scalar particle. In the JBW model we calculated α_{QED} and found the corresponding power of the Z_3 divergent factor which we denote g_{QED}.

28.1.5 Finding the Universe g_U

Now we perform the same calculation for universe vacuum polarization and find the g value, which we denote g_U, that corresponds to α_U. The value of g_U will be seen to lead to the power g in the universal scale factor almost exactly.

The universe eigenvalue function is[135]

$$F_2(\alpha_U) = F_1(\alpha_U) - [2/3 + \alpha_U/(2\pi) - (1/4)[\alpha_U/(2\pi)]^2] \quad (28.15)$$

For

$$\alpha_U \equiv \alpha_{QED}/4 = 0.001824338 \quad (28.16)$$

we found the eigenfunction value

$$F_2(\alpha_U = 0.001824338) = 5.10824 \times 10^{-12} \cong 0 \quad (28.17)$$

Examining $F_2(\alpha_U)$[136] as a function of g_U we found the value of g_U corresponding to α_U is

$$g_U = -0.00014525 \quad (28.18)$$

Thus the universe vacuum polarization is

$$\Gamma_U(p) = (p/\Lambda)^{2g_U} \quad (28.19)$$

The fourier transform is[137]

---

[135] We assume the universe eigenvalue function has the same form as the QED eigenvalue function.

[136] $F_2(\alpha_U)$ and $F_2(g_U)$ are alternate notations for the same function.

[137] Those who might object to fourier transforming to time t should remember that inside a Black Hole the "time-like" coordinate is the radius and the time variable t is comparable to a spatial coordinate. The possibility that the

$$a(t) = (1/2\pi) \int_0^\infty dp/p \exp(-ipt)\, \Gamma_U(p) \qquad (28.20)$$

$$= k\, (t/T)^{-2g_U} \qquad (28.21)$$

where k is a constant and where

$$1/T = \Lambda \qquad (28.22)$$

with Λ being the "momentum space" cutoff mass. Comparing eq. 28.1 and 28.21 we find

$$g = -2g_U$$
$$= 0.0002905 \qquad (28.23)$$

From eq. 28.21 for the power g of a(t) we see the universal scale factor g is

$$g = 0.000282377 \qquad (28.24)$$

Thus the value of g calculated from the universe vacuum polarization differs from the actual value of g by less than 3%. Given the approximate nature of our JBW calculation of vacuum polarization the agreement is remarkable.[138]

In addition we found the "fine structure constant" for the vector interaction to be given by eq. 4.22 resulting in

$$e_U = (4\pi\alpha_U)^{1/2} = 0.151411 \qquad (28.25)$$

---

universe is a Black Hole is not excluded. This fourier transform appears in Blaha (2019c) in eq. 25.25 with a typographic error—the division by p was omitted.

[138] And may be exact! The value of the Hubble Constant H in recent times varies from about 70 – 75 making the calculation of g also approximate. We chose an average value of 73.24 to obtain the value of g above. If we chose the current value for H to be 75.58 we would have $g = -2g_U$ exactly. Note: studies of binary black hole merger gravity waves have given a Hubble Constant of 75.2 km s^{-1} Mpc^{-1} (and earlier of 78 km s^{-1} Mpc^{-1}), and studies of light bent by distant galaxies give H = 72.5 km s^{-1} Mpc^{-1}. Thus the value H = 75.58 is not unreasonable. See section 22.1 for a summary of studies of H.

Thus we have shown the universe vacuum polarization $\Gamma_U(p)$ when transformed to time is the universal scale factor a(t) up to a constant. The evolution of our universe is set by universe vacuum polarization. Other 4D universes may be expected to be similar.

The above relation we have found between QED-like vacuum polarization and universe vacuum polarization (Dark Energy) appears to confirm our interpretation of universe Dark Energy as mainly a consequence of universe vacuum polarization due to a universe vector interaction.[139]

28.1.6 Dark Energy is Equivalent to Universe Vacuum Polarization

Dark Energy is elusive both on the experimental and theoretical levels. We know it exists through its effects on our universe. Yet interactions with matter have not been found. Thus it is somewhat of a phantom.

The existence of Dark Energy, which, clearly, strongly affects the evolution of the universe, means that the Einstein equation, usually regarded as central to universe evolution, is incomplete for that purpose. It does not specify the total energy density ρ_{tot}.

$$\dot{a}^2 - 8\pi G\rho_{tot}a^2/3 = -k \tag{28.26}$$

However we can obtain a "handle" on the total energy density by inserting our universal scale factor a(t) in the Einstein equation together with the known radiation density, matter density and Cosmological Constant Λ terms:

$$\rho_{tot}(t) \equiv \rho_{crit}\Omega_{tot}(t) = \rho_{crit}[\Omega_\Gamma(t) + \Omega_M(t) + \Omega_\Lambda + \Omega_T]$$

where the unknown part needed to makes the Einstein equation correct is the elusive Dark Energy $\rho_T(t)$

$$\rho_T(t) = \rho_{crit}\Omega_T(t) \tag{28.27}$$

---

[139] Rather like the discovery of the Ω^- particle in the 1960s confirmed Gell-Mann's SU(3) theory.

Then we can calculate energy density $\rho_{\text{Dark}}(t)$ as a function of time as well as related quantities as the following plots show. Figs. 29.1 through 29.5 display time plots of $\Omega_T(t)$.

28.1.7 Quasi-Free Universe Particles

Since $F_2 \cong 0$ by eq. 28.17 universe particles are very much like free particles since the vacuum polarization is zero except for a divergence due to the effect of the three subtracted terms displayed in eq. 28.15.

Universe particles are not totally free particles due to gravitation and Standard Model interactions such as electromagnetism. We treated the case of free universe particles in Blaha (2018e).

28.1.8 Doubling Relation Between Coupling Constants

The coupling constants that we have derived show a doubling whose fundamental significance remains to be understood.

INTERACTION	COUPLING CONSTANT[140]
Universe Interaction e_U	0.1514
QED $e_{QED} = (4\pi\alpha_{QED})^{1/2}$	0.303
Weak SU(2) g_W	0.619
Strong SU(3) g_S	1. 22

Figure 28.2 The interaction coupling constants show a regular doubling. A fundamental cause for doubling is not apparent.

28.2 Second Approach to the Universal Scale Factor

In the above proof we addressed only the small time (large momentum) behavior of the Universal Scale Factor exponent g. We now consider the other parameter h (which has no apparent relation to H or the constant h = 0.689.) We will show h has a

[140] M. Tanabashi *et al* (Particle Data Group), Phys. Rev. D**98**, 030001 (2018).

simple direct interpretation based on the Einstein equation under the assumption that the usually assumed energy density constants Ω_γ, Ω_m, and Ω_Λ may be time dependent:

$$\Omega_\gamma(t),\ \Omega_m(t),\ \Omega_\Lambda(t)$$

and the assumption that there is another ultra-energy density $\Omega_T(t)$.

28.2.1 Origin of h

The value of the parameter h is remarkably close to the standard value of the Hubble parameter expressed in eV. It is also remarkably close to t_{now}^{-1}. Thus

$$h = 1.49 \times 10^{-33}\ \text{eV} \cong H_0 = 68.9\ \text{km s}^{-1}\ \text{Mpc}^{-1} = 1.47 \times 10^{-33}\ \text{eV} \cong t_{now}^{-1} \tag{28.28}$$

We therefore provisionally approximate eq. 27.1 as

$$a(t) \cong (t/t_{now})^{g + H_0 t} \tag{28.29}$$

We will demonstrate that the H_0t exponent is due to the existence of an "explosive" phase with energy density $\rho_T(t) = \rho_{cr}\Omega_T(t)$ where ρ_{cr} is the critical energy, which is related to the Hubble parameter by $H_0^2 = 8\pi G\rho_{cr}/3$. Einstein's equation is:

$$\dot{a}^2 - H_0^2 a^2(t)[\,\Omega_\gamma/a^4(t) + \Omega_m/a^3(t) + \Omega_\Lambda + \Omega_T(t)] = -k \tag{28.30}$$

or

$$H(t) = \dot{a}/a = [H_0^2(\Omega_\gamma/a^4(t) + \Omega_m/a^3(t) + \Omega_\Lambda + \Omega_T) - k/a^2(t)]^{1/2}$$

where we assume the factors are functions of time: $\Omega_\gamma(t)$, $\Omega_m(t)$, $\Omega_\Lambda(t)$ and $\Omega_T(t)$. We keep $\Omega_\Lambda = 0.689$. Conservation laws of the type considered by Weinberg (1972) are not conserved in their simple form but deviate due to large pressures and changes in energy density due to "input/output" from the universe energy. In particular,

$$\Omega_m \neq \text{constant} \tag{28.31}$$

$$\Omega_\Lambda \neq \text{constant}$$
$$\Omega_T \neq \text{constant}$$

28.2.2 Overall Universe Expansion

We now consider the overall universe expansion defining

$$\rho_{tot}(t) \equiv \rho_{crit}\Omega_{tot}(t) = \rho_{crit}[\Omega_\gamma(t) + \Omega_m(t) + \Omega_\Lambda + \Omega_T(t)] \quad (28.32)$$

$$\dot{a}^2 - H_0^2 a^2(t)\Omega_{tot}(t) = 0 \quad (28.33)$$

We find the solution for $\Omega_{tot}(t)$ where

$$a_{tot}(t) = (t/t_{now})^{g + H_0 t} \quad (28.34)$$

is

$$\Omega_{tot}(t) = [1 + \ln(t/t_{now}) + g/(H_0 t)]^2 \quad (28.35)$$

We note that $\Omega_{tot}(t)$ can be approximately written as

$$\Omega_{tot}(t) = [1 + \ln(H_0 t) + g/(H_0 t)]^2 \quad (28.36)$$

by eq. 28.28. *Ω_{tot} increases as $(\ln t)^2$ as t gets large, and increases as $[g/(H_0 t)]^2$ as $t \to 0$.*

The maximum of Ω_{tot} occurs at $g = H_0 t_{max}$ or

$$t_{max} = g/H_0 = 1.32 \times 10^{14} \text{ sec} \quad (28.37)$$

The time t_{max}, which is approximate due to approximations in eq. 28.28, is relatively close to the minimum (Big Dip) of H at t = 8.71×10^{13} sec. At the Ω_{tot} maximum

$$\Omega_{tot}(t_{max}) = 37.7 \quad (38.38)$$

in contrast to $\Omega_\Lambda = 0.689$ showing the general dominance of the Dark Energy in $\Omega_{tot}(t)$ as discussed in chapter 29 later.

The minima of Ω_{tot} with values 0.0 occur at

$$t = 1.36 \times 10^{13} \text{ sec} \tag{28.39}$$
$$t = 1.55 \times 10^{17} \text{ sec} \tag{28.40}$$

Note both minima of Ω_{tot} are zero. Consequently Ω_T must be negative at those points since $\Omega_\gamma/a^4(t) + \Omega_m/a^3(t) + \Omega_\Lambda$ is presumably positive. Ω_T is much larger in magnitude than $\Omega_\gamma/a^4(t) + \Omega_m/a^3(t) + \Omega_\Lambda$ *except* at points in the vicinity of the minima. See Fig. 28.3 for Ω_{tot} details.

28.3 Universe Expansions and Contractions indicated by Ω_{tot}

As the universe expands we expect the total energy density Ω_{tot} to decline, and if the universe contracts we expect the energy density to increase. Fig. 28.4 shows the pattern of expansion and contraction of the universe over time. The universe shows a "Dead Bang Bounce"[141] in reaction to the vast expansion of the universe after the big Bang. After the bounce the universe expands again and Ω_{tot} declines. It is possible that another bounce could happen again in the future as Fig. 28.5 indicates. Perhaps the universe is subject to a series of lesser and lesser bounces with corresponding expansions and contractions. The Universal Scale Factor only applies up to the current time and shows only one bounce. It could be modified to handle additional bounces in the future as in Fig. 28.5. This would raise the possibility of an oscillating universe.

28.4 Roles of g and h of the Universal Scale Factor

It is clear from the above considerations that the small time behavior of a(t) is governed by g. Remarkably it corresponds to the small distance behavior of the derivation based on vacuum polarization.

It is also clear that the large time (recent) behavior of a(t) is determined by h, which analogously might be viewed as the infrared (large distance) behavior of the vacuum polarization. Thus time in the universe corresponds to distance in vacuum polarization.

[141] In Finance it is called a Dead Cat Bounce.

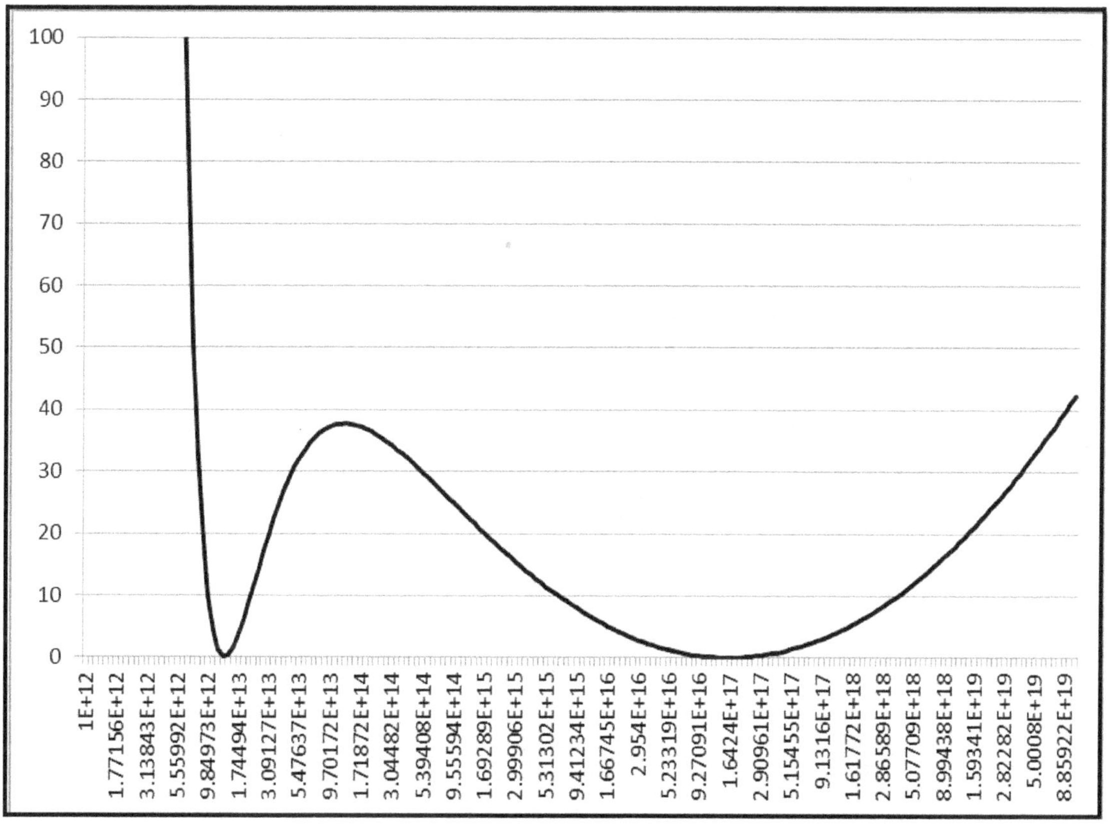

Figure 28.3 The energy density $\Omega_{tot}(t)$ of eq. 28.35 plotted as a function of time in seconds. The minima are at t = 1.36 $\times$ 10^{13} sec with Ω_{tot} = 0.0, and at t = 1.55 $\times$ 10^{17} sec with Ω_{tot} = 0.0. The maximum is at t = 1.32 $\times$ 10^{14} sec with Ω_{tot} = 37.7.

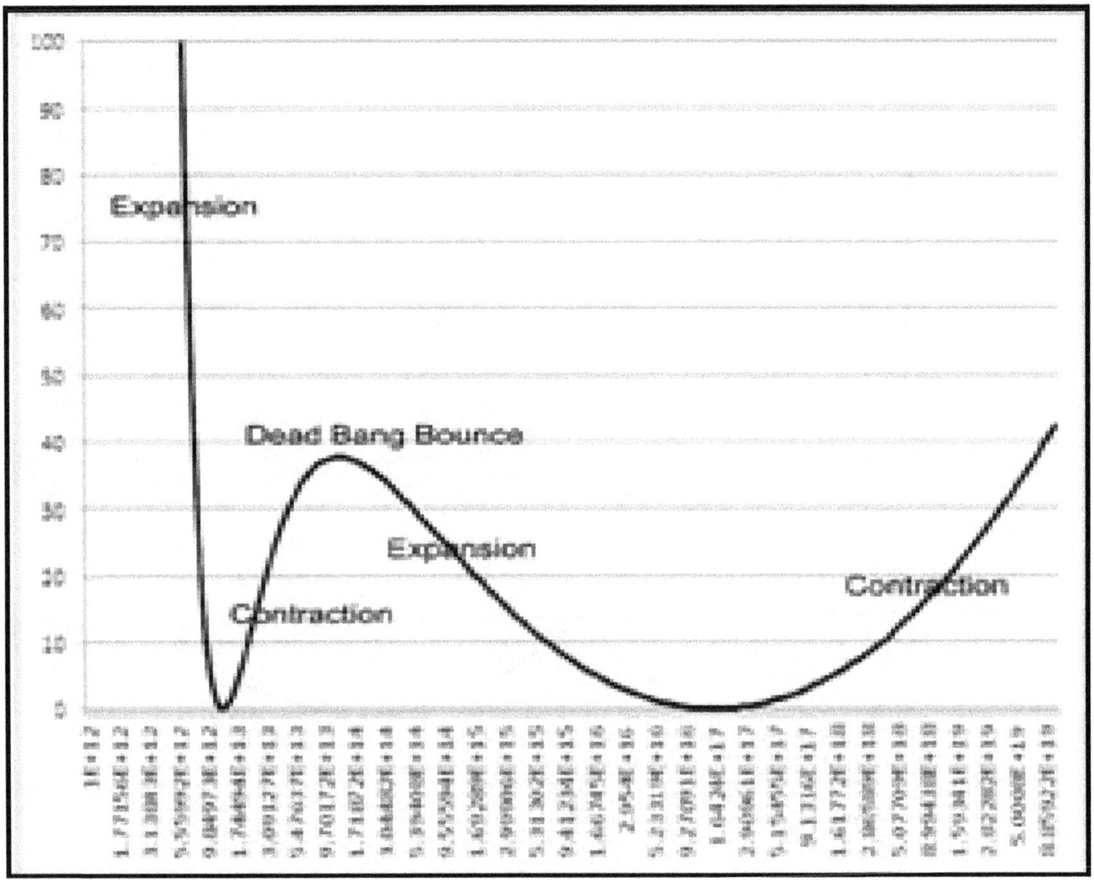

Figure 28.4 The expansions and contractions of the universe as indicated by the changes in total energy density. The diagram shows a "Dead Bang Bounce" in response to the vast expansion of the universe after the Big Bang.

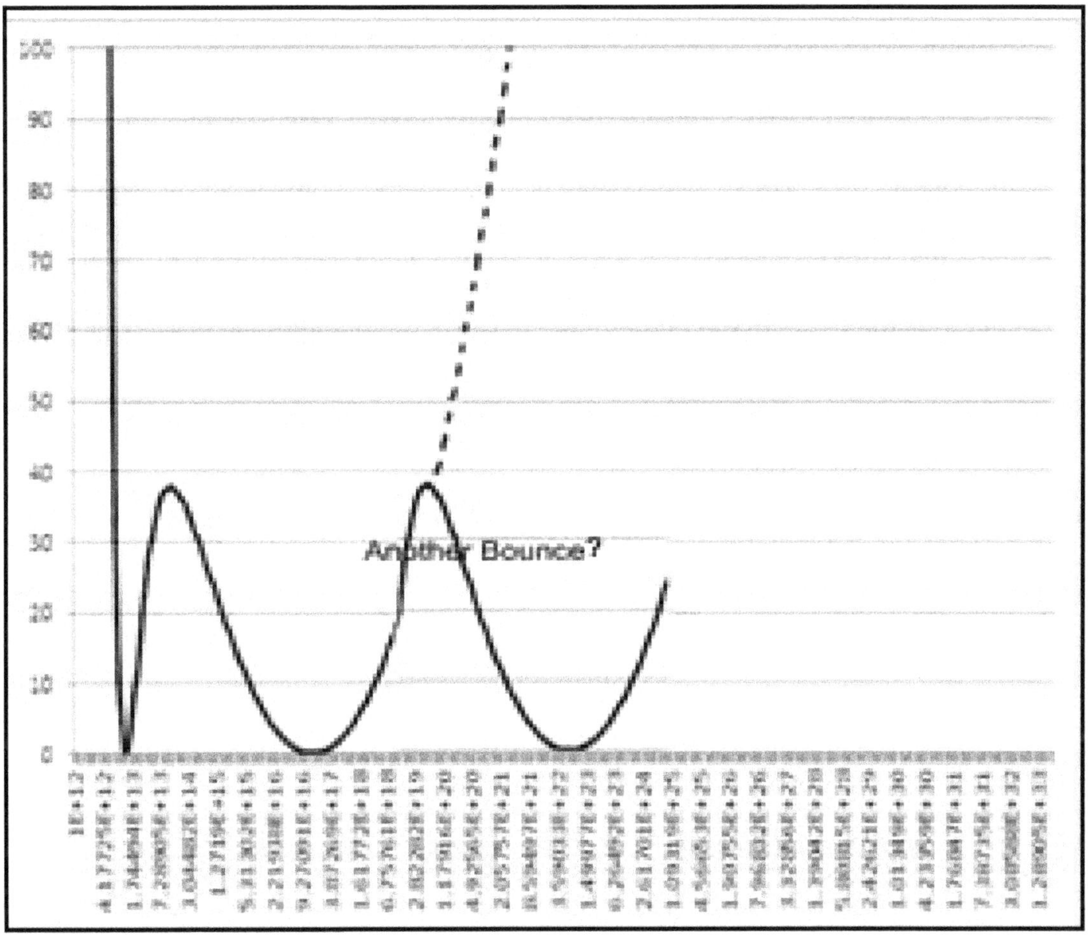

Figure 28.5 The expansions and contractions of the universe as indicated by the changes in total energy density with an illustrative future second bounce added. A second bounce raises the possibility of an oscillating universe. This The dotted line is that of the Universal Scale Factor plot.

28.5 Why $h \cong H_0 \cong 1/t_{now}$?

An important question that emerges from the above discussion is:

Why do we have the approximate equalities $h \cong H_0 \cong 1/t_{now}$

It appears $h \cong H_0$ is understandable to achieve consistency with Einstein equation in above calculation.

One may also understand $h \cong 1/t_{now}$ as simple agreement of the Universal Scale Factor with the available Hubble Constant data.

The resulting form of $\Omega_{tot}(t)$ is very much in agreement with what one might expect:

$$\Omega_{tot}(t) = [1 + \ln(H_0 t) + g/(H_0 t)]^2 \qquad (28.36)$$

Since the form of $\Omega_{tot}(t)$ (which is based on the Einstein equation) depends solely H_0 and the vacuum polarization derived constant g.

29. Energy Density, Pressure, Dark Energy, and Equation of State Implied by the Universal Scale Factor

The universal scale factor defined in chapter 27 implies the time dependence of the universe's Energy Density, Pressure, Dark Energy, and Equation of State. In this chapter we calculate these quantities and then plot their values as a function of time.

We start with the expressions for the universal scale factor and its derivatives:

$$a(t) = (t/t_{now})^{g + ht} \tag{27.1}$$

with

$$da(t)/dt = a[g/t + h + h \ln(t/t_{now})] \tag{29.1}$$

and

$$d^2a(t)/dt^2 = da/dt\ [g/t + h + h \ln(t/t_{now})] + a(h - g/t)/t$$

29.1 Dark Energy

The Einstein equation

$$\dot{a}^2 - 8\pi G\rho_{tot}a^2/3 = -k + \Lambda\ a^2c^2/3 \tag{29.3}$$

enables us to determine the dark energy energy) $\Omega_T(t)$ beyond the cosmological constant, using the energy density:

$$\rho_{tot}(t) \equiv \rho_{crit}\Omega_{tot}(t) = \rho_{crit}[\Omega_\gamma(t) + \Omega_m(t) + \Omega_\Lambda + \Omega_T(t)] \tag{29.4}$$

Thus

$$\Omega_T(t) = (H(t)^2 + c^2k/a^2)/H_0^2 - \Omega_\gamma/a^4 - \Omega_m/a^3 - \Omega_\Lambda \tag{29.5}$$

where

$$H_0^2 = 8\pi G\rho_{crit}/3$$

The quantity Ω_T is the excess of energy beyond that specified by Ω_γ, Ω_M, and Ω_Λ in the conventional expressions for the mass-energy density. The calculation of Ω_T which is in part energy density, (and possibly in part a Megaverse energy influx), is based on eq. 29.3 using the Einstein equation in the form

$$H(t) = (da/dt)/a(t) = [H_0^2 \rho_{tot}(t)/\rho_{cri} - c^2k/a^2(t)]^{1/2}$$

Figs. 29.1, 29.2, 29.4 and 29.5 plot $\Omega_T(t)$ and its derivative. It is clearly much larger than $\Omega_H(t)$ (Fig. 29.2) showing why the standard approaches to calculating a(t) using $\Omega_H(t)$ are faulty. The ratio Ω_H/Ω_T is appreciable only after 380,000 years. See Fig. 29.14.

$$\Omega_H = \Omega_\gamma(t) + \Omega_m(t) + \Omega_\Lambda \qquad (29.6)$$

29.2 Energy Density

The energy density is given by eq. 29.5. It is plotted in Figs. 29.6, 29.7 and 29.8. Fig. 29.6 shows a drop below zero starting at t = 0.12 sec. ($\log_{10}$ t = -0.92) signaling very low values for ρ. There is a minimum at $t = 1.2 \times 10^{16}$ sec. in Fig. 29.7.

Note also the drop below zero in Fig. 29.6 beginning at $t = 5 \times 10^{-16}$ sec. signaling the start of very small values for ρ.

The derivative of total energy density $d\rho_{tot}/dt$ is plotted in Fig. 29.8.

29.3 Pressure

The Friedmann-Lemaître-Robertson-Walker metric equations yield the pressure

$$p = -(c^2\rho_{crit}/(3H_0^2))[2(d^2a/dt^2)/a + (da/dt)^2/a^2 + kc^2/a^2 - \Omega_\Lambda] \qquad (29.7)$$

using eqs. 29.1 and 29.2 above.

An alternate approach to determining the pressure p uses the energy conservation equation

$$d(\rho_{tot}R^3)/dR = -3pR^2 \qquad (29.8)$$

where

$$R = k^{-1/2}\, a(t) \qquad (29.6)$$

and gives

$$p = - (da/dt)^2 \rho_{tot} - a\, da/dt\, d\rho_{tot}/dt \qquad (29.7)$$

$$p = -a(t)^2 (H(t)^2 \rho_{tot} + 1/3\, H(t)\, d\rho_{tot}/dt\,)/[(2.85 \times 10^{37})c^2] \qquad (29.8)$$

where $k = 5.56 \times 10^{-57}$ cm^{-2}, ρ is the total energy-mass density, p is the pressure, and the denominator is required by the dimensions to obtain p in gm/cm^3. See Fig. 29.9, 29.10 and 29.11.

In Fig. 29.9 note the low values of $\log_{10}(-p(t)) \approx -115$ beginning at $\log_{10}$(t sec.) = 12.1 ($t = 1.2 \times 10^{12}$ sec.). The pressure is always negative indicating a pressure for expansion. At $t = 1.19 \times 10^{-165}$ sec. we find $\log_{10}(-p(t)) = 593$ or $p(t) = -10^{593}$ gm/cm^3.

In Fig 29. 11 the maximum pressure in the range of 380,000 years to the present is -2.62×10^{-124} gm/cm^3 at $t = 1.0 \times 10^{17}$ sec.

29.4 Derivative of Dark Energy

The derivative of the Dark Energy, which we identify as $d\Omega_T/dt$ is

$$d\Omega_T/dt = H(t)[-2g/(H_0^2 t^2) + 2h/(H_0^2 t) - 2c^2 k/a^2 + 4\Omega_\gamma/a^4 + 3\Omega_M/a^3] \qquad (29.9)$$

We find that it is well-approximated by

$$d\Omega_T/dt \approx H(t)[-2g/(H_0^2 t^2) + 2h/(H_0^2 t) - 2c^2 k/a^2\,] \qquad (29.10)$$

Note the dip below zero in Fig. 29.5 beginning at $t = 1.2 \times 10^{10}$ sec. – somewhat earlier than the Big Dip.

29.5 Dominance of Dark Energy

Similarly we find eq. 29.5 is well approximated by

$$\Omega_T(t) \approx (H(t)^2 + c^2 k/a^2)/H_0^2 \qquad (29.3')$$

due to the relative smallness of Ω_H. Dark Energy dominates.

The total energy density is

$$\rho_{tot}(t) \equiv \rho_{crit}\Omega_{tot}(t) = \rho_{crit}[\Omega_\gamma(t) + \Omega_m(t) + \Omega_\Lambda + \Omega_T(t)] \quad (29.11)$$

See Fig. 29.6. It is well approximated by

$$\rho_{tot}(t) \approx \rho_{crit}\Omega_T(t) \quad (29.12)$$

29.6 Equation of State

The equation of state

$$w = p/\rho$$

is plotted in Figs. 29.12 and 29.13 as a function of time. Fig. 29.12 shows an enormous range of values from the Big Bang to the present. Fig. 29.13 shows that w is of the order of 10^{-95} for most of the interval from 380,000 years to the present. At its "peak" Fig. 29.13 shows $w = -1.36 \times 10^{-95}$ at the time $t = 1. \times 10^{17}$ sec.

Since w is approximately zero in these time intervals it can be viewed as describing cold dust or gas.

Since quintessence is viewed as indicated by $w \neq -1$, the theory has quintessence.

29.7 Deceleration Parameter q

The deceleration parameter q is plotted in Figs. 29.15 and 29.16.

$$q = -\, a d^2a/dt^2/(da/dt)^2 \quad (29.13)$$

It is proportional to the second derivative of the universal scale factor.

If $q < 0$ then it indicates accelerating expansion of the universe. If $q > 0$ then it indicates a decelerating universe.

Figs. 29.15 and 29.16 for q both have a pronounce maximum and minimum. The maximum occurs at $t = 1.2 \times 10^{13}$ sec. The minimum occurs at 1.6×10^{17} sec. To the left of the maximum (early times) $q > 0$ indicating decelerating expansion. To the right of the maximum at $t > 1.7 \times 10^{14}$ sec. we find $q < 0$ indicating accelerating expansion The accelerating expansion started 13.78 billion years ago – "just" after the Big bang. At present q = -2.0 – accelerating expansion – as suggested by astrophysical experiments. In the future, at $t = 8.2 \times 10^{17}$ sec. we found q = -1.2. The accelerating expansion will continue.

29.8 Comments on Universal Scale Factor Quantities

The complete universe "life" history presented in the following plots is consistent with a declining pressure, a declining density, and a declining pressure from the Big Bang phase consistent with our physical expectations. The Big Dip and subsequent could be due in part to a sharp influx of energy (Fig. 29.2) possibly from the Megaverse.

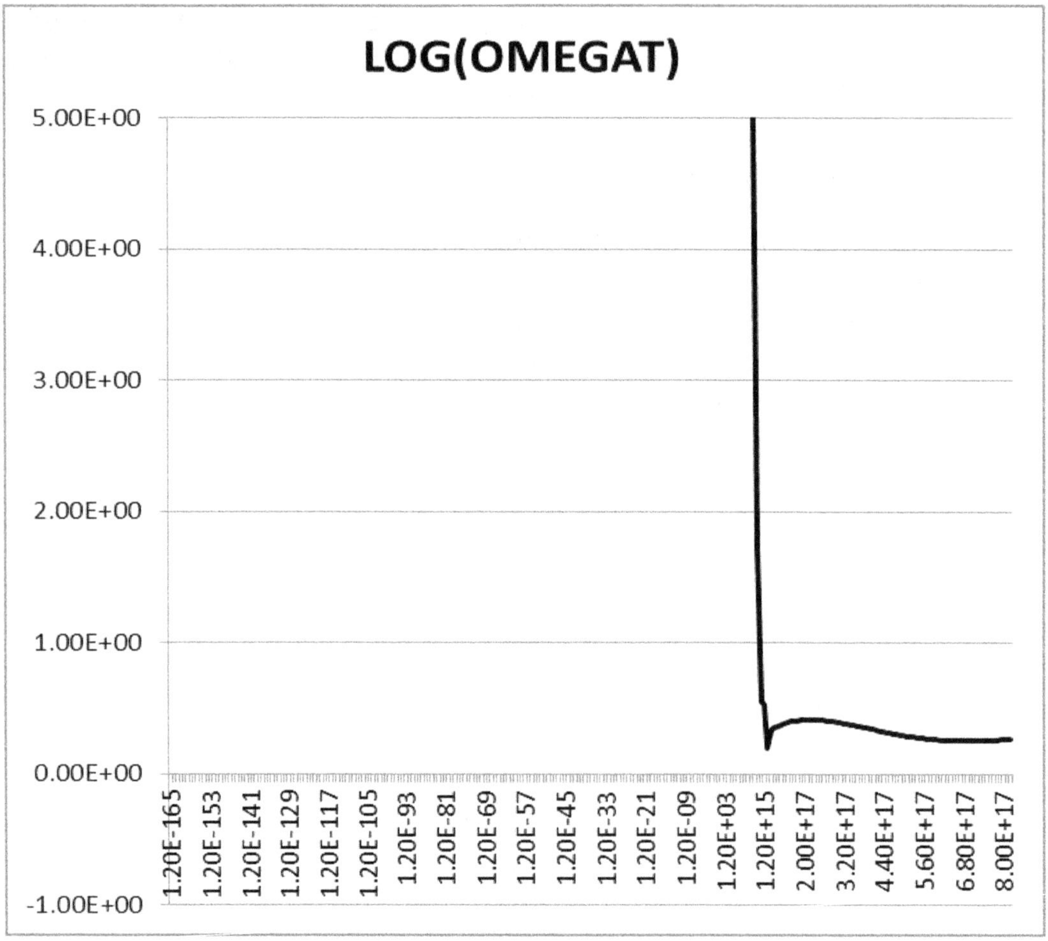

Figure 29.1 Log $\Omega_T(t)$ plotted vs. log time in seconds from 1.19×10^{-165} sec. to the 8.2×10^{17} sec. (almost double the present time). Note the Big Dip in $\Omega_T(t)$ at about $t = 8.71 \times 10^{13}$ sec. followed by a rise then a decline. At $t = 1.19 \times 10^{-165}$ sec. we found (not shown) log $\Omega_T(t) \approx 358$, an enormous value, corresponding to the level of vacuum energy found in quantum field theory. It declines to the plotted data shown above.

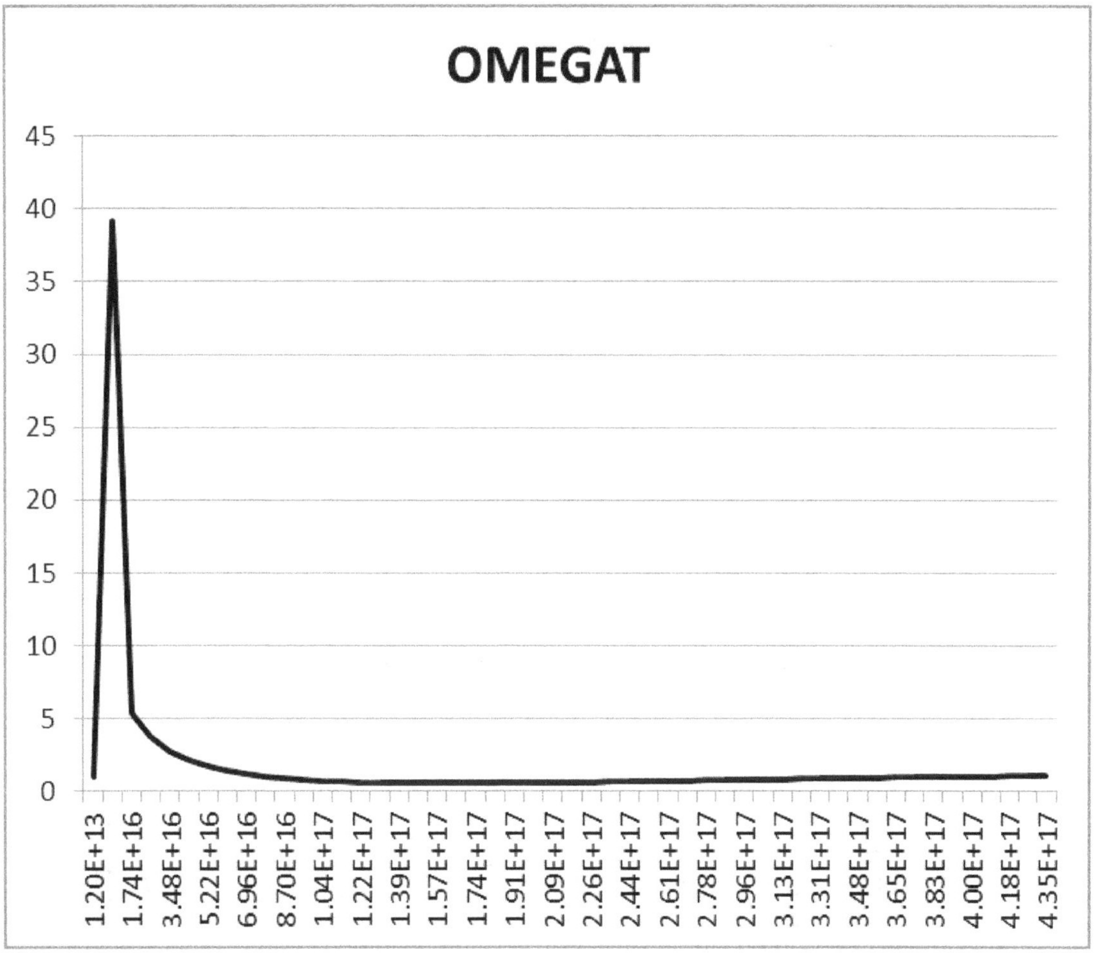

Figure 29.2 $\Omega_T(t)$ plotted vs. time in sec. from the year 380,000 to the present. Note the peak at t = 8.71×10^{13} sec. suggesting an influx into the universe at the transition to the matter-dominated phase, and then a decline followed by a raise. The peak value of $\Omega_T(t)$ is 39.2.

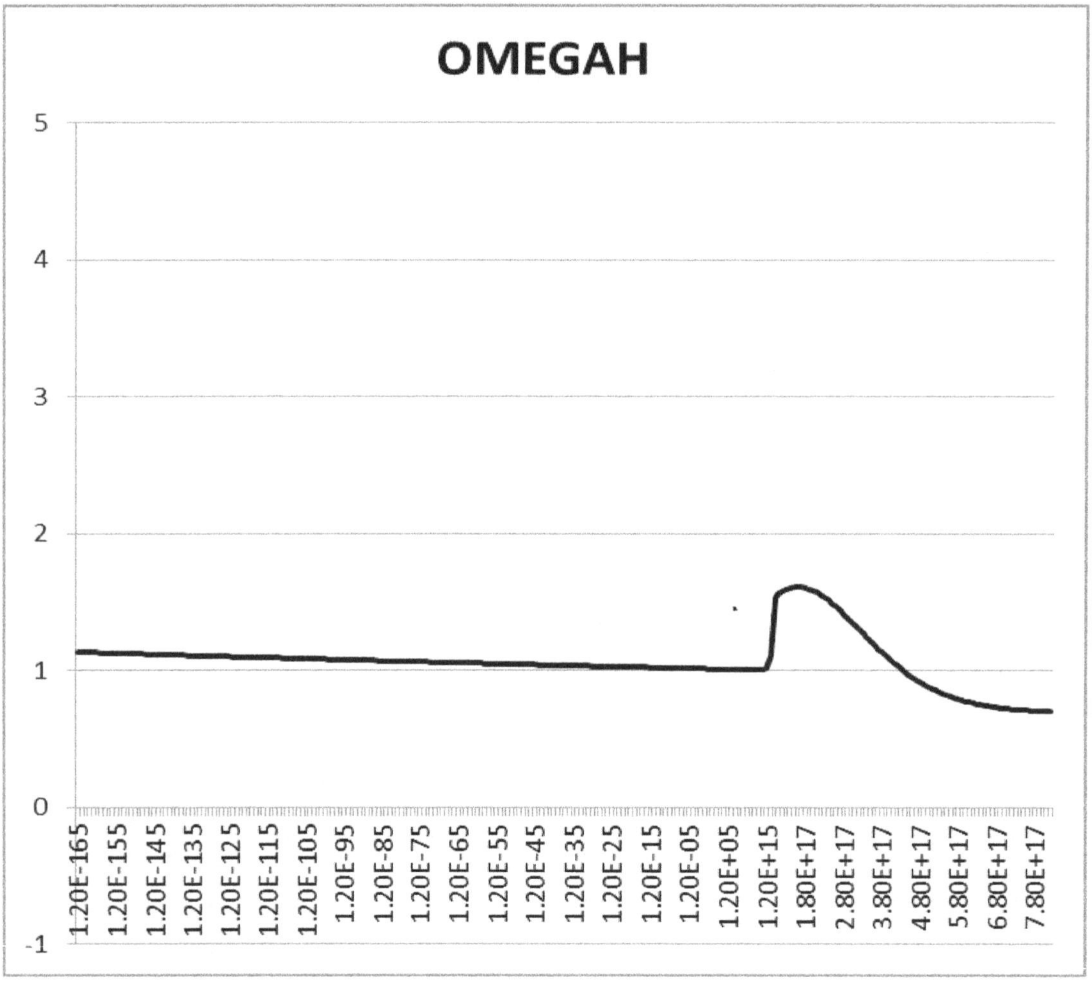

Figure 29.3 $\Omega_H(t)$ plotted vs. log time in seconds from 1.19×10^{-165} sec. to the 8.2×10^{17} sec. (almost double the present time). Note the start of a rise at $t = 1.2 \times 10^{13}$ sec. which coincides with the start of the appearance of atoms at t_T = 380,000 years = 1.19837×10^{13} sec.

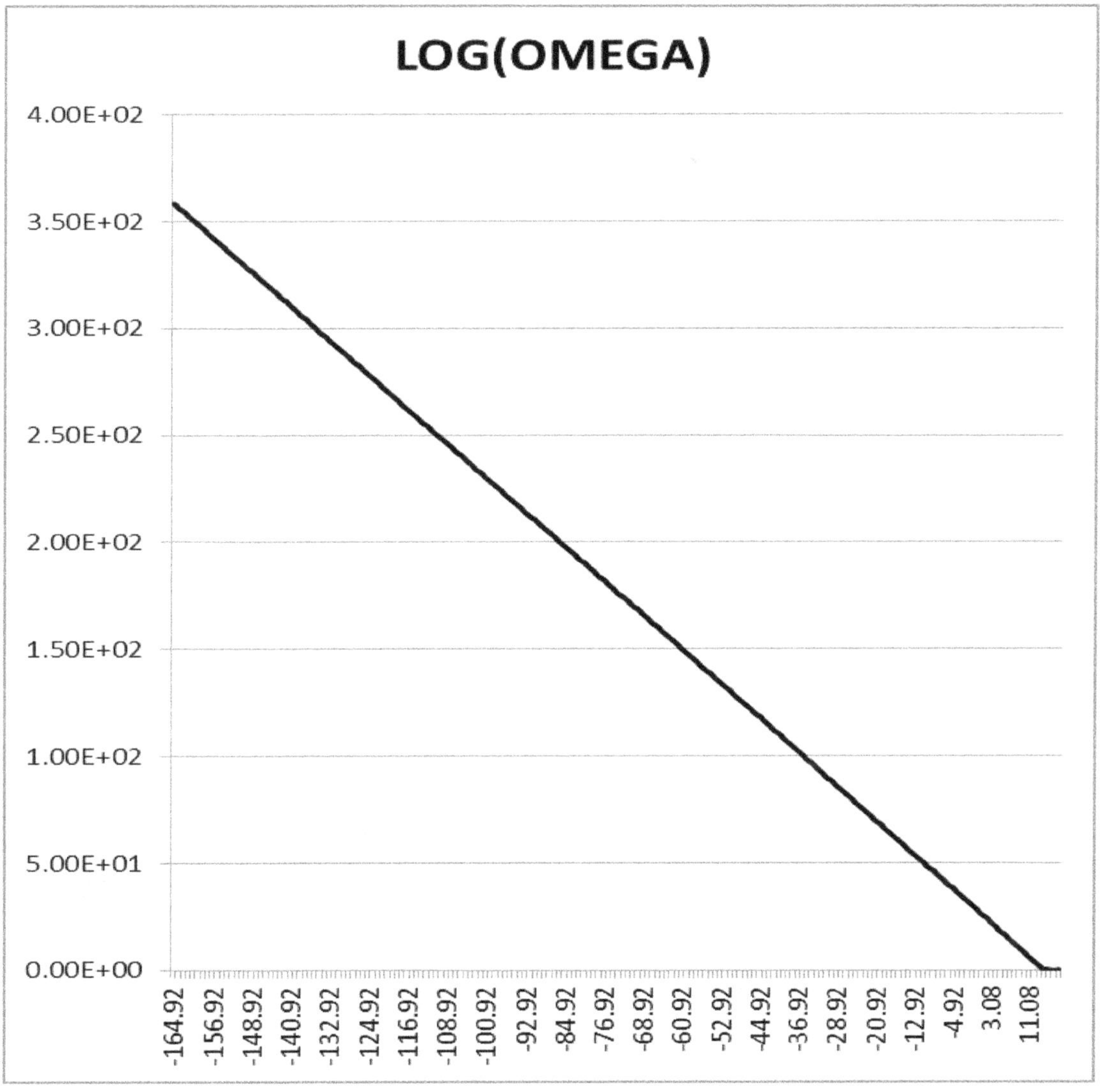

Figure 29.4 $\text{Log}_{10}\ \Omega_T(t)$ plotted vs. $\log_{10}$ t sec. from the Big Bang metastate to the present. Note the "dip" below zero of $\Omega_T(t)$ at $\log_{10}$ t = 11.

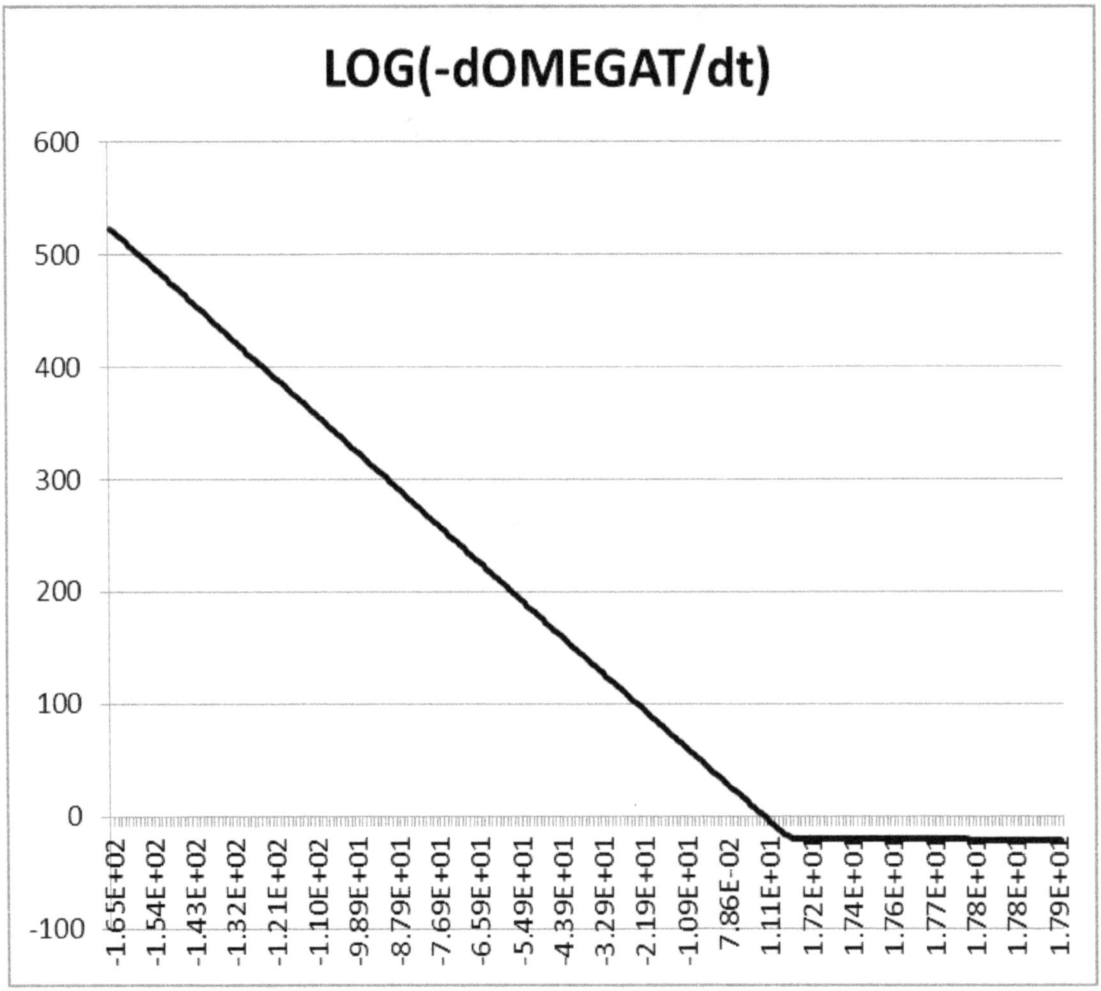

Figure 29.5 $\mathrm{Log}_{10}(-d\Omega_T(t)/dt)$ plotted vs. $\log_{10}(t)$ sec. from the Big Bang metastate $t = 1.19 \times 10^{-165}$ sec. to the future: $t = 8.2 \times 10^{17}$ sec. Note dip below zero beginning at $t = 1.2 \times 10^{10}$ sec. – somewhat earlier than the Big Dip.

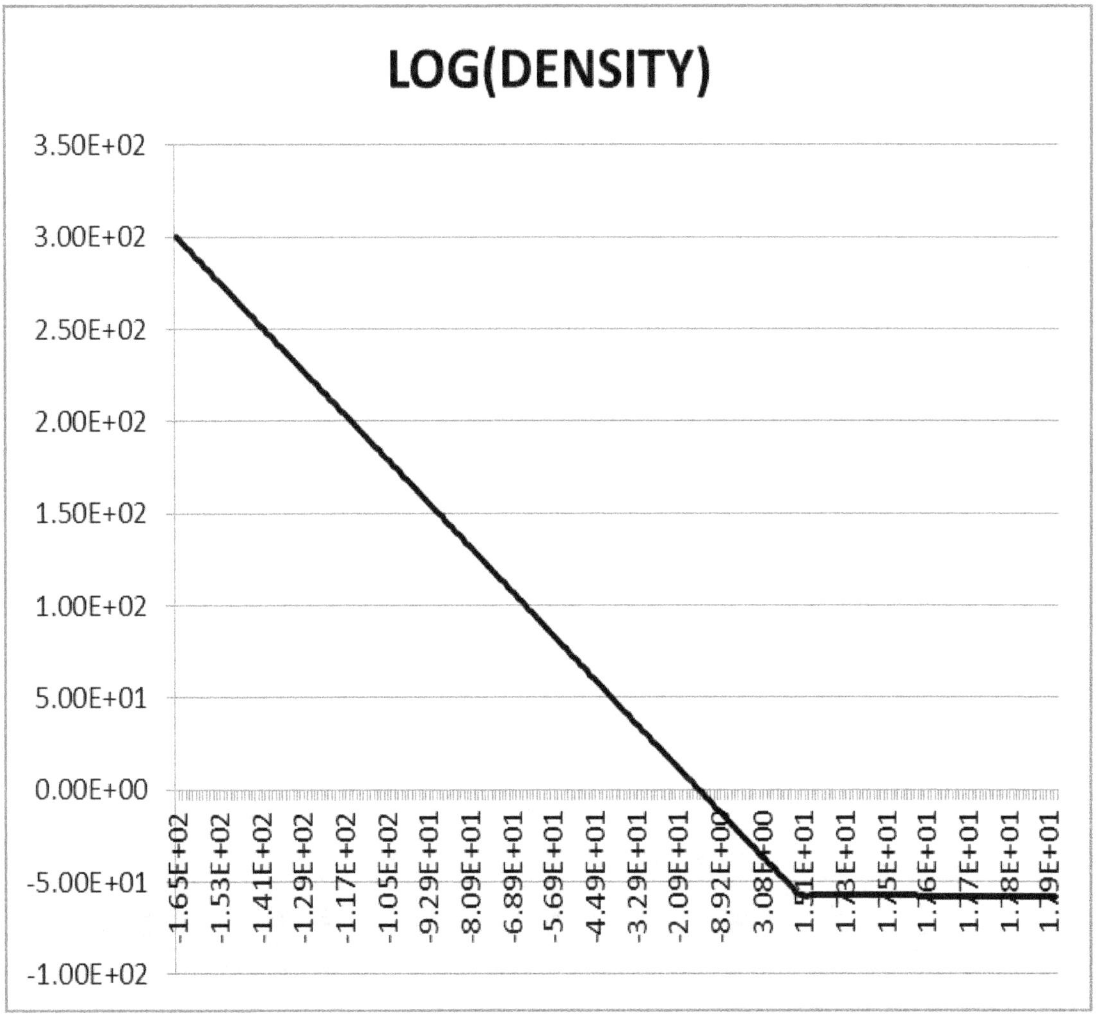

Figure 29.6 Log Density: $Log_{10}(\rho_{tot}(t)\ g/cm^{-3})$ plotted vs. $log_{10}(t)$ from the Big Bang metastate at $t = 1.19 \times 10^{-165}$ sec. to the future: $t = 8.2 \times 10^{17}$ sec. Note the drop below zero beginning at $t = 5 \times 10^{-16}$ sec. signaling the start of very small values for ρ.

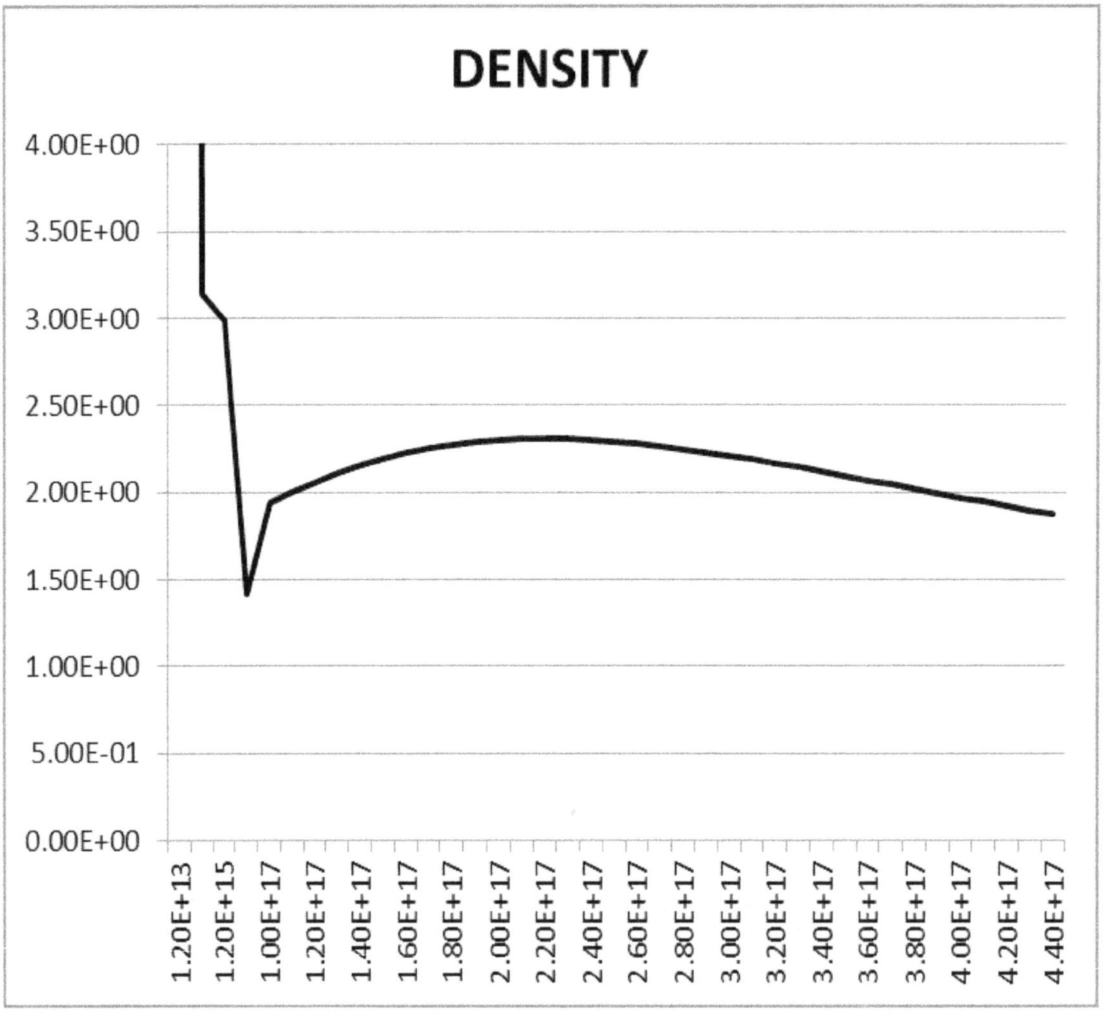

Figure 29.7 Energy density ρ(t) g/cm^{-3} $\times$ 10^{29} plotted vs. t from t = 1.19 $\times$ 10^{13} sec. (380,000 years) to the present 4.35 $\times$ 10^{17} sec. Note the dip at t = 1.2 $\times$ 10^{16} sec. with a small value for ρ = 1.42 $\times$ 10^{-29} g/cm^{-3}.

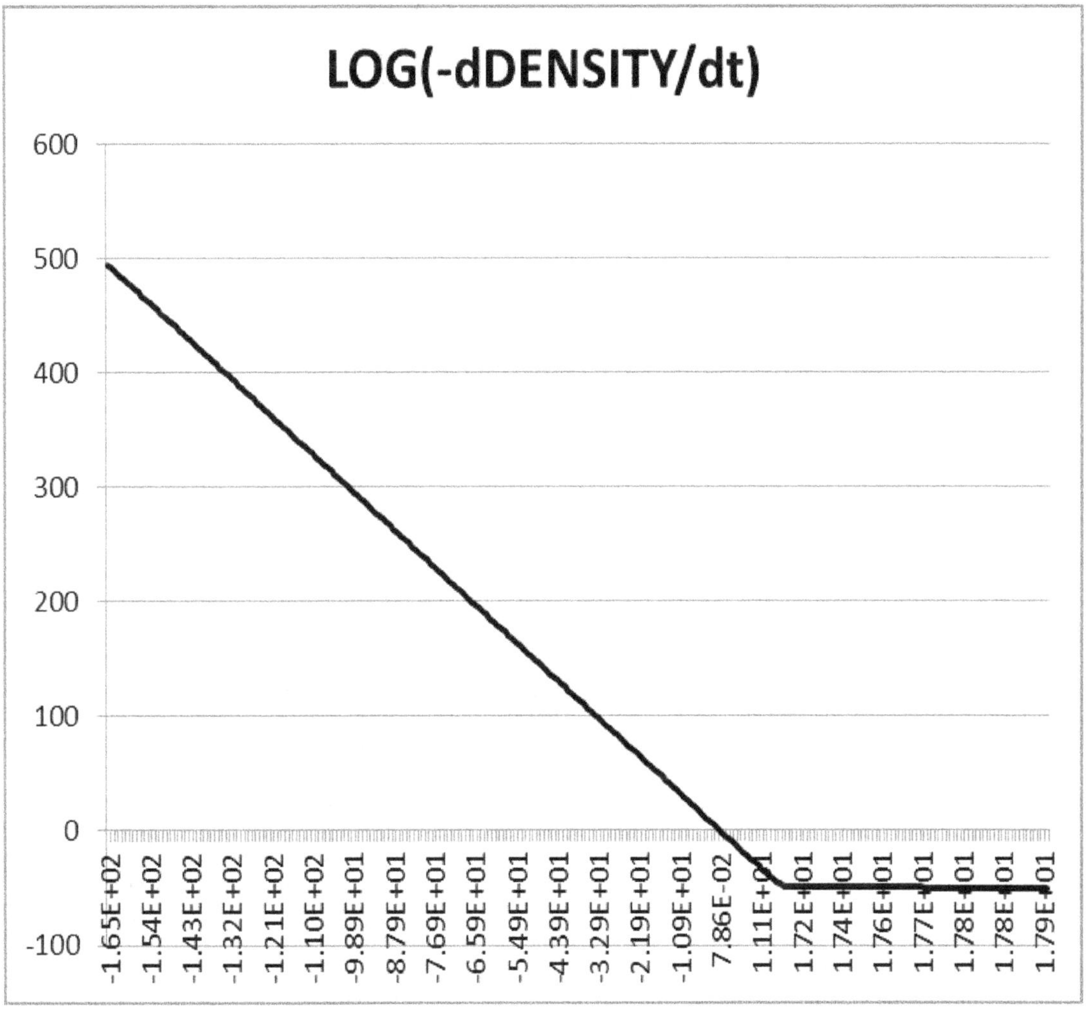

Figure 29.8 $\text{Log}_{10}(-d\rho(t)/dt \text{ g s}^{-1} \text{ cm}^{-3})$ plotted vs. $\log_{10}(t)$ from the Big Bang metastate at $t = 1.19 \times 10^{-165}$ sec. to the future: $t = 8.2 \times 10^{17}$ sec. The plot goes below zero at about $\log_{10}(t \text{ sec.}) = 7.8 \times 10^{-2}$ ($t = 1.198$ sec.) where $\log_{10}(-d\rho/dt \text{ g s}^{-1} \text{ cm}^{-3}) = -0.76$.

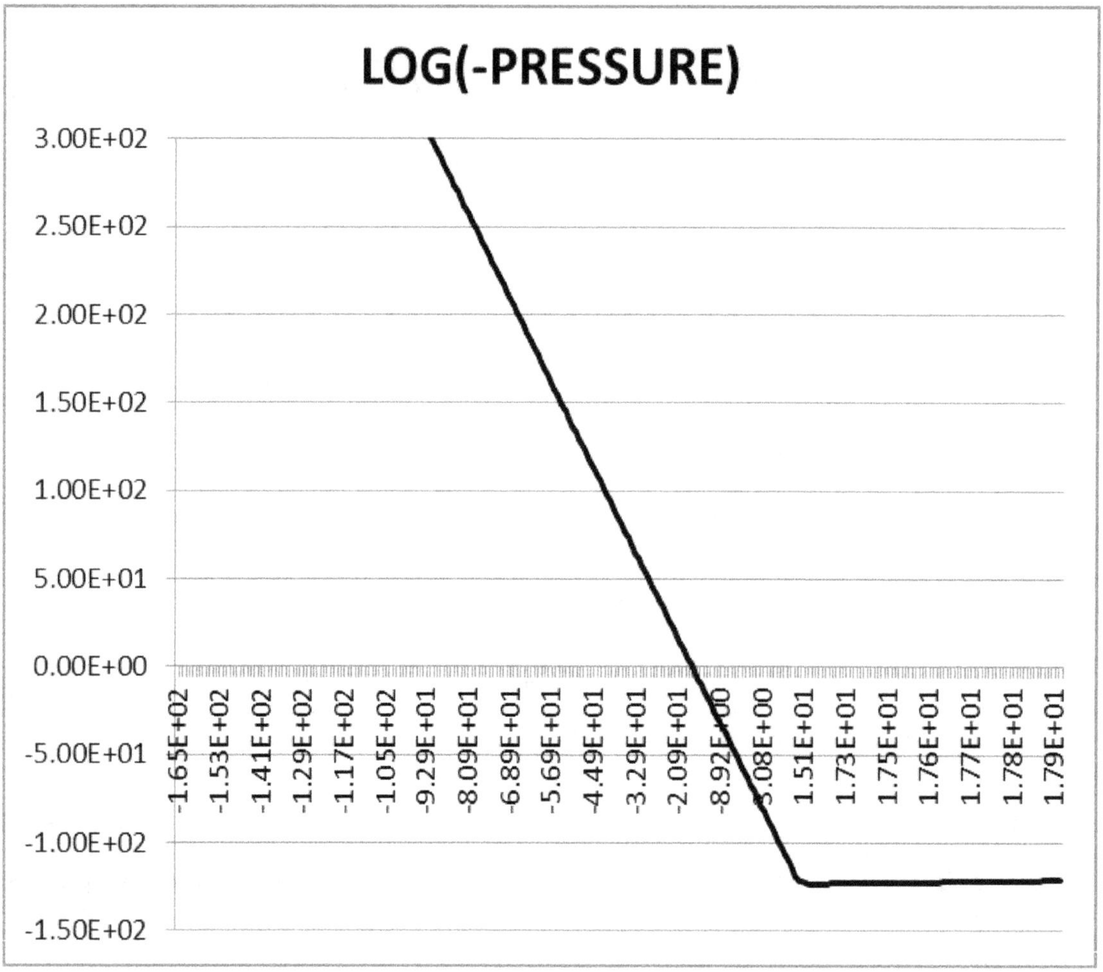

Figure 29.9 Pressure: $\text{Log}_{10}(-p(t))$ plotted vs. $\log_{10}(t)$ from the Big Bang metastate at $t = 1.19 \times 10^{-165}$ sec. to the future time $t = 8.2 \times 10^{17}$ sec. Note the low values of $\log_{10}(-p(t)) \approx -115$ beginning at $\log_{10}(t \text{ sec.}) = 12.1$ ($t = 1.2 \times 10^{12}$ sec.). The pressure is always negative indicating a pressure for expansion. At $t = 1.19 \times 10^{-165}$ sec. we find $\log_{10}(-p(t)) = 593$ or $p(t) = -10^{593}$.

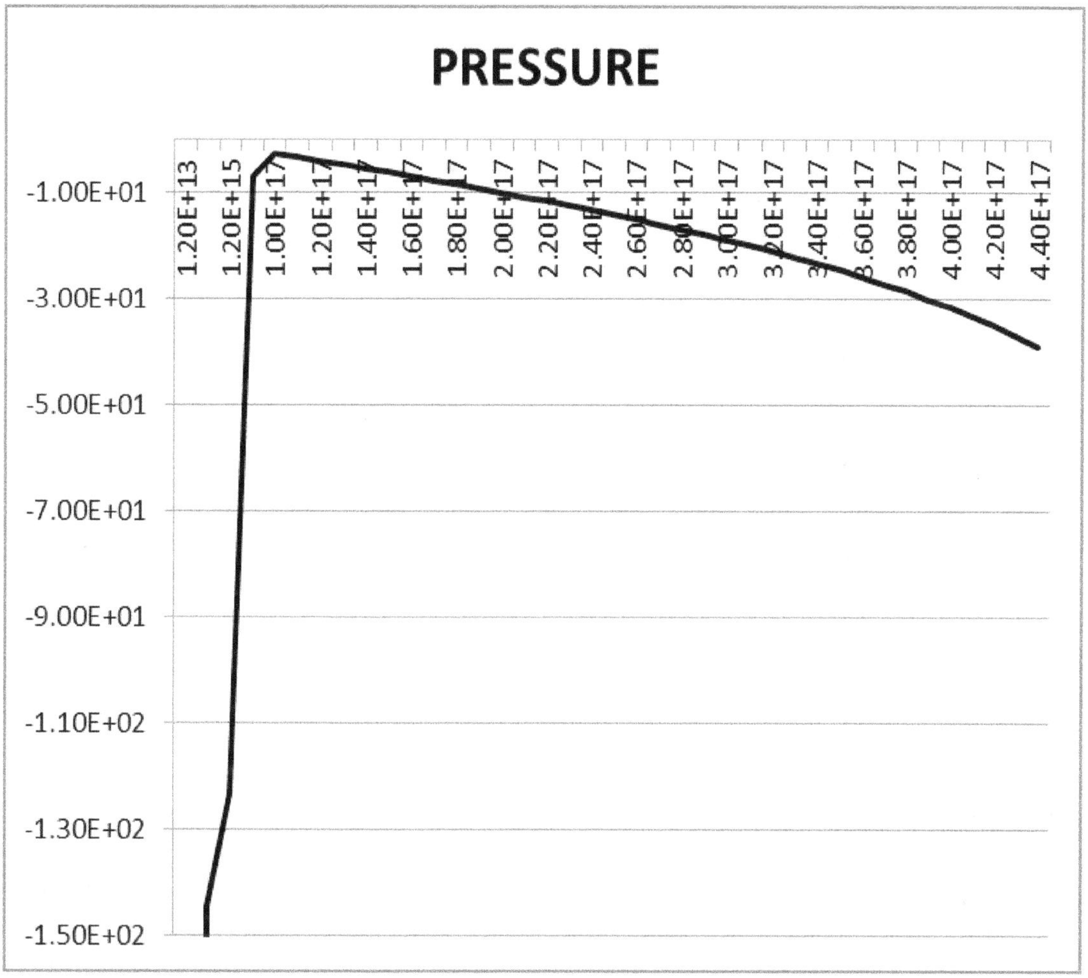

Figure 29.10 Negative Pressure: $p(t) \times 10^{124}$ plotted vs. t sec. from $t = 1.19 \times 10^{13}$ sec. (380,000 years) to the present 4.35×10^{17} sec.

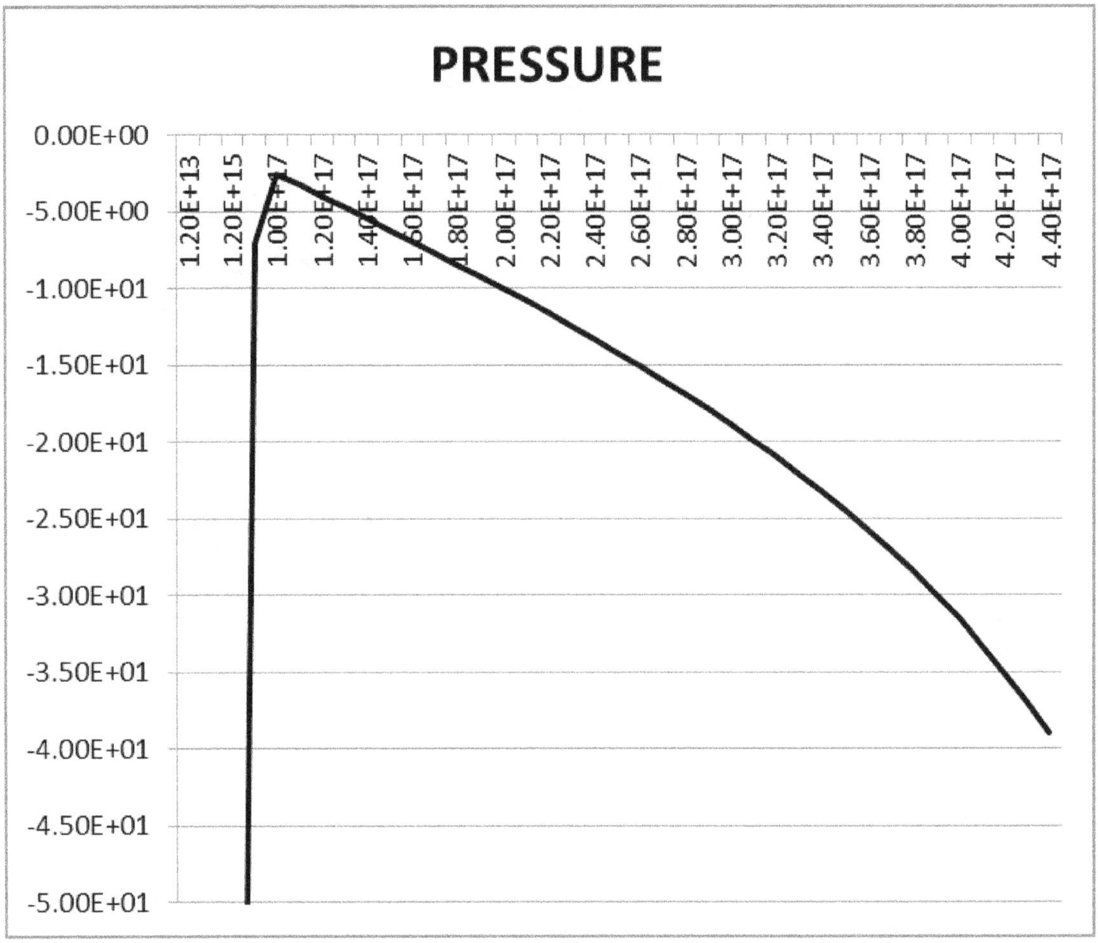

Figure 29.11 Negative Pressure: p(t) × 10^{124} plotted vs. t sec. from t = 1.19 × 10^{13} sec. (380,000 years) to the present 4.35 × 10^{17} sec. The maximum is -2.62 × 10^{-124} at t = 1.0 × 10^{17} sec. A closer view than Fig. 29.10.

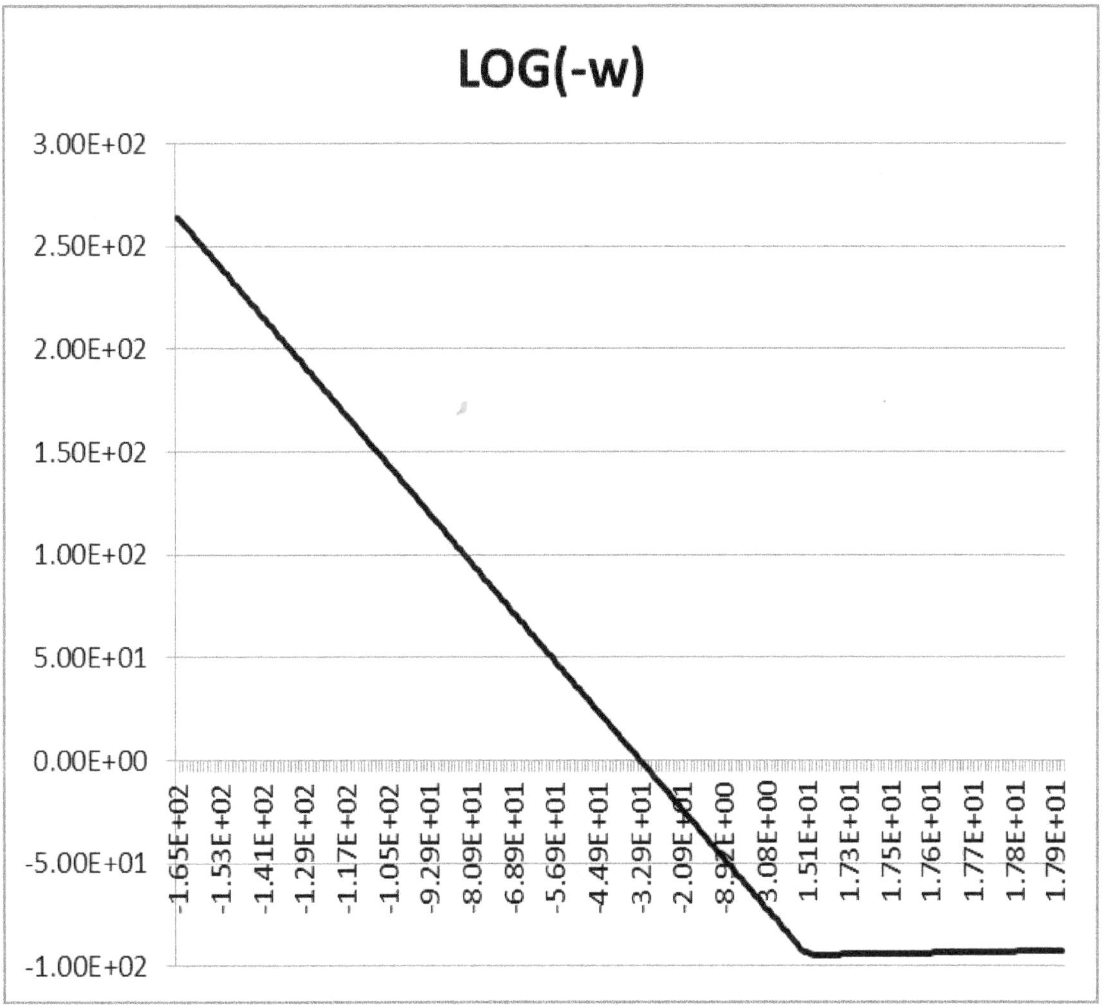

Figure 29.12 The equation of state as a function of time $\log_{10}(-w)$ plotted vs. $\log_{10}(t)$ from the Big Bang metastate at $t = 1.19 \times 10^{-165}$ sec. to the future time $t = 8.2 \times 10^{17}$ sec. Note: $w \neq -1$. Quintessence!

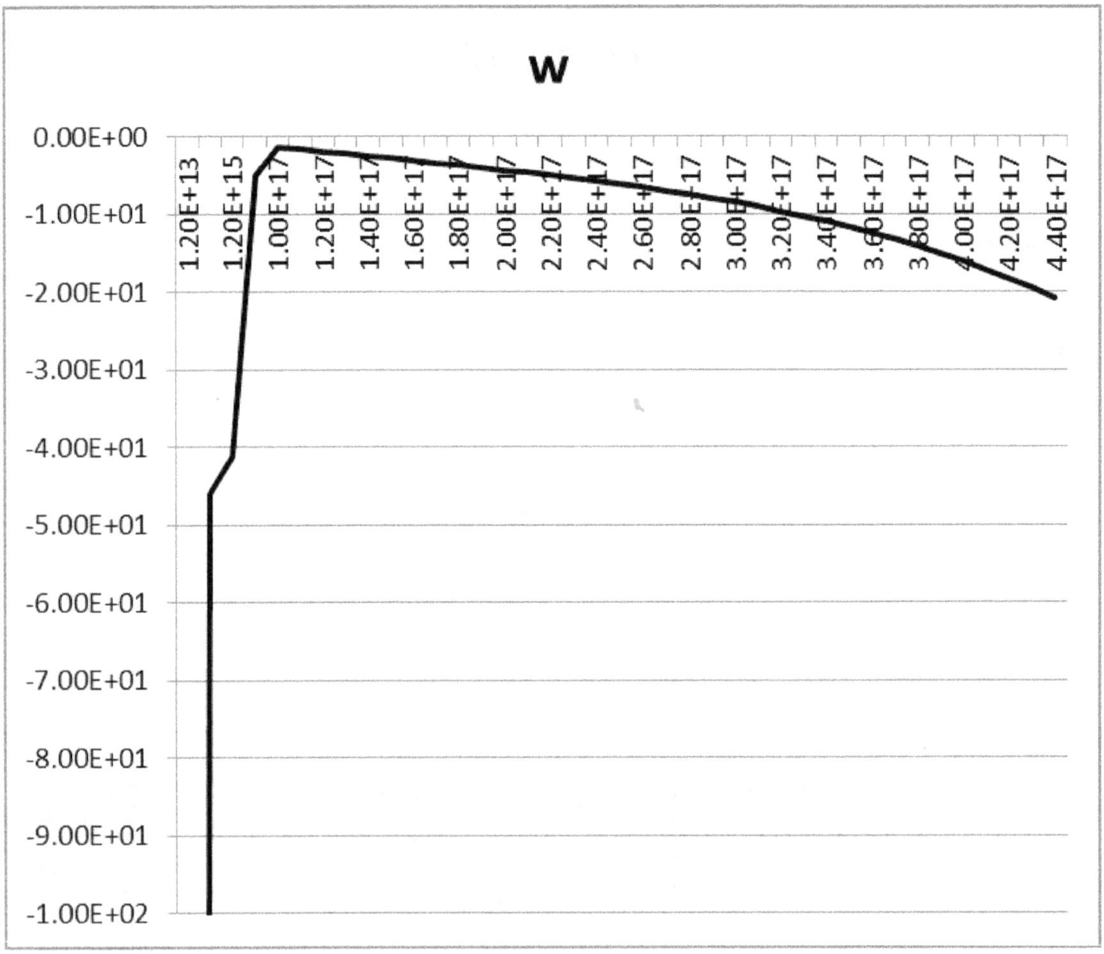

Figure 29.13 The equation of state $w \times 10^{95}$ as a function of time plotted vs. t sec. from $t = 1.19 \times 10^{13}$ sec. (380,000 years) to the present time. Note the peak of $w = -1.36 \times 10^{-95}$ at $t = 1. \times 10^{17}$ sec. Note: $w \neq -1$. Quintessence!

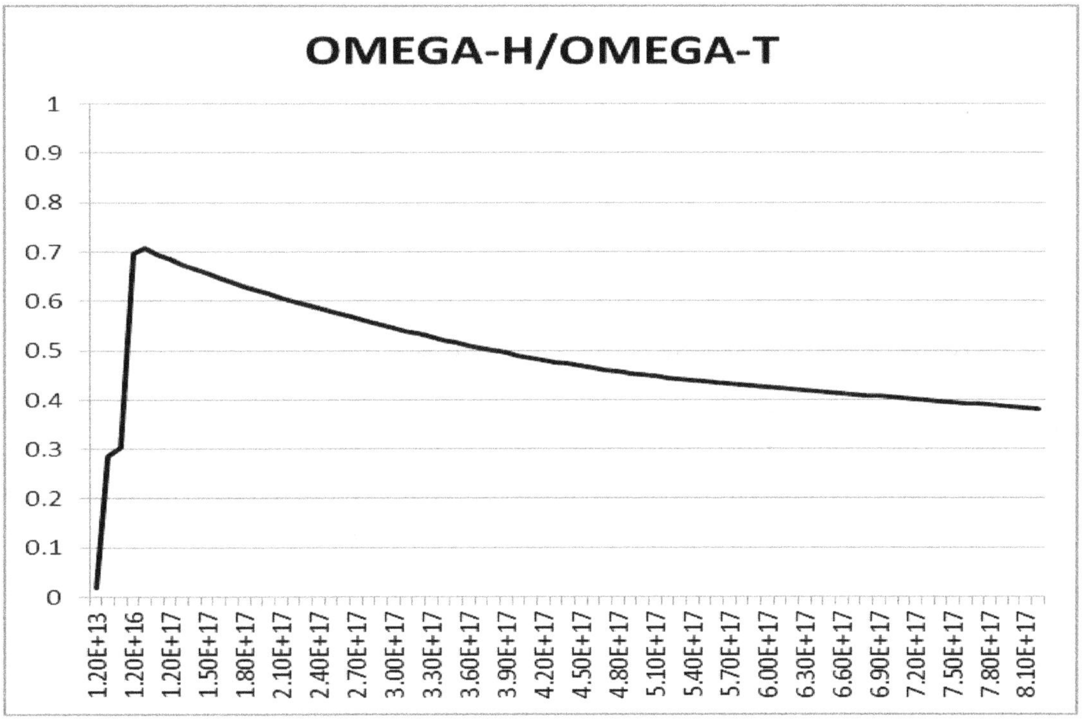

Figure 29.14 The ratio Ω_H/Ω_T showing the dominance of Dark Energy Ω_T from t = 1.19×10^{13} sec. (380,000 years) to the future time t = 8.2×10^{17} sec. Prior to 380,000 the Dark Energy was greater by many orders of magnitude. The peak value of Ω_H/Ω_T is at t = 1.2×10^{16} sec.

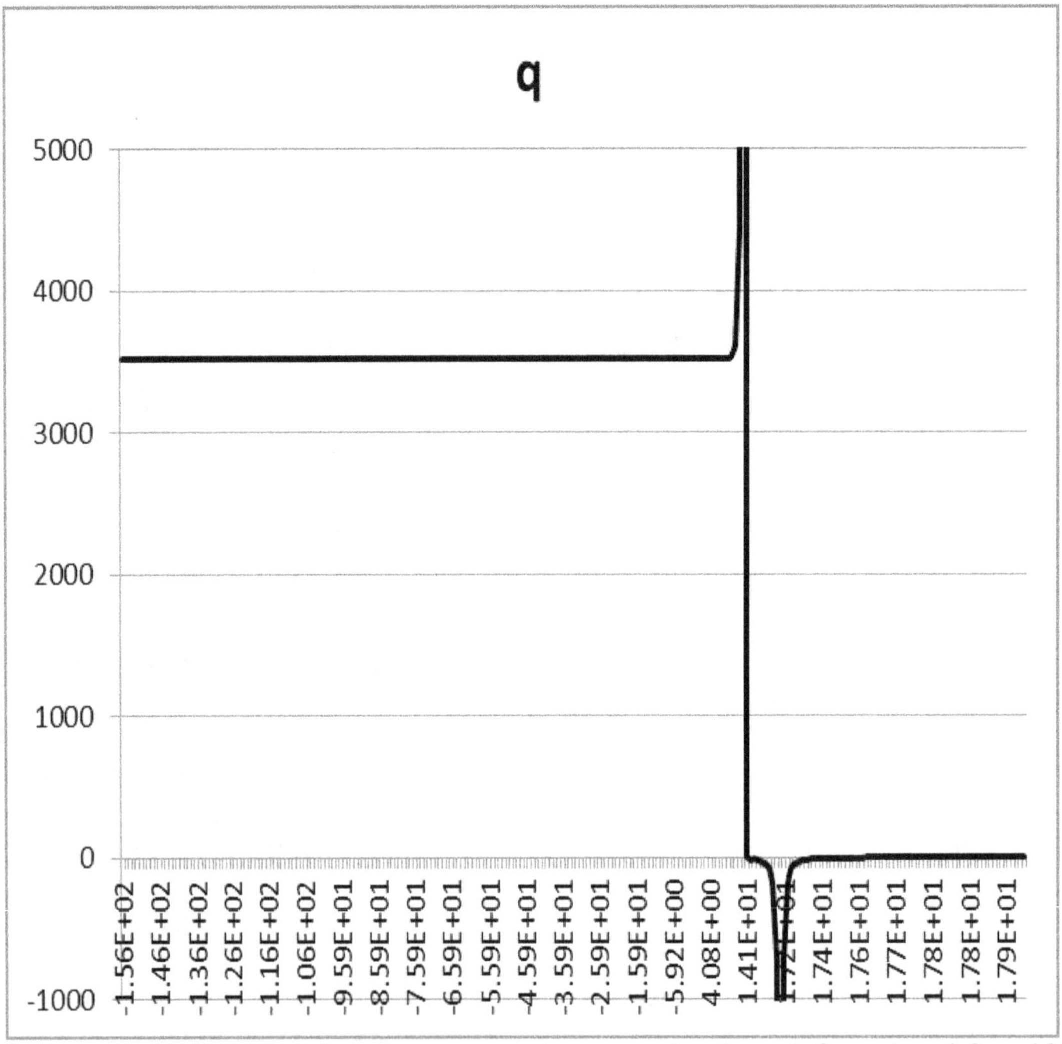

Figure 29.15 The deceleration parameter q vs. log(t sec) from $t = 1.2 \times 10^{-156}$ sec. to the future $t = 8.2 \times 10^{17}$ sec. The peak occurs at $t = 1.2 \times 10^{13}$ sec. The minimum occurs at 1.6×10^{17} sec. Note: $q > 0$ for $t < 1.7 \times 10^{14}$ sec.

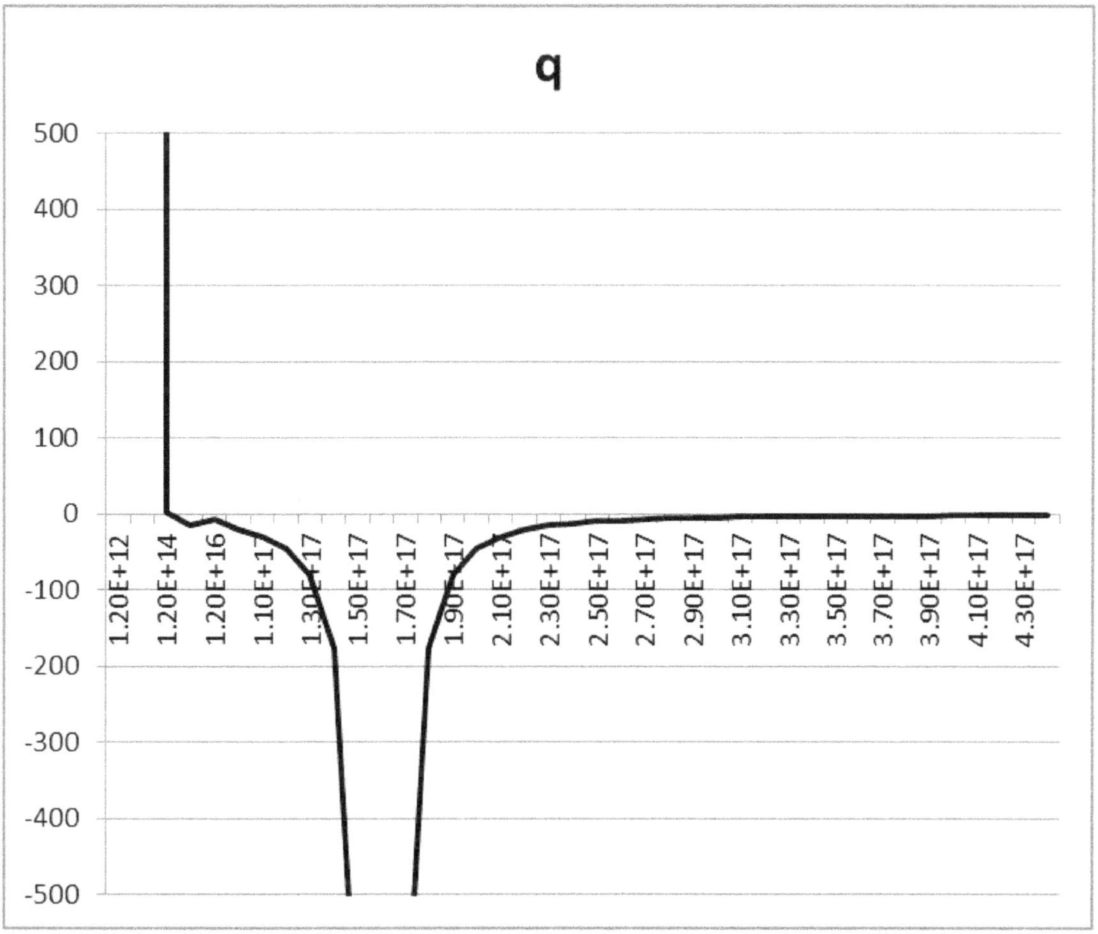

.

Figure 29.16 The deceleration parameter q vs. log(t sec) from t = 1.2×10^{12} sec. to t = 4.4×10^{17} sec. The peak occurs at t = 1.2×10^{13} sec. The dip occurs at 1.6×10^{17} sec. To the left of the peak q > 0 indicating decelerating expansion. To the right of the peak for t > 1.7×10^{14} sec. we find q < 0 indicating accelerating expansion starting 13.78 billion years ago – "just" after the Big bang. q = -2.0 presently. In the future, at t = 8.2×10^{17} sec. we found q = -1.2.

30. Some Implications of the Universal Scale Factor

In this chapter we consider some implications of our model Quantum Big Bang and Quantum Vacuum Universe based on the time development of relevant parameters presented in chapters 27 through 29.

30.1 Superclusters and Voids

In our study of the Quantum Big Bang[142] we found that the early Big Bang Metastate universe had a wide variation in the Hubble Constant. At its center with radial distance $r = 0$, we found $H = 1.79 \times 10^{218}$ km s^{-1} Mpc^{-1} while at its periphery $H = 1.14 \times 10^{126}$ km s^{-1} Mpc^{-1}. Thus we see the "explosion" of the energy-mass density of the central region into the outer regions. The result would appear to be a "wave" of mass-energy "emptying" the center and creating a bulge towards the outer region. Thus we anticipate that the primordial universe had variations in mass-energy density favoring the creation of a wave of high mass-energy density and an inner region of low mass-energy. The result is precursors of voids and supercluster bubbles.

In the later Quantum Vacuum Universe period we found a rapid universe contraction to 70% of its previous size. Subsequently the universe expanded – also fairly rapidly. The contraction effectively squeezed the outer portions of the universe again creating a wave of mass-energy. The expanding wave then (as in the case of water waves) developed foam (bubbles) of mass-energy of high density that evolved into the superclusters and voids that we find today.

The number of superclusters is estimated to be ten million. Superclusters seem to contain of the order of 10^5 galaxies or more. Thus the combination of the Quantum

---

[142] See Blaha (2019e).

Big Bang period and the Quantum Vacuum period explain the presence of superclusters and voids found in today's universe.

It appears we have an understanding of the large scale evolution of the universe. Fig. 30.1 presents the large scale pattern of universe evolution.

q rises to a peak value of 408,273 at $t = 1.2 \times 10^{13}$ sec. indicating decelerating expansion.

Dark Energy Ω_T rises to a peak of 39.2 at $t = 8.71 \times 10^{13}$ sec. causing H^2 to simultaneously become very large (Einstein equation), although H (Fig. 22.6) is negative and large with a minimum of -445 km s^{-1} Mpc^{-1}.

For $t > 1.7 \times 10^{14}$ sec. we find $q < 0$ indicating accelerating expansion as shown in H

ρ_{tot} declines to a minimum of 1.42×10^{-29} g/cm^{-3} at $t = 1.2 \times 10^{16}$ sec. after which it rises.

At $t = 1.0 \times 10^{17}$ sec. w and –pressure both peak. Note $w \approx 0$ suggesting a cold dust or gas. The pressure is also quite small. Both peaks are due to the increase in ρ_{tot} after reaching its minimum.

$q > 0$ for $t < 1.7 \times 10^{14}$ sec. (decelerating). q reaches a minimum at $t = 1.6 \times 10^{17}$ sec.

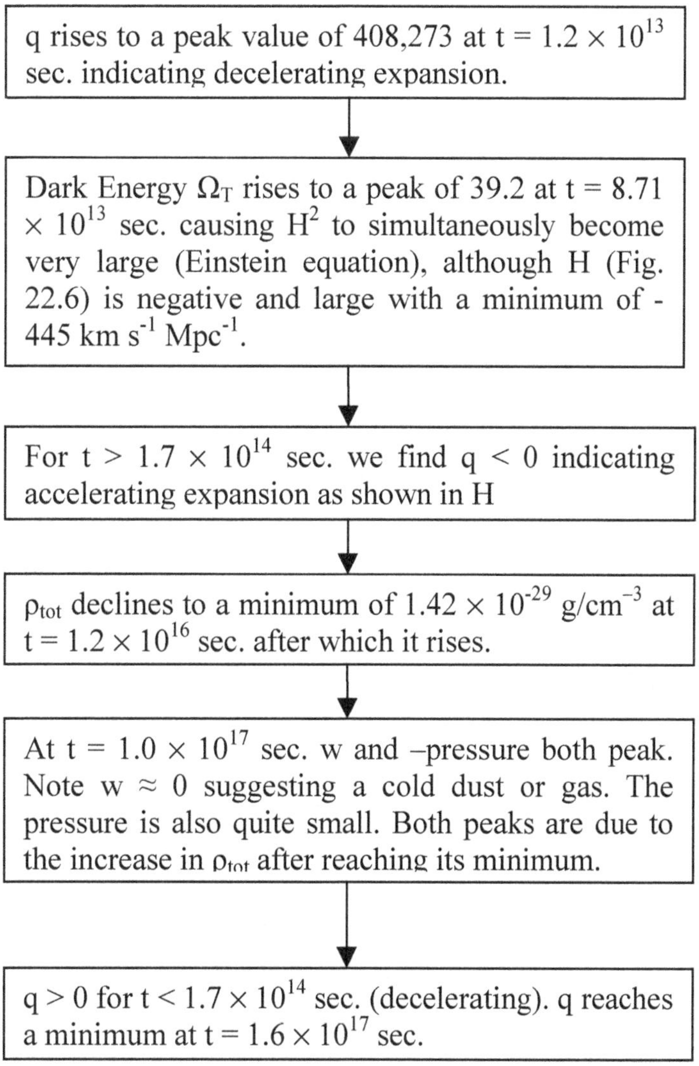

Figure 30.1 Scenario Based on figures of chapters 27-29. The figures in chapters 27 through 29 give the evolution in time of the universe.

31. Evidence for Entities Beyond Our Universe

It is easy to assert the existence of other universes. In this chapter we present theoretical and experimental support for a space of universes that we have called the *Megaverse*. We have discussed Megaverse features in earlier books including Blaha (2017c). We present some details on the Megaverse in chapters that follow. A method to escape our universe is presented in Blaha (2018e) and in earlier books by the author.

31.1 Theoretical and Experimental Support

Why are we not content with one universe given its enormous size and variety? It appears that there are important theoretical reasons, and some important experimental observations, that suggest that there is more than our universe 'out there.'

In this chapter[143] we will discuss theoretical reasons and experimental suggestions of a larger space—that we call the *Megaverse*—that contains our universe and, most likely, other universes. The existence of a Megaverse resolves several theoretical issues and may address some important astronomical puzzles that have appeared in recent years.

The theoretical issues, which have been subjects of discussion for many years, are:

1. The need for a 'clock' to measure 'time' knowing that it is to some extent relative and local.
2. The need for a 'quantum observer' to complete the understanding of quantum gravity as described by the Wheeler-DeWitt equation and in other efforts to develop a quantum gravity.
3. The need for other universes to provide theoretical measuring platforms for quantities beyond the charge and mass of the universe. We think here of the

[143] Most of this chapter appears in Blaha (2015a) and in earlier books by the author.

other quantum numbers of particles and particle number operators such as Baryon number.

4. The need for an ultimate source of mass and inertia in our universe.

In Blaha (2015a) and earlier books we have suggested that there are weighty reasons to believe that other universes exist.[144] The existence of other universes is a solution to these problems.

These problems have a source in Quantum Gravity and the interpretation of the Wheeler-DeWitt equation in particular. See chapter 26 for a discussion of the Wheeler-DeWitt equation and its implications. We now consider the issues raised above.

31.1.1 Universe Clocks

Asynchronous Logic provides the equivalent of a clock for the synchronization of processes within large electrical systems such as VLSI chips. Similarly there is a need for a universal clock for our universe. As DeWitt[145] points out in his studies of quantum gravity,

'"The variables … [of the quantized Friedmann model] because of their lack of hermiticity, are not rigorously observable and hence cannot yield a measure of proper time which is valid under all circumstances. … . It is for this reason that we may say that "time" is only a phenomenological concept … If the principle of general covariance is truly valid then the quantum mechanics of everyday usage with its dependence on the Schrödinger equations … is only a phenomenological theory. For the only "time" which a covariant theory can admit is an intrinsic time defined by the contents of the universe itself. Any intrinsically defined time is necessarily non-Hermitean, which is equivalent to saying that there exists no clock, whether geometrical or material, which can yield a measure of time which is operationally valid under *all* circumstances, and hence there exists no operational method for determining the Schrödinger state function with arbitrarily high precision."

---

[144] In Blaha (2013a), before the Higgs particle was discovered at CERN we suggested an alternate mechanism was possible if a sister universe existed (making the existence of other universes a reasonable possibility. The Higgs discovery makes the sister universe mechanism unlikely.

[145] DeWitt, B. S., Phys. Rev. **160**, 1113 (1987).

The lack of a clock within our universe invalidates quantum mechanics in principle and Quantum Gravity in particular. DeWitt concludes, "Thus [quantum gravity] will say nothing about time unless a clock to measure time is provided."

Unruh[146] also has an issue with the source of time:

"One of the key problems is that of time. We see and experience the world in terms of time. We see things grow, develop, and change. However, time does not enter into the Euclidean formulation of quantum gravity directly. In the usual Hamiltonian formulation, the Hamiltonian for quantum gravity is made up of densities which are the generators, not only of spatial coordinate transformations, but also of temporal coordinate transformations. The content of four of Einstein's equations is that some generators are zero. Thus all wave functions are invariant under all spatial and all temporal coordinate transformations. There is nothing in the wave function or the amplitudes which refers to the coordinate t, or the corresponding points of the manifold in any way. How then do we recover the indubitable and ubiquitous experience we have of time? The standard answer is that our experience of time is actually an experience of different correlations between physical quantities in the world. Time is replaced by the readings of clocks. I know that time has changed, not through any direct experience with time, but because the hands of my watch have changed.

Although the implementation of this idea is actually extremely difficult in practice, and although I personally believe that one should formulate one's quantum theory of gravity so as to contain time explicitly, let us nevertheless pursue the consequences of this idea of time as defined internally, as the "reading" of a dynamic variable. For an observer inside the theory, his "time" is not the coordinate t. Rather his time is some one of the given dynamic variables of the theory: y or P. Thus although the coupling to the baby universes via the effective action S is independent of the coordinates t or x, that does not mean that the observer inside the theory will experience the interactions as being independent of time. For him and/or her, time is one of the dynamic variables and so it can depend on the various dynamic variables of the theory, even if it does not depend on the time coordinate t. In general one would expect the

---

[146] Unruh, W. G., Phys. Rev. D **40**, 1053 (1989).

observer to see what looks to him like a time-dependent interaction with the baby universes. At one time, some one of the baby universes may couple strongly to the large universe, while at some other time, another of the baby universes will couple more strongly."

In Blaha (2015a) and earlier books, we suggested the existence of other universes provides a 'clock' in principle for our universe. And being universes, these other universes are excellent clocks. DeWitt points out,

"Because every clock has a "one-sided" energy spectrum, its ultimate accuracy must necessarily be inversely proportional to its rest mass. When the whole universe is cast in the role of a clock, the concept of time can of course be made fantastically accurate (at least in principle) … "

Setting a mass scale using other universes, also sets[147] a time scale and resolves the issue of a clock for our universe. *In principle the existence of other universes validates the role of time in the Copenhagen interpretation of Quantum Mechanics.*

31.1.2 Quantum Observer

Attempts to create a quantum gravity theory have to confront the need for an *Observer* in any quantum theory within the context of the Copenhagen interpretation. DeWitt points out,

"The Copenhagen view depends on the assumed a priori existence of a classical level to which all questions of observation may ultimately be referred. Here, however, the whole universe is the object of inspection; there is no classical vantage point, and hence the interpretation question must be re-argued from the beginning. While we do not wish to stress this point unduly, since, after all, the Friedmann model ignores the vast complexities of the real universe, it is nevertheless clear that the quantum theory of space-time must ultimately force a deviation from the traditional Copenhagen doctrine."

---

[147] For example the Planck time value is set by the Planck mass.

And Unruh states

"One of the key features in the interpretation of such transition amplitudes, or wave functions, is the idea that we, as observers are also a part of the Universe as a whole. We, as physical observers, must be describable from within the theory and not as observers external to the theory as in usual quantum mechanics. In usual quantum mechanics, the interpretation is usually given in terms of observers that are outside of the theory. There one makes a split, with the quantum world at one side of the split, and the observer on the other. von Neumann argued that the predictions of quantum mechanics, at least under certain assumptions, are independent of the exact location of that split, but Bohr argued adamantly for the necessity of such a split (classical observers and quantum world). *There is a great difficulty in setting up such a split for physical observers contained within and influenced by a quantum universe,* [italics added] and for the Universe as a whole, especially including gravity, one cannot argue that the predictions will be independent of where one puts the split. Since all energies interact gravitationally, and our observations are surely energetic phenomenon, the treatment of the energetics of observation as classical would lead to different predictions than if they were treated quantum mechanically. One is therefore forced to devise an interpretation of quantum mechanics in which the observer is part of the quantum system, rather than outside the quantum system.

This means that the interpretation of these transition amplitudes becomes somewhat non-intuitive. One must ask what the system looks like from within, from the viewpoint of an observer who is part of that world, rather than being able to interpret them directly in terms of probabilities for observations made by an external observer."

While the O*bserver* question is addressed by a number of authors, the proposed answers are not entirely convincing. *The existence of other universes provides macroscopic Quantum Observers for our universe.* And our universe provides a macroscopic quantum observer for other universes. Thus the quantum observer issue is resolved.

These considerations lead us to view the existence of other universes as a critical solution to the above problems.

31.1.3 The Higgs Mechanism is Explainable by Extra Dimensions

The Higgs Mechanism 'explains' (generates) fermion and boson masses. However the Higgs potential contains a quadratic term with a constant with the dimensions of [mass]. In a sense the Higgs Mechanism trades one mass for another. From where do the Higgs potentials' masses come?

A further explanation is needed is to determine the origin of the "dimensionful" mass terms in the Higgs' particle equations themselves. At present little if any thought has been given to the origin of these terms. We suggested that, excluding a "*deus ex machina*" source, the only known way to generate these mass terms in the Higgs' equations is through the separation of equations technique of differential equations. This technique requires additional parameters which can only be the coordinates of *extra unknown dimensions*. The best example of the generation of mass terms appears in the Schwarzschild solution of General Relativity where a separation constant, often denoted M, appears that has the dimension of [mass].

Thus extra space-time dimensions would resolve the origin of Higgs potentials' masses. Given extra dimensions it is reasonable to expect that these extra dimensions contain universes. Thus the Megaverse!

31.1.4 Possible Accretion of Megaverse Matter to Fuel Expansion of Our Universe

If matter is distributed outside of universes in the Megaverse, and if this matter can be accreted to universes by gravitational attraction, then the apparent increasing expansion of our universe may be due to this accretion. In chapter 14 of Blaha (2017c) we presented a model in which this possibility is realized. If true, then we would have tangible evidence of the residence of our universe in the Megaverse.

31.1.5 Asynchronous Logic is a Requirement of Universes

By establishing Asynchronous Logic principles[148] as the basis for the existence of universes and for setting the number of dimensions in each universe – four; and basis of fermion particles - qubes – we have found deeper principles of organization for the

---

[148] The basis of this section is described in detail in Blaha (2015a). That book places Physics within a logical framework that is a possible deeper ground for fundamental Physics theory.

foundations of physics. The principles built on this foundation serve to enable the coordination of complex physical processes.

Usually we look at particle processes primarily from a space-time perspective: particles collide and produce new particles. We primarily think of the incoming and outgoing particles in a collision. However, considering the set of fundamental particles – and the particle transforming interactions in themselves – neglecting space-time and momentum considerations – leads us to view particles as constituting an alphabet and their interactions as a type of computer grammar.[149] Then the Asynchronicity Principles enable us to bring in space-time in a way that gives us the maximum complexity with the most minimal assumptions. As Leibniz[150] points out our universe has maximal complexity with minimal assumptions. (See chapter 1.)

31.1.6 The Meaning of Total Quantities of a Universe

The 'external' properties of a universe are normally questioned—for the simple reason that it is assumed that there is no 'outside' of our universe. For example, Misner (1973) asserts:[151]
'There is no such thing as "the energy (or angular momentum, or charge) of a closed universe," according to general relativity, and this for a simple reason. To weigh something one needs a platform on which to stand to do the weighing.'

Misner *et al* presumes no such platform exists. If there is but one closed universe as most currently believe then one cannot measure any totals of a closed universe (which ours may be to be). Yet if we take a more general view that our universe is only one of many then it becomes possible to measure total mass, charge, angular momentum, baryon number, and many other quantities of interest. Indeed, the existence of other universes (within the encompassing Megaverse) opens the door to an understanding of time, mass, energy, and all the other quantities necessary to develop a dynamical theory of universes.

---

[149] This conceptual approach was first described in Blaha (1998) who went on to characterize our universe as one enormous word evolving in time.
[150] See Rescher (1967).
[151] Pp. 457 - 458.

Our new 'rotations of interactions' formalism (described previously) enables us to rotate measurable quantities. These quantities (quantum numbers) furnish a set of totals for our universe such as baryon numbers (normal and Dark), lepton numbers, angular momentum and so on that characterize our universe.

Later we will also see that one can then treat universes as 'particles', and develop 'universe dynamics', which might explain knotty problems such as the Big Bang and its precursor (if any). We will do this in subsequent chapters after first considering the possible structure of universes in general in the Megaverse.

31.2 Possible Experimental Evidence for the Megaverse

At first glance it would seem impossible to produce evidence for the existence of other universes. However there are subtle means by which we can 'sense' experimentally 'nearby' universes should they exist. The mechanism would appear to be gravitational effects exerted on objects within our universe by unseen objects of enormous mass. Currently there appears to be three experimental suggestions of the existence of 'nearby' universes and one theoretical argument based on an influx of mass-energy from the Megaverse that may cause the expansion of our universe.

31.2.1 Great Attractors

One potential support is the discovery of the Great Attractor (at the center of the Laniakea Galaxy Supercluster), and the more massive Shapley Attractor (centered in the Shapley Supercluster)[152]. These attractors contain massive numbers of galaxies and are drawing galaxies over a distance of millions of light years towards them.

If another universe(s) is 'near' our universe it could act as a 'gravitational magnet' and draw galaxies within our universe towards it to form one or more superclusters which could then act as attractors. Thus attractors might indirectly reveal the presence of other nearby universes—contrary to the expected large scale uniformity of the universe. The only other apparent source of superclusters is chance. Chance seems an unsatisfactory possibility in the present case.

---

[152] Tully, R. Brent; Courtois, Helene; Hoffman, Yehuda; Pomarède, Daniel, "The Laniakea Supercluster of galaxies". Nature (4 September 2014). 513 (7516): 71–73; arXiv:1409.0880.

31.2.2 Bright Bumps in Universe Suggesting Collision with Another Universe

A recent study[153] of the residual brightness of parts of the accessible universe found that bright patches appeared if a model of the CMB (Cosmic Microwave Background) with gases, stars and dust was 'subtracted' from the PLANCK map of the entire sky. After the subtraction one would expect only noise spread throughout the sky. However, bright patches were seen in a certain range of frequencies. These anomalies are thought to be a result of our universe colliding with another object – presumably another universe in the Megaverse.

31.2.3 Cold Spot in Universe Suggesting Collision with Another Universe

Another recent study[154] of a huge cold region of the universe spanning billions of light years revealed that this region is not a relatively empty region but rather is similar to in its distribution of galaxies to the rest of the universe. Previous the Cold Spot (an area where cosmic microwave background radiation – the leftover Big Bang radiation is weak – making it significantly colder (0.00015C colder) than the average temperature of the universe.)

An analysis of 7,000 galaxy redshifts using new high-resolution data has now shown that the Cold Spot is similar to the rest of the universe. The Durham University group suggested that the Cold Spot might have been caused by a collision between our universe and another Universe. They further suggested that there is only a 1 in 50 chance that it could be explained by standard cosmology. could produce this feature

Thus we have another important piece of circumstantial evidence in favor of other universes and thus the Megaverse.

31.2.4 Megaverse Energy-Matter Infusion into Our Universe

In chapter 14 of Blaha (2017c) we presented a model for an influx of mass-energy from the Megaverse to support the Bondi-Gold-Hoyle-Narlikar Steady State Cosmology, which was originally based on the 'continuous creation of mass-energy' by Hoyle and Narliker. This model explains why the value of Ω makes the universe close

[153] Ranga-Ram Chary, arXiv.org:/1510.00126 (2015).
[154] T. Shanks et al, Durham University (Australia), Monthly Notices of the Royal Astronomical Society, 2016 .

to flat. If this model is correct then we would have concrete support for a Megaverse with a low mass-energy density leaking mass-energy into our universe. *More generally, it suggests that universes are surfaces of high mass-energy density in a Megaverse of low mass-energy density – with a ratio of mass-energy densities of the other of* 10^{30}.

31.2.5 Conclusion

We conclude that data is beginning to emerge favoring multiple universes and a physical Megaverse in support of the theoretical justifications presented earlier.

31.3 Historical Trend Towards Larger Space-Time Structures

Looking back through the history of Mankind's view of the universe we see a clear progression to a larger and larger view. Before the 16th century the earth was the universe. In the 16th century Giordano Bruno (and possibly others) suggested that the stars were suns with many worlds circling them. So our view of the universe expanded to include stars.

Then over time it was noticed that nebulae existed in space. The astronomer, Edwin Hubble, studied the Andromeda nebula with the 'new' Mt. Wilson telescope and in 1929 announced that it was a galaxy composed of stars. Now the universe was conceptually similar to our current view.

Now we seem to have significant theoretical considerations and some suggestive experimental data that lead us to consider the possibility that our universe is not alone—that our universes is but one of many universes in a space we call the Megaverse. This book (and Blaha (2017c)) pulls together much of our earlier work and adds new insights into the nature of the Megaverse. We shall take the Megaverse as fact, extrapolate the form of our universe into the form of other possible universes, and develop a fairly detailed theory of the Megaverse and its resident universes. We will consider escaping our universe into the Megaverse—mindful that such travel will not happen until the *very* distant future. There are many technical bridges to cross before we can travel to other universes.

Given the vastness of our universe one might ask Why travel? We considered reasons in some detail in Blaha (2017c). For now, it suffices to say, Because they are there. Mankind has always grown and prospered through exploration and exploitation of new territories. Indeed the eminent Historian, Arnold Toynbee, stated that new 'turf' is the source of growth in all civilizations. Eventually, in the very distant future, we may need the new turf in the Megaverse.

31.4 Other Megaverses? Will It Ever End?

Does the trend to larger and larger expanses of space suggest that our Megaverse may be but one of many duplicate Megaverses of the same number of dimensions (but different orientations)? One cannot decide this question in our present, or likely near future, state of knowledge.

However, based on our estimate of the dimension of the Megaverse, and its basis in the geometry and group structure of its interactions, it is reasonable to conjecture duplicate Megaverses would have the same number of dimensions as our Megaverse, and have universes with the same number of dimensions as our universe, and thus have a Physics in each other Megaverses' universes similar to the Physics of our universe.

This scenario would appear to be unlikely in the author's view as it appears uneconomical and gives rise to the question How could a plethora of Megaverses arise?

Another scenario, in which Megaverses appear within Megaverses like the toy Chinese nested boxes, also appears unlikely. For it would require a chain of Megaverses of ever increasing dimension, and raise the question of its origin—a question that would never be answerable. Nor would the nested set of Megaverses be experimentally accessible.

So we are content with one Megaverse.

32. The Nature of Other Universes: SuperUniverses in the Megaverse

Having concluded the discussion of the Unified SuperStandard Theory from Complex General Relativity and Quantum Field Theory using the five axioms of chapter 1 in this volume and Blaha (2018e), and the discussion of the expansion of our universe based on vacuum polarization with its eigenvalue function, we note that other four dimensional (4D) universes are likely to have the same Physics (and thus Chemistry and Biology) as our universe.

In chapter 26 we raised the possibility of such universes within a larger space, the Megaverse.[155] We called these universes SuperUniverses. They have the same general features as our universe although they may differ in energy, size and distribution of galaxies..

We suggest all four-dimensional universes of sufficient size are SuperUniverses. Universes of higher (or lower) dimension may exist in the Megaverse and may also have a variant form of Unified SuperStandard Theory appropriate to their dimensionality.

The line of reasoning motivating SuperUniverse similarity follows.:

32.1 The Unified SuperStandard Theory and Cosmology Determine the Universe's Features

The overall physical theory of our universe has the features:

1. It is described by Complex General Relativity.

---

155 We use the term Megaverse for clarity because there are a number of different variants of Multiverses.

2. The particles, and particle interactions, in our universe are those of the Unified SuperStandard Theory.

3. In our universe we found the known Standard Model coupling constants, which were found to good approximation in our approach, are *self-determined* in Quantum Field Theory by vacuum polarization calculations. Since the Standard Model interactions are responsible for the vast majority of the features of Physics, Chemistry and Life we find the universe's detailed structure is set by the Unified SuperStandard Theory which embodies the Standard Model.

4. Our universe has an expansion scale factor and Hubble Constant determined by vacuum polarization. The Dark Energy dominance in our universe is also determined by vacuum polarization as are the Big Dip, galaxies, Superclusters, and voids.

Thus we find a "complete" description of our universe in the Unified SuperStandard Theory and our cosmological theory based on vacuum polarization.

32.2 Unified SuperStandard Theory Features are the Same in all 4D Universes – SuperUniverses

The complete self-contained description of universe would apply to any 4D universe due to its fundamental basis. Consequently, if there are other universes as we suggested earlier,[156] then any 4D universe should be similar to ours with the same Physics and Chemistry and Biology. Life in other universes is possible but can be expected to take different forms. We have called these universes *SuperUniverses*.

SuperUniverses can differ in some ways that do not change our overall conclusions: In some SuperUniverses a) anti-particles may dominate; b) left-right symmetry may differ; c) Life may favor either dextrorotary or levorotary molecules.

---

[156] See Blaha (2018e) and earlier books for discussions of the Megaverse and universe particle quantum field theory..

Appendix A. Subgroups of the Complex Lorentz Group

Appendices C – F specify the subgroups of the Complex Lorentz Group in some detail within the framework of complex Lorentz transformations. The subgroups are the U(1) and SU(2) subgroups corresponding to the ElectroWeak interactions, the SU(3) subgroup corresponding to the Strong interactions, and another pair of subgroups, U(1) and SU(2), corresponding to the Dark ElectroWeak interactions. These groups have a covering group SU(7) that we map to the corresponding set of particle interactions.

Appendix B. Reality Group of Complex General Relativity

We have seen that Complex Special Relativity is the basis of flat space-time phenomena. Flat space-time coordinates are complex-valued in general. The real-valued coordinates that we experience in everyday life are the result of our measuring instruments: clocks and rulers. Real-valued coordinates are generated from complex-valued coordinates by Reality group transformations.

If flat space-time is governed by Complex Special Relativity then it is clear that curved 'space-time' is governed by Complex General Relativity. Here again there is a Reality group the General Relativistic Reality group – a U(4) group – that maps complex-valued General Relativity coordinates to real-valued curved coodinates.[157] There is a corresponding U(4) Internal Symmetry Reality group that we call the *Species group*. This group rotates fermions as described in chapter 3.

We can isolate the General Relativistic Reality group by factoring complex General Relativistic coordinate transformations into parts that consist of a real-valued General Coordinate transformation and complex-valued coordinate transformations. It will be apparent that the General Relativistic Reality group emerges in this discussion. We begin by defining the tetrad notation.

B.1 Tetrad (Vierbein) Formalism

The *vierbein* formalism begins with the Equivalence Principle that allows us to define an inertial coordinate system in the neighborhood of any point Z in space-time. We will use the notation $\varsigma^{\alpha}(Z)$ to denote the inertial coordinates at Z. We define a tetrad or vierbein as

[157] Much of this chapter appears in Blaha (2016h) and (2017a).

$$v^{\alpha}{}_{\mu}(x) = (\partial \varsigma^{\alpha}(x)/\partial x^{\mu})_{x=Z} \tag{B.1}$$

and, in a neighborhood of Z, we can invert the relation between ς and x to define an inverse

$$w^{\mu}{}_{\alpha}(x) = (\partial x^{\mu}(\varsigma)/\partial \varsigma^{\alpha})_{x=X} \tag{B.2}$$

such that

$$\begin{aligned} w^{\mu}{}_{\alpha}(x)v^{\alpha}{}_{\nu}(x) &= \delta^{\mu}{}_{\nu} \\ w^{\mu}{}_{\beta}(x)v^{\alpha}{}_{\mu}(x) &= \delta^{\alpha}{}_{\beta} \end{aligned} \tag{B.3}$$

In real-valued General Relativity all *tetrads* are real-valued. In Complex General Relativity a *tetrad* $v^{\alpha}{}_{\mu}(x)$ is complex-valued.

The metric at a curved space-time point X is defined in terms of *tetrads* as

$$\begin{aligned} g_{\rho\sigma}(x) &= \eta_{\alpha\beta}\, v^{\alpha}{}_{\rho}(x)v^{\beta}{}_{\sigma}(x) \\ g^{\rho\sigma}(x) &= \eta^{\alpha\beta}\, w^{\rho}{}_{\alpha}(x)w^{\sigma}{}_{\beta}(x) \end{aligned} \tag{B.4}$$

The inverse of a *tetrad* transformation can also be expressed as

$$w_{\beta}{}^{\nu}(x) = v_{\beta}{}^{\nu}(x) = \eta_{\beta\alpha}g^{\nu\mu}(x)v^{\alpha}{}_{\mu}(x)$$

Then a *tetrad* and its inverse satisfy

$$v^{\alpha}{}_{\mu}(x)v_{\beta}{}^{\mu}(x) = \delta^{\alpha}{}_{\beta} \tag{B.5}$$

and

$$v^{\alpha}{}_{\mu}(x)v_{\alpha}{}^{\nu}(x) = \delta^{\nu}{}_{\mu}$$

There are two general types of space-time transformations that can be performed on a tetrad.

1. A complex-valued (possibly real-valued) General Relativistic coordinate transformation:

$$v'^{\alpha}{}_{\mu}(x) = \partial x^{\nu}/\partial x'^{\mu}\, v^{\alpha}{}_{\nu}(x)$$

2. A complex-valued, local *Lorentzian transformation*

$$v'^{\beta}{}_{\mu}(x) = \Lambda(x)^{\beta}{}_{\alpha}\, v^{\alpha}{}_{\mu}(x)$$

where $\Lambda(x)^{\beta}{}_{\alpha}$ is an element of a subset of the local Complex Lorentz Group.

The local Lorentzian transformations $\Lambda(x)^{\beta}{}_{\alpha}$ consist of local Lorentz transformations that are real-valued, and complex-valued Lorentz transformations. Both types of transformations satisfy the orthogonality condition:

$$\eta_{\alpha\beta}\Lambda^{\alpha}{}_{\rho}(x)\Lambda^{\beta}{}_{\sigma}(x) = \eta_{\rho\sigma} \tag{B.6}$$

Thus the *tetrad* partakes of both local (position dependent) General Relativistic transformations and local Lorentzian transformations.

B.2 Complex General Relativistic Transformations

The General Relativistic Reality group interaction emerges from complex General Relativistic transformations. We can separate elements of the set of all complex General Coordinate transformations into a product of two factors: a real-valued General Coordinate transformation and a complex-valued General Coordinate transformation. The set of complex factors can be further factored into those that satisfy

$$\Lambda(\omega, \mathbf{u})^{T} G \Lambda(\omega, \mathbf{u}) = G \tag{B.7}$$

and those that do not. We then see that the set of those that do not satisfy the above equation form a curved space representation of the U(4) group under 'multiplication' of transformations.

The elements of the set of real and complex General Coordinate transformations whose flat complex space-time limit satisfy the above equation form the elements of the Complex Lorentz group.[158]

We thus find the set of all 4-dimensional complex, curved space General coordinate transformations can be visualized as in Fig. B.1. The next section describes the interplay of the three parts displayed in Fig. B.1.

B.3 Structure of Complex General Coordinate Transformations

Complex General Coordinate transformations can be uniquely factored into products of two terms, which will later be further factored into three factors. They have the form

$$\partial x''^{\nu}(x)/\partial x^{\mu} = U(x'')^{\nu}{}_{\beta}\, \partial x'^{\beta}(x)/\partial x^{\mu} \tag{B.8}$$

where

$$x''^{\nu}(x) = U(x'')^{\nu}{}_{\beta} x'^{\beta}$$
$$x'^{\mu}(x) = U^{-1\mu}{}_{b}(x'')\, x''^{b}$$

where $U(x')^{\nu}{}_{\beta}$ is complex and where $\partial x'^{\beta}(x)/\partial x^{\mu}$ is a purely real General Coordinate transformation.

We define

$$U\,(x'')^{\mu}{}_{\nu} = w^{\mu}{}_{a}(x'')[\exp(i\, \textstyle\sum_{k} g_k \Phi_k(x'')\tau_k)]^{a}{}_{b}\, v^{b}{}_{\nu}(x'') \tag{B.9}$$
$$U^{-1}(x'')^{\mu}{}_{\nu} = w^{\mu}{}_{a}(x'')[\exp(-i\textstyle\sum_{k} g_k \Phi_k(x'')\tau_k)]^{a}{}_{b}\, v^{b}{}_{\nu}(x'')$$

where τ_k is a hermitean U(4) generator matrix, where the constants g_k are real, and where the Φ_k are real. The uniqueness of the factorization follows from the Reality group (and U(4)) property that any complex 4-vector can be uniquely mapped to any specified real 4-vector.

---

[158] It is this part of curved space-time General Relativity that becomes the flat space-time Complex Lorentz group, which leads to the SU(3)⊗SU(2)⊗U(1)⊗SU(2)⊗U(1) Standard Model Reality group.

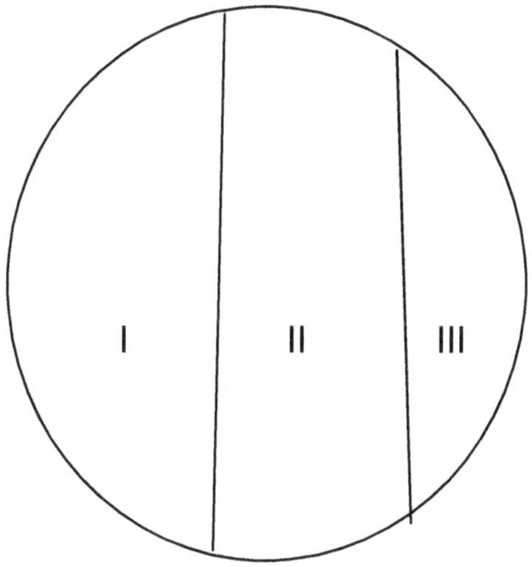

Figure B.1. A visualization of the set of General Coordinate transformations separated into real-valued General coordinate transformations (part I), complex transformations that satisfy $\Lambda(\omega, \mathbf{u})^T G\Lambda(\omega, \mathbf{u}) = G$ (part II), and complex transformations that do not satisfy $\Lambda(\omega, \mathbf{u})^T G\Lambda(\omega, \mathbf{u}) = G$ (part III). Part I and part II combine in the limit of flat space-time to form the Complex Lorentz group. Parts II and III elements form a U(4) group that we call the General Relativistic Reality group.

Given the factorization above it becomes possible to separate the affine connection correspondingly.

B.4 Complex Affine Connection – General Relativistic Reality Group

The structure of a complex general coordinate transformation enables us to calculate its affine connection for later use in determining the covariant derivative, and the dynamic equations. First the affine connection for the transformation to real-valued x' coordinates from inertial coordinates ς^ρ is

$$\Gamma^{\sigma}{}_{\lambda\mu}\,(x') = \partial x'^{\sigma}/\,\partial\varsigma^{\rho}\;\partial^{2}\varsigma^{\rho}/\,\partial x'^{\lambda}\partial x'^{\mu} \tag{B.10}$$

Next the Reality group transformation has the affine connection

$$\Gamma^{\sigma}{}_{\lambda\mu}\,(x'') = \partial x''^{\sigma}/\,\partial\varsigma^{\rho}\;\partial^{2}\varsigma^{\rho}/\,\partial x''^{\lambda}\partial x''^{\mu}$$

which can be re-expressed as

$$\begin{aligned}\Gamma^{\sigma}{}_{\lambda\mu}\,(x'') &= \partial x''^{\sigma}/\,\partial x'^{\beta}\;\partial x'^{\beta}(\varsigma)/\partial\varsigma^{\rho}\;\partial/\partial x''^{\mu}[\partial\,\varsigma^{\rho}/\,\partial x'^{\alpha}\;\partial\, x'^{\alpha}/\,\partial x''^{\lambda}] \\ &= \partial x''^{\sigma}/\,\partial x'^{\beta}\;\partial x'^{\alpha}/\partial x''^{\lambda}\;\partial x'^{\gamma}/\partial x''^{\mu}\;\Gamma^{\beta}{}_{\alpha\gamma}(x') + \partial x''^{\sigma}/\,\partial x'^{\beta}\;\partial^{2}x'^{\beta}/\,\partial x''^{\lambda}\partial x''^{\mu}\end{aligned} \tag{B.11}$$

Next substituting the General Relativistic Reality group transformation

$$x''^{\nu}(x) = U(x'')^{\nu}{}_{\beta}x'^{\beta}$$
$$x'^{\mu}(x) = U^{-1}(x'')^{\mu}{}_{\beta}\,x''^{\beta}$$

together with

$$\partial x''^{\sigma}/\,\partial x'^{\beta} = \partial[U(x'')^{\sigma}{}_{\alpha}x'^{\alpha}]/\partial x'^{\beta} = U(x'')^{\sigma}{}_{\beta} + x'^{\alpha}\,\partial U(x'')^{\sigma}{}_{\alpha}/\partial x'^{\beta}$$

$$\partial x'^{\sigma}/\partial x''^{\beta} = \partial[U^{-1}(x'')^{\sigma}{}_{\alpha}x''^{\alpha}]/\partial x''^{\beta} = U^{-1}(x'')^{\sigma}{}_{\beta} + x''^{\alpha}\,\partial U^{-1}(x'')^{\sigma}{}_{\alpha}/\partial x''^{\beta}$$

we find the second term above (eq. B.11) is the Reality fields affine connection

$$\Gamma_{R}{}^{\sigma}{}_{\lambda\mu}(x'') = \partial[U(x'')^{\sigma}{}_{\alpha}x'^{\alpha}]/\partial x'^{\beta}\;\partial\,\{\,\partial[U^{-1}(x'')^{\beta}{}_{\alpha}x''^{\alpha}]/\,\partial x''^{\lambda}\}/\partial x''^{\mu}$$

and so we find the affine connections are approximately additive. Thus approximately

$$\Gamma^{\sigma}{}_{\lambda\mu}(x'') = \Gamma_{GR}{}^{\sigma}{}_{\lambda\mu}(x') + \Gamma_{R}{}^{\sigma}{}_{\lambda\mu}(x'')$$

if $x''^{\sigma} \simeq x'^{\sigma}$ where $\Gamma_{GR}{}^{\sigma}{}_{\lambda\mu}(x')$ is the real General Relativistic affine connection.

A complex transformation of types II and III in Fig. B.1 has the form:

$$U(x'')^{\mu}{}_{\nu} = w^{\mu}{}_{a}(x'')[\exp(i \sum_k \Phi_k(x'')\tau_k)]^{a}{}_{b}\, l^{b}{}_{\nu}(x'')$$
$$U^{-1}(x'')^{\mu}{}_{\nu} = w^{\mu}{}_{a}(x'')[\exp(-i\sum_k \Phi_k(x'')\tau_k)]^{a}{}_{b}\, l^{b}{}_{\nu}(x'')$$

where τ_k is a U(4) generator matrix. Its infinitesimal transformation is approximately

$$U(x'')^{\nu}{}_{\beta} \approx \delta^{\nu}{}_{\beta} + i \sum_k \Phi_k(x'')[\tau_k]^{\nu}{}_{\beta} \tag{B.12}$$

$$U^{-1}(x'')^{\nu}{}_{\beta} \approx \delta^{\nu}{}_{\beta} - i \sum_k \Phi_k(x'')[\tau_k]^{\nu}{}_{\beta}$$

using the *vierbein* flat space-time limits

$$w^{\mu}{}_{a}(x'') \approx \delta^{\mu}{}_{a}$$
$$l^{b}{}_{\nu}(x'') \approx \delta^{b}{}_{\nu}$$

where

$$\Phi_k(x) = \int^{x} dy_\lambda\, A_{Rk}{}^{\lambda}(y) \tag{B.13}$$

Then

$$\Gamma_R{}^{\sigma}{}_{\lambda\mu} = -\tfrac{1}{2}i\{\sum_k A_{Rk}(x'')_{\mu}[\tau_k]^{\sigma}{}_{\lambda} + \sum_k A_{Rk}(x'')_{\lambda}[\tau_k]^{\sigma}{}_{\mu}\} \tag{B.14}$$
$$= A_R{}^{\sigma}{}_{\mu\lambda} + A_R{}^{\sigma}{}_{\lambda\mu}$$

(summed over k) with the matrix $A_R{}^{\sigma}{}_{\mu\lambda}$ given by

$$A_R{}^{\sigma}{}_{\mu\lambda} = -\tfrac{1}{2}i\sum_k A_{Rk\mu}[\tau_k]^{\sigma}{}_{\lambda} \tag{B.15}$$

with $A_R{}^{\sigma}{}_{\mu\lambda}$ transformable to matrix row and column numbers

$$A_{R\text{flat}}{}^{\mu a}{}_{b} = A_{R\text{flat}k}{}^{\mu}[\tau_k]^{\sigma}{}_{\lambda}\delta_{\sigma}{}^{a}\delta^{\lambda}{}_{b}$$

using the flat space-time vierbein values. So $A_{R\text{flat}}{}^{a}{}_{\mu b}$ may be written in matrix form as

$$A_{R_{flat\mu}} = -\tfrac{1}{2}i\sum_k A_{R_{flatk\mu}}\tau_k \qquad \text{(B.16)}$$

In the flat space-time limit the $A_{Rk}{}^{\lambda}(y)$ becomes the Coordinate Species group U(4) gauge fields $A_{R_{flatk}}{}^{\lambda}(y)$.

The relevant *quadratic* $A_R{}^{\sigma}{}_{\mu\lambda}$ terms from eq. 11.22 that are needed to find the dynamic equation for the gauge fields $A_{R_{flat}}{}^{i}{}_{\mu}$ are contained in

$$\mathcal{L}_A = \text{Tr}\,\sqrt{g}[M\partial_\nu R^1{}_{\sigma\mu}\partial^\nu R^{2\sigma\mu} + aR^1{}_{\sigma\mu}R^{2\sigma\mu} + bg^{\sigma\mu}(R^1{}_{\sigma\mu} + R^2{}_{\sigma\mu}) + 1/4(g_{\mu\nu} + g^2{}_{\mu\nu})T^{\mu\nu}] \qquad \text{(B.17)}$$

We can let

$$R^i{}_{\sigma\mu} = R^{i\beta}{}_{\sigma\beta\mu} \equiv \partial_\mu(A_R{}^{i\beta}{}_{\sigma\beta} + A_R{}^{i\beta}{}_{\beta\sigma}) - \partial_\beta(A_R{}^{i\beta}{}_{\sigma\mu} + A_R{}^{i\beta}{}_{\mu\sigma}) \qquad \text{(B.18)}$$

for i = 1, 2. In the flat space-time limit we chose the Landau gauge

$$\partial_\mu A_{R_{flat}}{}^{i\mu a}{}_{b} = 0 \qquad \text{(B.19)}$$

As a result

$$R^i{}_{\sigma\mu} \equiv \partial_\mu(A_R{}^{i\beta}{}_{\sigma\beta} + A_R{}^{i\beta}{}_{\beta\sigma}) \qquad \text{(B.20)}$$

Using

$$A_{R_{flat}}{}^{i\mu\sigma}{}_{\lambda} = A_{R_{flat}}{}^{i\mu}[\tau_k]^a{}_b\delta^\sigma{}_a\delta_\lambda{}^b \qquad \text{(B.21)}$$
$$A_{R_{flat}}{}^{i}{}_{\mu} = -\tfrac{1}{2}i\sum_k A_{R_{flat}}{}^{i}{}_{k\mu}$$

and taking the trace in eq. B.17 we obtain

$$\mathcal{L}_A = \text{Tr}\,\sqrt{g}[8M\partial_\nu\partial_\mu A_{R_{flat}}{}^1{}_\sigma\partial^\nu\partial^\mu A_{R_{flat}}{}^{2\sigma} + 8a\partial_\mu A_{R_{flat}}{}^1{}_\sigma\partial^\mu A_{R_{flat}}{}^{2\sigma} + 1/4(g_{\mu\nu} + g^2{}_{\mu\nu})T^{\mu\nu}] \qquad \text{(B.22)}$$

in the flat space-time limit. Eq. B.17 needs to take account of the complex nature of ($g_{\mu\nu}$ + $g^2{}_{\mu\nu}$) until transformed by the infinitesimal form of the complex Reality transformation:

$$(g_{\beta\alpha} + g^2{}_{\beta\alpha})' \rightarrow U(x'')_\beta{}^\mu (g_{\mu\nu} + g^2{}_{\mu\nu})U^{-1}(x'')^\nu{}_\alpha$$
$$= (\delta_\beta{}^\mu + i \textstyle\sum_k \Phi_k(x'')[\tau_k]_\beta{}^\mu)(g_{\mu\nu} + g^2{}_{\mu\nu})(\delta^\nu{}_\alpha - i \sum_k \Phi_k(x'')[\tau_k]^\nu{}_\alpha)$$
$$\cong (g_{\beta\alpha} + g^2{}_{\beta\alpha}) + i\{\textstyle\sum_k \Phi_k(x'')[\tau_k]_\beta{}^\mu - i \sum_k \Phi_k(x'')[\tau_k]^\nu{}_\alpha)(g_{\mu\nu} + g^2{}_{\mu\nu})$$
$$\cong (g_{\beta\alpha} + g^2{}_{\beta\alpha}) + i\textstyle\sum_k \Phi_k(x'')\{[\tau_k]_{\beta\alpha} - [\tau_k]_{\beta\alpha}\} \qquad (B.23)$$

Approximating $\Phi_k(x'')$ with an infinitesimal line we find

$$\textstyle\sum_k \Phi_k(x) \cong \delta x_\lambda A_{Rflat}{}^{1\lambda}(x) \qquad (B.24)$$

by eq. B.13. Thus

$$1/4(g_{\mu\nu} + g^2{}_{\mu\nu})T^{\mu\nu} \cong \tfrac{1}{4}\,[(g_{\beta\alpha} + g^2{}_{\beta\alpha}) + i\,\delta x_\lambda A_{Rflat}{}^{1\lambda}(x)\{[\tau_k]_{\beta\alpha} - [\tau_k]_{\beta\alpha}\}]T^{\alpha\beta}$$

Applying the canonical Euler-Lagrange method we obtain the dynamical equations (using integration by parts to handle higher order derivative terms):[159]

$$\Box^2 A_{R_{flat}}{}^1{}_\sigma + (a/M)\Box A_{R_{flat}}{}^1{}_\sigma + i\delta x_\sigma\{[\tau_k]_{\beta\alpha} - [\tau_k]_{\beta\alpha}\}T^{\alpha\beta}/(32M) = 0 \qquad (B.25)$$
$$\Box^2 A_{R_{flat}}{}^2{}_\sigma + (a/M)\Box A_{R_{flat}}{}^2{}_\sigma = 0 \qquad (B.26)$$

Since the $A_{R_{flat}}$ gauge field is gravitational in nature it exists, as eq. B.25 shows, as a type of gravitational interaction whose source is the energy-momentum tensor. Following the standard derivation of the gravitational potential we find the Coulomb interaction of $A_{R_{flat}}{}^{10}$.

B.5 PseudoQuantization of Affine Connections

Having obtained the form of the general affine connection we now PseudoQuantize them for later use in our unification program. We define

---

[159] It is possible that the Reality transformation also depends on $A_{R_{flat}}{}^2{}_\sigma$. Then eq. A.26 would have an energy-momentum tensor term as well. Consequently there would be an additional interaction of the same form as in eq. A.27.

$$R^{1\beta}{}_{\sigma\nu\mu} = \partial_\mu H^{1\beta}{}_{\sigma\nu} - \partial_\nu H^{1\beta}{}_{\sigma\mu} + H^{1\gamma}{}_{\nu\sigma} H^{1\beta}{}_{\gamma\mu} - H^{1\gamma}{}_{\mu\sigma} H^{1\beta}{}_{\gamma\nu}$$
$$R^{2\beta}{}_{\sigma\nu\mu\rho} = \partial_\mu H^{2\beta}{}_{\sigma\nu} - \partial_\nu H^{2\beta}{}_{\sigma\mu} + H^{2\gamma}{}_{\nu\sigma} H^{2\beta}{}_{\gamma\mu} - H^{2\gamma}{}_{\mu\sigma} H^{2\beta}{}_{\gamma\nu} + \quad \text{(B.27)}$$
$$+ H^{1\gamma}{}_{\nu\sigma} H^{2\beta}{}_{\gamma\mu} - H^{1\gamma}{}_{\mu\sigma} H^{2\beta}{}_{\gamma\nu} + H^{2\gamma}{}_{\nu\sigma} H^{1\beta}{}_{\gamma\mu} - H^{2\gamma}{}_{\mu\sigma} H^{1\beta}{}_{\gamma\nu}$$

where

$$H^{\sigma}{}_{\nu\mu} = \Gamma_{GR}{}^{\sigma}{}_{\nu\mu} + \Gamma_{GR}{}^{2\sigma}{}_{\nu\mu} + \Gamma_R{}^{1\sigma}{}_{\nu\mu} + \Gamma_R{}^{2\sigma}{}_{\nu\mu} \quad \text{(B.28)}$$

and where $\Gamma_{GR}{}^{\sigma}{}_{\nu\mu}$ and $\Gamma_{GR}{}^{2\sigma}{}_{\nu\mu}$ are affine connections for real-valued General Relativity, and $\Gamma_R{}^{1\sigma}{}_{\nu\mu}$ and $\Gamma_R{}^{2\sigma}{}_{\nu\mu}$ are affine connections for a complex-valued set of transformations embodying a U(4) gauge group that combine with real-valued General Relativistic transformations to yield Complex General Relativistic transformations.

The affine connection is most often viewed as a derived quantity—part of the derivation of the curvature tensor in General Relativity. It is typically derived from manipulations of the metric $g_{\mu\nu}$. However, the affine connection can also be viewed as a set of independent fields that become related to the metric via dynamic equations.

Some years ago A. Einstein and H. Weyl[160] pointed out that the metric and the affine connection should be treated as independent quantities and subject to independent arbitrary infinitesimal variations:

> "In contrast to Einstein's original "metric" conception in terms of the $g_{\nu\mu}$ there was later developed, by Eddington, by Einstein himself, and recently by Schrödinger, an affine field theory operating with the components $\Gamma^{\sigma}{}_{\nu\mu}$ of an affine connection. But in 1925 Einstein also advocated a "mixed" formulation by means of a lagrangian in which both the $g_{\nu\mu}$ and the $\Gamma^{\sigma}{}_{\nu\mu}$ are taken as basic field quantities and submitted to independent arbitrary infinitesimal variations.[161] In certain respects this seems to be the most natural procedure."

Following this approach we have introduced the above affine connections for use in the construction of our unification of particle interactions.

---

[160] H. Weyl, Phys. Rev. 77, 699 (1950).
[161] A. Einstein, Sitzungsber., Preuss. Akad. Der Wissensch. (1925), p. 414.

Appendix C. Coordinate Reality Group Analogue to ElectroWeak Doublets

C.0 Introduction

In appendix H we establish the four species of fermions based on Complex Lorentz group (L_C) boosts. In this appendix[162] we will introduce the coordinate space *analogue* of the ElectroWeak interactions based on Complex Lorentz group considerations. We begin by generalizing the free Dirac equation to a 2×2 matrix of Dirac-like equations that have a larger group covariance. This matrix equation is applied to a doublet consisting of a normal Dirac particle wave function and a tachyon wave function. We will identify these doublets as *analogues* of ElectroWeak lepton doublets.

Then starting in section C.5 we consider a generalized 2×2 equation matrix (covariant under the L_C group) for doublets of *complexon* particles with complex 3-momenta consisting of an up-type complexon and a down-type tachyonic complexon. Because of an inherent SU(3) symmetry we will identify these doublets as analogues of quark ElectroWeak doublets. SU(3) symmetry leads us to identify each complexon quark in a doublet as a color SU(3) triplet analogue.

C.1 Transformations of Dirac and Tachyon Equations

A Left-handed boost of the Dirac equation transforms the Dirac equation into the spin ½ tachyon equation, and vice versa:

$$S_L(\Lambda_L(\omega, \mathbf{u}))\psi(x) \rightarrow \psi_T'(x') \qquad \text{(C.1a)}$$
$$S_L(\Lambda_L(\omega, \mathbf{u}))\psi_T(x) \rightarrow \psi'(x')$$

---

162 This chapter is extracted from Blaha (2007b).

Also, noting the appearance of a γ^5, we see

$$S_L(\Lambda_L(\omega, \mathbf{u}))(\gamma^\mu\partial/\partial x^\mu - m)S_L^{-1}(\Lambda_L(\omega, \mathbf{u})) = (i\gamma^\mu\partial/\partial x'^\mu - m) \quad \text{(C.1b)}$$
$$S_L(\Lambda_L(\omega, \mathbf{u}))\gamma^5(i\gamma^\mu\partial/\partial x^\mu - m)\gamma^5 S_L^{-1}(\Lambda_L(\omega, \mathbf{u})) = (\gamma^\mu\partial/\partial x'^\mu - m)$$

where

$$x'^\mu = i\Lambda_L{}^\mu{}_\nu(\omega, \mathbf{u})x^\nu \quad \text{(C.1c)}$$
$$\partial/\partial x'^\mu = -i\Lambda_L{}^\nu{}_\mu(\omega, \mathbf{u})\partial/\partial x^\nu$$

with

$$x' = E(\mathbf{v})x = i\Lambda_L(\mathbf{v})x$$

Eqs. C.1a – C.1c imply

$$S_L(\Lambda_L(\omega, \mathbf{u}))(\gamma^\mu\partial/\partial x^\mu - m)\psi_T(x) = (i\gamma^\mu\partial/\partial x'^\mu - m)S_L(\Lambda_L(\omega, \mathbf{u}))\psi_T(x)$$
$$= (i\gamma^\mu\partial/\partial x'^\mu - m)\psi'(x') \quad \text{(C.1d)}$$

and

$$S_L(\Lambda_L(\omega, \mathbf{u}))\gamma^5(i\gamma^\mu\partial/\partial x^\mu - m)\psi(x) = (\gamma^\mu\partial/\partial x'^\mu - m)S_L(\Lambda_L(\omega, \mathbf{u}))\gamma^5\psi(x)$$
$$= (\gamma^\mu\partial/\partial x'^\mu - m)\psi_T'(x') \quad \text{(C.1e)}$$

where

$$\psi'(x') = S_L(\Lambda_L(\omega, \mathbf{u}))\psi_T(x) \quad \text{(C.1f)}$$

and

$$\psi_T'(x') = S_L(\Lambda_L(\omega, \mathbf{u}))\gamma^5\psi(x) \quad \text{(C.1g)}$$

Note the Dirac equation is not a left-handed Complex Lorentz covariant.

C.2 Doublet Extended Dirac Equations

We will now consider the issue of generalizing the Dirac equation so that the extended equation is covariant under both Lorentz transformations and Left-handed Complex Lorentz transformations.

The only obvious method to obtain an extended Dirac equation that is covariant under Complex Lorentz transformations is to define an 8×8 matrix generalization. Let

$$đ(x) = \begin{bmatrix} (\gamma^{\mu}\partial/\partial x^{\mu} - m) & 0 \\ 0 & (i\gamma^{\mu}\partial/\partial x^{\mu} - m) \end{bmatrix} \tag{C.2}$$

be an 8 × 8 matrix operator with the 4 × 4 matrix elements shown, and let

$$\Psi(x) = \begin{bmatrix} \psi_T(x) \\ \psi(x) \end{bmatrix} \tag{C.3}$$

be an 8 component column vector composed of a Dirac field and a tachyon field. Then the extended free Dirac equation is

$$đ(x)\Psi(x) = 0 \tag{C.4}$$

We now define the 8 × 8 Left-handed Complex Lorentz transformation

$$S_{L8}(v) = \begin{bmatrix} 0 & S_L(\Lambda_L(v))\gamma^5 \\ S_L(\Lambda_L(v)) & 0 \end{bmatrix} \tag{C.5}$$

with inverse transformation

$$S_{L\,L8}{}^{-1}(\Lambda_L(v)) = \begin{bmatrix} 0 & S_L{}^{-1}(\Lambda_L(v)) \\ \gamma^5 S_L{}^{-1}(\Lambda_L(v)) & 0 \end{bmatrix} \tag{C.6}$$

Note: we use the notations $S_L(\Lambda_L(v))$ and $S_L(\Lambda_L(\omega, \mathbf{u}))$ interchangeably. Applying S_{L8} to eq. C.4 yields

$$0 = S_{L8}(\Lambda_L(v))đ(x)\Psi(x) = đ(x')\Psi'(x') \tag{C.7}$$

where

$$\Psi'(x') = \begin{bmatrix} S_L\gamma^5\psi(x) \\ S_L\psi_T(x) \end{bmatrix} = \begin{bmatrix} \psi_T'(x') \\ \psi'(x') \end{bmatrix} \tag{C.8}$$

Thus the extended Dirac equation is covariant under generalized Left-handed Complex Lorentz transformations such as eqs. C.5-C.6. Covariance requires the tachyon and the Dirac particles must have the same absolute value for the mass which is the qubel mass in the free fermion case.

It is easy to show that the extended Dirac equation eq. C.4 is also covariant under conventional Lorentz transformations in the 8 × 8 representation:

$$S_8(\Lambda(v)) = \begin{bmatrix} S(\Lambda(v)) & 0 \\ 0 & S(\Lambda(v)) \end{bmatrix} \tag{C.9}$$

with inverse

$$S_8^{-1}(\Lambda(v)) = \begin{bmatrix} S^{-1}(\Lambda(v)) & 0 \\ 0 & S^{-1}(\Lambda(v)) \end{bmatrix} \tag{C.10}$$

and non-diagonal Lorentz transformations:

$$S_{8A}(\Lambda(v)) = \begin{bmatrix} 0 & S(\Lambda(v)) \\ S(\Lambda(v)) & 0 \end{bmatrix} \tag{C.11}$$

with inverse transformation

$$S_{8A}{}^{-1}(\Lambda(v)) = \begin{bmatrix} 0 & S^{-1}(\Lambda(v)) \\ S^{-1}(\Lambda(v)) & 0 \end{bmatrix} \tag{C.12}$$

Under a conventional Lorentz transformation we find

$$0 = S_8(\Lambda(v))đ(x)\Psi(x) = đ(x')\Psi'(x') \tag{C.13}$$

$$0 = S_{8A}(\Lambda(v))đ(x)\Psi(x) = đ(x')\Psi'(x')$$

The lagrangian density that corresponds to our 8-dimensional construction is

$$\mathcal{L}_8 = \overline{\Psi}(x)đ(x)\Psi(x) \tag{C.14}$$

where

$$\overline{\Psi}(x) = \Psi^\dagger\Gamma^0 \tag{C.15}$$

and

$$\Gamma^0 = \begin{bmatrix} i\gamma^0\gamma^5 & 0 \\ 0 & \gamma^0 \end{bmatrix} \tag{C.16}$$

The action

$$I = \int d^4x\mathcal{L}_8 \tag{C.17}$$

is invariant under Lorentz transformations S_8 and S_{8A}.

Then the Hamiltonian density for the 8-dimensional theory is

$$\mathcal{H}_8(x) = \begin{bmatrix} i\psi_T^\dagger\gamma^5(\boldsymbol{\alpha}\cdot\nabla + \beta m)\psi_T & 0 \\ 0 & \psi^\dagger(-i\boldsymbol{\alpha}\cdot\nabla + \beta m)\psi \end{bmatrix} \tag{C.18}$$

C.3 Non-Invariance of the Extended Free Action under a Left-handed Extended Lorentz Transformation

The action eq. C.17 is not invariant under Left-handed Complex Lorentz transformations. The fundamental cause of this non-invariance is the three dimensional nature of space. In the case of Dirac particles one can define a Lorentz invariant action because time is one-dimensional. Thus one can use $\psi^\dagger\gamma^0 = \bar{\psi}$ to form the Dirac field lagrangian and action. A key factor in Lorentz invariance is the relation between the inverse and hermitean conjugate of the spinor boost operator

$$\gamma^0 S^{-1}\gamma^0 = S^\dagger \tag{C.19}$$

In the case of the tachyon lagrangian and action, Left-handed Complex Lorentz invariance is not possible because the tachyonic equivalent to eq. C.19 is

$$S_L^{-1}(\Lambda(\mathbf{v}))\boldsymbol{\gamma}\cdot\mathbf{p}/|\mathbf{p}| = i\gamma^0 S_L^\dagger(\Lambda(\mathbf{v})) \tag{C.20}$$

where $\mathbf{p} = m\gamma_s\mathbf{v}$. The appearance of $\boldsymbol{\gamma}\cdot\mathbf{p}/|\mathbf{p}|$ in eq. C.20 precludes the invariance of the free tachyon action.

We will now show the effect of a Left-handed Complex Lorentz transformation (eqs. C.5 and C.6) on the lagrangian density eq. C.1c. The two non-zero parts of the lagrangian density $\mathcal{L}_8$ (eq. C.14) are

$$\mathcal{L}_1 = \psi_T{}^{\dagger} i\gamma^0\gamma^5(\gamma^\mu\partial/\partial x^\mu - m)\psi_T(x) \tag{C.21}$$

and

$$\mathcal{L}_2 = \psi^{\dagger}\gamma^0(i\gamma^\mu\partial/\partial x^\mu - m)\psi(x) \tag{C.22}$$

where † represents complex conjugation. The effect of the transformation, eqs. C.5-C.6, on these terms is

$$\begin{aligned}\mathcal{L}_1{}' &= \psi_T{}^{\dagger} i\gamma^0\gamma^5 S_L{}^{-1}S_L(\gamma^\mu\partial/\partial x^\mu - m)\, S_L{}^{-1}S_L\psi_T(x)\\ &= \psi_T{}^{\dagger} i\gamma^0\gamma^5 S_L{}^{-1}(i\gamma^\mu\partial/\partial x'^\mu - m)S_L\psi_T(x)\\ &= -\psi_T{}^{\dagger} S_L{}^{\dagger}\gamma^5(\boldsymbol{\gamma}\cdot\mathbf{p}/|\mathbf{p}|)(i\gamma^\mu\partial/\partial x'^\mu - m)S_L\psi_T(x)\\ &= \psi'^{\dagger}(x')(\boldsymbol{\gamma}\cdot\mathbf{p}/|\mathbf{p}|)\gamma^5(i\gamma^\mu\partial/\partial x'^\mu - m)\psi'(x')\end{aligned} \tag{C.23}$$

and

$$\begin{aligned}\mathcal{L}_2{}' &= \psi^{\dagger}\gamma^0\gamma^5 S_L{}^{-1}S_L\gamma^5(i\gamma^\mu\partial/\partial x^\mu - m)\gamma^5 S_L{}^{-1}S_L\gamma^5\psi(x)\\ &= \psi^{\dagger}\gamma^0\gamma^5 S_L{}^{-1}(\gamma^\mu\partial/\partial x'^\mu - m)S_L\gamma^5\psi(x)\\ &= i\psi^{\dagger}\gamma^5 S_L{}^{\dagger}(\boldsymbol{\gamma}\cdot\mathbf{p}/|\mathbf{p}|)(\gamma^\mu\partial/\partial x'^\mu - m)S_L\gamma^5\psi(x)\\ &= i\psi_T{}'^{\dagger}(x')(\boldsymbol{\gamma}\cdot\mathbf{p}/|\mathbf{p}|)(\gamma^\mu\partial/\partial x'^\mu - m)\psi_T{}'(x')\end{aligned} \tag{C.24}$$

using eqs. C.20, C.1f and C.1g, where $\psi'(x')$ is a solution of the Dirac equation obtained by Left-handed Complex Lorentz boosting (by $\mathbf{v} = \mathbf{p}/(\gamma m)$) of a tachyon field and where $\psi_T{}'(x')$ is a solution of the tachyon equation obtained by Left-handed Complex Lorentz boosting (by $\mathbf{v} = \mathbf{p}/(\gamma m)$) of a Dirac field. Eqs. C.23-C.24 clearly show that $\mathcal{L}_8$ is *not* invariant under Left-handed Complex Lorentz transformations.

Consequently the action of eq. C.17 is only invariant under inhomogeneous Lorentz transformations. *This state of affairs is actually an advantage when we derive internal symmetry features of the Superstandard Model because it will be seen to prevent any interplay between unbroken internal symmetry ElectroWeak SU(2) rotations and Left-handed Complex Lorentz transformations.*

C.4 The Diracian Dilemma – To what do Left-handed Extended Lorentz Boost Particles Correspond? Answer: Left-handed Leptons

The development of this 8-dimensional formalism, and in particular, the "bi-spinor" wave function consisting of a Dirac spinor and and a tachyon spinor, raises the question, "Is there a particle interpretation for the "bi-spinor" wave function?" Dirac faced a similar issue in 1928-1930 with the negative energy states of the Dirac equation. He developed "hole theory" which eventually led to the interpretation of holes in the sea of filled negative energy states as *positrons*. We now face the same problem: with what pairs of particles do we identify the doublets consisting of a Dirac particle and a tachyon?

The obvious natural interpretation of these 8-spinors is ElectroWeak isodoublets such as:

$$\Psi_\ell(x) = \begin{bmatrix} \psi_{\ell T} \\ \\ \psi_\ell \end{bmatrix} \sim \begin{bmatrix} \nu \\ \\ e \end{bmatrix} \qquad \text{(C.25)}$$

are for leptons to have "e" represent a charged lepton and ν represent a neutrino. With this interpretation we can introduce SU(2) gauge interactions and develop one-generation, leptonic Weak theory naturally.

C.5 To what do Complexons Correspond? Quarks

We have identified two of the four types of spin ½ fermions as leptons. The remaining two types of spin ½ fermions – complexons – ψ_C and ψ_{CT} seem to naturally correspond to quarks since their equations of motion and wave functions have a natural SU(3) symmetry as we pointed out earlier. We therefore associate an analogue color SU(3) symmetry with these two types of spin ½ complexons. The Electroweak doublet of quarks then is

$$\Psi_q^{\ a}(x) = \begin{bmatrix} \psi_C^{\ a} \\ \\ \psi_{CT}^{\ a} \end{bmatrix} \sim \begin{bmatrix} u^a \\ \\ d^a \end{bmatrix} \tag{C.26}$$

where u is an "up" type quark and d is a "down" type quark.[163]

The rationale for constructing quark doublets is the same as in the leptonic case: We wish to define a generalization of the "Dirac-like" equations of motion that is covariant under L_C boosts.

C.6 Quark Doublets

We assume that quark doublets consist of a complexon[164] and a tachyonic complexon[165] and to this extent they mirror lepton doublets. In this section we will develop a generalized free complexon equation and describe its features.

---

[163] While the lepton situation is clear in the sense that charged leptons cannot be tachyons since their masses are known (Thus only tachyonic neutrinos are the only currently allowed possibility.), the quark situation is somewhat unclear. We have provisionally chosen the "down" type of quark (d, s, and b) as tachyonic. The association of bound states of these quarks such as the K^0 and B^0 systems which are known to have CP violation, and the CP violation engendered by tachyons, encourages this interpretation.

In addition, $W^{\pm}$ charge asymmetry in $p\bar{p}$ collisions indicate the d sea in a proton is greater than the u sea (K. Abe et al, PRL **74**, 850 (1995)) as does the asymmetry of Drell-Yan production in deep inelastic scattering on p and n targets (A. Baldit et al, Phys. Lett. **B332**, 244 (1994)). These results are to be expected since there is no mass gap for a d tachyon sea while there is a mass gap for a u Dirac particle sea. Complexon quarks may explicate the discrepancies between theory and experiment in the spin structure functions of the parton model for nucleons.

[164] **An "ordinary" complexon can "exceed the speed of light" just like a tachyonic complexon because a complexon has a complex valued velocity enabling it to evade the real-valued singularity at v = c.**

[165] The global SU(3) symmetry of complexons makes their identification with quarks reasonable. However, the complexon theory that we develop and use for quark dynamics in the Standard Model is <u>not</u> required. Our SuperStandard Theory could use Dirac fermion dynamics for the up-type quarks and tachyon dynamics for down-type quarks. Then the (broken) Left-handed Extended Lorentz group would be the basic space-time group rather than L_C. We choose to use complexon dynamics for quarks because they have an internal SU(3)-like structure suggestive of color SU(3). More importantly, their spin dynamics is different and thus may resolve the differences between theory and experiment for the deep inelastic parton spin-dependent structure functions. Nevertheless, quarks could be

C.6.1 Summary of L_C Boosts to Generate Spin ½ Equations

We begin by recapitulating L_C boost features for coordinates and spinors:[166]

$$\Lambda_C(\mathbf{v_c}) = \exp[i\omega\hat{\mathbf{w}}\cdot\mathbf{K}] \tag{H.61}$$
$$\omega = (\omega_r^2 - \omega_i^2 + 2i\omega_r\omega_i\,\hat{\mathbf{u}}_r\cdot\hat{\mathbf{u}}_i)^{1/2} \tag{H.62}$$
$$\hat{\mathbf{w}} = (\omega_r\hat{\mathbf{u}}_r + i\omega_i\hat{\mathbf{u}}_i)/\omega \tag{H.63}$$
$$\hat{\mathbf{w}}\cdot\hat{\mathbf{w}} = \hat{\mathbf{u}}_r\cdot\hat{\mathbf{u}}_r = \hat{\mathbf{u}}_i\cdot\hat{\mathbf{u}}_i = 1 \tag{H.64a}$$
$$\mathbf{v_c} = \hat{\mathbf{w}}\tanh(\omega) \tag{H.64b}$$

The corresponding L_C spinor boost for $m^2 > 0$ particles with complex 3-momenta from chapter 2 of Blaha (2018e) is

$$\begin{aligned} S_C(\omega, \mathbf{v_c}) &= \exp(-i\omega\sigma_{0k}\hat{w}_k/2) = \exp(-\omega\gamma^0\boldsymbol{\gamma}\cdot\hat{\mathbf{w}}/2) \\ &= \cosh(\omega/2)I + \sinh(\omega/2)\gamma^0\boldsymbol{\gamma}\cdot\hat{\mathbf{w}} \end{aligned} \tag{2.99}$$

with inverse transformation

$$\begin{aligned} S_C^{-1}(\omega, \mathbf{v_c}) &= \gamma^2\gamma^0K^{-1}S_C^{\dagger}K\gamma^0\gamma^2 = \gamma^2\gamma^0 S_C^{\,T}\gamma^0\gamma^2 = \exp(\omega\gamma^0\boldsymbol{\gamma}\cdot\hat{\mathbf{w}}/2) \\ &= \cosh(\omega/2)I - \sinh(\omega/2)\gamma^0\boldsymbol{\gamma}\cdot\hat{\mathbf{w}} \end{aligned} \tag{2.100}$$

The Dirac-like complexon equation resulting from the boost is

$$[i\gamma^0\partial/\partial t + i\boldsymbol{\gamma}\cdot(\nabla_r + i\nabla_i) - m]\psi_C(t, \mathbf{x}_r, \mathbf{x}_i) = 0 \tag{2.123}$$

where $\mathbf{x} = \mathbf{x}_r - i\mathbf{x}_i$. The subsidiary condition is

$$\nabla_r\cdot\nabla_i\,\psi_C(t, \mathbf{x}_r, \mathbf{x}_i) = 0 \tag{2.123a}$$

---

similar to leptons in this regard and form a doublet of a Dirac fermion and an ordinary tachyon. Whether quarks are complexons or not is an experimental question!

[166] **The equation numbering of this subsection C.6.1 follows that of Blaha (2007b).**

The L_C coordinate boost that leads to $m^2 < 0$ tachyonic complexons with complex 3-momenta is

$$\Lambda_{CL}(\mathbf{v_c}) \equiv \Lambda_{CL}(\omega, \hat{\mathbf{w}}) = \exp[i(\omega + i\pi/2)\hat{\mathbf{w}}\cdot\mathbf{K}] \tag{2.185}$$

where

$$\omega = (\omega_r^2 - \omega_i^2)^{1/2} \tag{H.62}$$
$$\hat{\mathbf{w}} = (\omega_r\hat{\mathbf{u}}_r + i\omega_i\hat{\mathbf{u}}_i)/\omega \tag{H.63}$$
$$\hat{\mathbf{w}}\cdot\hat{\mathbf{w}} = \hat{\mathbf{u}}_r\cdot\hat{\mathbf{u}}_r = \hat{\mathbf{u}}_i\cdot\hat{\mathbf{u}}_i = 1 \tag{H.64a}$$
$$\mathbf{v_c} = \hat{\mathbf{w}}\tanh(\omega + i\pi/2) = \hat{\mathbf{w}}\,\mathrm{cotanh}(\omega) \tag{2.189}$$
$$\omega_L = \omega + i\pi/2$$

The L_C spinor boost for tachyonic complexons is

$$S_{CL}(\Lambda_{CL}(\omega, \hat{\mathbf{w}})) = \exp(-i\omega_L\sigma_{0i}\hat{w}_i/2) = \exp(-\omega_L\gamma^0\boldsymbol{\gamma}\cdot\hat{\mathbf{w}}/2)$$
$$= \cosh(\omega_L/2)I + \sinh(\omega_L/2)\gamma^0\boldsymbol{\gamma}\cdot\hat{\mathbf{w}} \tag{2.193}$$

The resulting Dirac-like tachyonic complexon equation is

$$[\gamma^0\partial/\partial t + \boldsymbol{\gamma}\cdot(\nabla_r + i\nabla_i) - m]\psi_{CL}(t, \mathbf{x_r}, \mathbf{x_i}) = 0 \tag{2.202}$$

with the subsidiary condition

$$\nabla_r\cdot\nabla_i\,\psi_{CL}(t, \mathbf{x_r}, \mathbf{x_i}) = 0 \tag{2.204}$$

C.6.2 L_C Boosts between Complexons and Tachyonic Complexons

An L_C spinor boost of a complexon can change it into a tachyonic complexon and vice versa:

$$S_{CL}(\Lambda_{CL}(\omega, \hat{\mathbf{w}}))\psi_C(x) \rightarrow \psi_{CT}'(x') \tag{C.27}$$
$$S_{CL}(\Lambda_{CL}(\omega, \hat{\mathbf{w}}))\psi_{CT}(x) \rightarrow \psi_C'(x')$$

Similarly the differential operator used in the equations of motion can also be transformed.

$$S_{CL}(\Lambda_{CL}(\omega, \hat{\mathbf{w}}))(\gamma^{\mu}D_{\mu} - m)S_{CL}^{-1}(\Lambda_{CL}(\omega, \hat{\mathbf{w}})) = (i\gamma^{\mu}D'_{\mu} - m) \quad \text{(C.28)}$$
$$S_{CL}(\Lambda_{CL}(\omega, \hat{\mathbf{w}}))\gamma^5(i\gamma^{\mu}D_{\mu} - m)\gamma^5 S_{CL}^{-1}(\Lambda_{CL}(\omega, \hat{\mathbf{w}})) = (\gamma^{\mu}D'_{\mu} - m)$$

where

$$x'^{\mu} = i\Lambda_{CL}{}^{\mu}{}_{\nu}(\omega, \mathbf{u})x^{\nu} \quad \text{(C.29)}$$
$$D'_{\mu} = -i\Lambda_{CL}{}^{\nu}{}_{\mu}(\omega, \mathbf{u})D_{\nu}$$

or in matrix form

$$X' = E_{CL}(\omega, \hat{\mathbf{w}})X \equiv i\Lambda_{CL}(\omega, \hat{\mathbf{w}})X \quad \text{(C.30)}$$

Eqs. C.27 – C.29 imply

$$S_{CL}(\Lambda_{CL}(\omega, \hat{\mathbf{w}}))(\gamma^{\mu}D_{\mu} - m)\psi_{CT}(x) = (i\gamma^{\mu}D'_{\mu} - m)S_{CL}(\Lambda_{CL}(\omega, \hat{\mathbf{w}}))\psi_{CT}(x)$$
$$= (i\gamma^{\mu}D'_{\mu} - m)\psi_C'(x') \quad \text{(C.31)}$$

and

$$S_{CL}(\Lambda_{CL}(\omega,\hat{\mathbf{w}}))\gamma^5(i\gamma^{\mu}D_{\mu} - m)\psi_C(x) = (\gamma^{\mu}D'_{\mu} - m)S_{CL}(\Lambda_{CL}(\omega,\hat{\mathbf{w}}))\gamma^5\psi_C(x)$$
$$= (\gamma^{\mu}D'_{\mu} - m)\psi_{CT}'(x') \quad \text{(C.32)}$$

where

$$\psi_C'(x') = S_{CL}(\Lambda_{CL}(\omega, \hat{\mathbf{w}}))\psi_{CT}(x) \quad \text{(C.33)}$$

and

$$\psi_{CT}'(x') = S_{CL}(\Lambda_{CL}(\omega, \hat{\mathbf{w}}))\gamma^5\psi_C(x) \quad \text{(C.34)}$$

Thus neither complexon dynamical equation is L_C covariant.

C.6.3 Doublet Dynamical Equation for Complexons

We will now consider the issue of generalizing the complexon dynamical equations so that the generalized equation is covariant under both Lorentz transformations and L_C boosts.

The only obvious method to obtain a generalized equation that is covariant under L_C boosts is to define an 8×8 matrix generalization. Let

$$đ_C(x) = \begin{bmatrix} (i\gamma^\mu D_\mu - m) & 0 \\ 0 & (\gamma^\mu D_\mu - m) \end{bmatrix} \tag{C.35}$$

be an 8 × 8 matrix operator with the 4 × 4 matrix elements shown, and let

$$\Psi_C(x) = \begin{bmatrix} \psi_C(x) \\ \psi_{CT}(x) \end{bmatrix} \tag{C.36}$$

be an 8 component column vector composed of a complexon field and a tachyonic complexon field. Then the generalized complexon equation is

$$đ_C(x)\Psi_C(x) = 0 \tag{C.37}$$

We now define the 8 × 8 Left-handed L_C boost transformation

$$S_{CL8} \equiv S_{CL8}(\Lambda_{CL}(\omega, \hat{\mathbf{w}})) = \begin{bmatrix} 0 & S_{CL}(\Lambda_{CL}(\omega, \hat{\mathbf{w}})) \\ S_{CL}(\Lambda_{CL}(\omega, \hat{\mathbf{w}}))\gamma^5 & 0 \end{bmatrix} \tag{C.38}$$

with inverse transformation

$$S_{CL8}^{-1} \equiv S_{CL8}^{-1}(\Lambda_{CL}(\omega, \hat{\mathbf{w}})) = \begin{bmatrix} 0 & \gamma^5 S_{CL}^{-1}(\Lambda_{CL}(\omega, \hat{\mathbf{w}})) \\ S_{CL}^{-1}(\Lambda_{CL}(\omega, \hat{\mathbf{w}})) & 0 \end{bmatrix} \tag{C.39}$$

Applying S_{CL8} to eq. C.37 yields

$$0 = S_{CL8}đ_C(x)\Psi_C(x) = đ_C(x')\Psi_C'(x') \tag{C.40}$$

where

$$\Psi_C'(x') = \begin{bmatrix} S_{CL8}\psi_{CT}(x) \\ S_{CL8}\gamma^5\psi_C(x) \end{bmatrix} = \begin{bmatrix} \psi_C'(x') \\ \psi_{CT}'(x') \end{bmatrix} \tag{C.41}$$

Thus the generalized complexon equation is covariant under L_C boosts. Covariance requires the complexon, and the tachyonic complexon, must have the same absolute value for the mass.

It is easy to show that the generalized complexon equation is also covariant under conventional Lorentz transformations represented as 4 × 4 diagonal blocks in an 8 × 8 matrix representation. (The demonstration is analogous to eqs. C.9 – C.13.)

The lagrangian density that corresponds to our 8-dimensional construction is

$$\mathcal{L}_{C8} = \overline{\Psi}_C(x)\not{d}_C(x)\Psi_C(x) \tag{C.42}$$

where

$$\overline{\Psi}_C(x) = \Psi_C^\dagger|_{\mathbf{x_i} = -\mathbf{x_i}}\,\Gamma_C^0 \tag{C.43}$$

and

$$\Gamma_C^0 = \begin{bmatrix} \gamma^0 & 0 \\ 0 & i\gamma^0\gamma^5 \end{bmatrix} \tag{C.44}$$

The action

$$I = \int d^4x\,\mathcal{L}_{C8} \tag{C.45}$$

is invariant under Lorentz transformations S_8 and S_{8A} (eqs. C.9 – C.12).

The Hamiltonian density for the 8-dimensional theory is

$$\mathcal{H}_{C8}(x) = \begin{bmatrix} \psi_C^\dagger(-i\boldsymbol{\alpha}\cdot\nabla_C + \beta m)\psi_C & 0 \\ 0 & i\psi_{CT}^\dagger\gamma^5(\boldsymbol{\alpha}\cdot\nabla_C + \beta m)\psi_{CT} \end{bmatrix} \quad \text{(C.46)}$$

where the spatial vector part of D^μ is complex

$$\nabla_C = \mathbf{D} \quad \text{(C.47)}$$

C.6.4 Non-Invariance of the Generalized Free Complexon Action under an L_C Boost

The action C.45 is not invariant under L_C boosts. The reason is similar to that of section C.3 for the "leptonic" type of particle: there is no simple relation between the hermitean conjugate of an L_C spinor boost and its inverse (a situation similar to eq. C.19 for the Dirac boost case).

Consequently the action of eq. C.45 is only invariant under inhomogeneous Lorentz transformations. *This state of affairs is again an advantage when we derive features of the Superstandard Model because it prevents any interplay between unbroken internal symmetry ElectroWeak SU(2) rotations and* L_C *transformations in the complexon (quark) sector.*

Appendix D. Coordinate Reality Group Analogue for ElectroWeak SU(2)⊗U(1) due to Real Superluminal Velocities

In the preceding appendix we developed a fermion doublet analogue framework for the ElectroWeak SU(2) interactions. We now turn to develop the ElectroWeak interactions analogue with an SU(2)⊗U(1) group structure, from the geometry of Complex Lorentz transformations.

In the discussions up to this point we have not considered imaginary (and more generally complex) coordinates resulting from a superluminal Lorentz transformation.[167] In this appendix we show that the coordinates generated from real-valued coordinates are complex-valued in general and require us to introduce another transformation that maps complex coordinates to real coordinates.[168] This transformation, which we will call a *Coordinate Reality Group* transformation, will be of significance because it has an SU(2)⊗U(1) group symmetry. It emerges when we consider superluminal transformations but is not required for ordinary sublight Lorentz transformations. This new SU(2)⊗U(1) symmetry is the analogue of the SU(2)⊗U(1) internal symmetry of the ElectroWeak sector of The Superstandard Model.

We introduce this new transformation by reconsidering the previous simple example wherein one coordinate system is traveling at a speed v in the x direction with respect to the "laboratory" system. See Fig. D.1.

The (left-handed[169]) Lorentz transformation is given by eq. D.1, and the coordinates in the two reference frames are related by eq. D.2.

---

[167] Superluminal transformations are a subset of Complex Lorentz transformations.

[168] The complex coordinates resulting from a superluminal transformation are physically viewed as real-valued by an observer in the new coordinate system. The apparent complexity of the coordinates resulting from a superluminal transformation are an artifact of the transformation.

[169] The right-handed Lorentz transformation case is analogous.

$$\Lambda_L(\omega, \mathbf{u} = (1,0,0)) = \begin{bmatrix} i\gamma_s & -i\beta\gamma_s & 0 & 0 \\ -i\beta\gamma_s & i\gamma_s & 0 & 0 \\ 0 & 0 & 1 & 0 \\ 0 & 0 & 0 & 1 \end{bmatrix} \quad \text{(D.1)}$$

implementing the coordinate transformation:

$$X' = \Lambda_L(\omega, \mathbf{u} = (1,0,0))X$$

or

$$\begin{aligned} t' &= i\gamma_s(t - \beta x) \\ x' &= i\gamma_s(x - \beta t) \\ y' &= y \\ z' &= z \end{aligned} \quad \text{(D.2)}$$

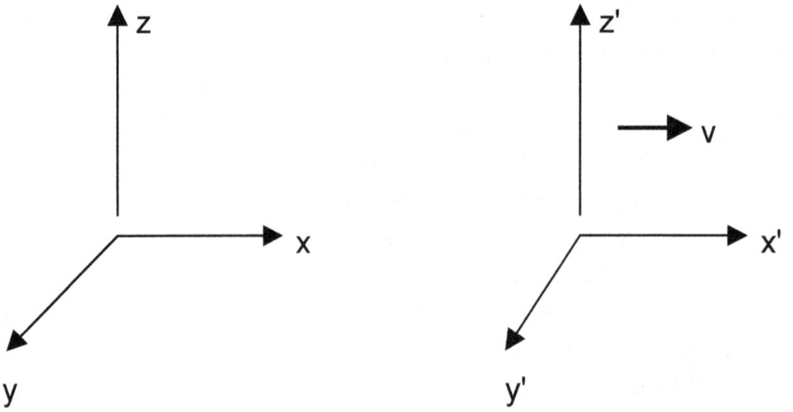

Figure D.1. Depiction of two coordinate systems. The "primed" coordinate system is moving with velocity v in the positive x direction with respect to the "unprimed" coordinate system. We choose parallel axes for convenience.

We now define a Coordinate Reality group transformation $\Pi_L(\mathbf{u})$ that maps the real coordinates of the unprimed reference frame to real coordinates in the primed reference frame.

$$\Pi_L(\mathbf{u}) = \begin{bmatrix} -i & 0 & 0 & 0 \\ 0 & -i & 0 & 0 \\ 0 & 0 & 1 & 0 \\ 0 & 0 & 0 & 1 \end{bmatrix} \qquad \text{(D.3)}$$

where **u** is the unit vector corresponding to the direction of **v** (the positive x direction in this example). Using $\Pi_L(\mathbf{u})$ we obtain an overall transformation from real coordinates to real coordinates in the case considered:

$$X'' = \Pi_L(\mathbf{u})\Lambda_L(\omega, \mathbf{u} = (1,0,0))X$$

or

$$\begin{aligned} t'' &= \gamma_s(t - \beta x) \\ x'' &= \gamma_s(x - \beta t) \\ y'' &= y \\ z'' &= z \end{aligned} \qquad \text{(D.4)}$$

where $\gamma_s = (\beta^2 - 1)^{-\frac{1}{2}}$. An observer in the primed reference frame would consider his/her time to be real when measured on a clock, and distances along the x axis to be real when measured with a ruler. Thus eq. D.4 makes good sense physically because in any reference frame, observers measure real distances and real times. For this reason we will call combined transformations of the type of eq. D.4 – from real coordinates to real coordinates – *physical* superluminal transformations for real-valued velocities.

It is important to note that $\Pi_L(\mathbf{u})$ is position dependent in general for more complicated Λ_L transformations and so the Coordinate Reality group is a local group of the Yang-Mills type. This is clear from eqs. D.1 and D.2. We will see the Coordinate Reality group is the analogue of local (Yang-Mills) SU(2)⊗U(1) ElectroWeak symmetry.

This simple example generalizes to arbitrary relative real velocities **v**. First we note that the Lorentz transformation for a velocity **v** that is a rotation of the velocity in the x-direction ($\mathbf{v} = |\mathbf{v}|R\mathbf{u}$ where R is the relevant rotation matrix) has the form

$$\Lambda_L(\omega, \mathbf{v}) = \mathfrak{R}(\mathbf{v}/v, \mathbf{u})\Lambda_L(\omega, \mathbf{u} = (1,0,0))\mathfrak{R}^{-1}(\mathbf{v}/v, \mathbf{u}) \qquad \text{(D.5)}$$

where $\mathfrak{R}(\mathbf{v}/v, \mathbf{u})$ is a rotation from the velocity direction **u** to direction **v**/v.

The original transformation (eq. D.2) can be written as

$$\Pi_L(\mathbf{u})\Lambda_L(\omega, \mathbf{u} = (1,0,0)) = \Pi_L(\mathbf{u})\mathfrak{R}^{-1}(\mathbf{v}/v, \mathbf{u})\Lambda_L(\omega, \mathbf{v})\mathfrak{R}(\mathbf{v}/v, \mathbf{u}) \quad \text{(D.6)}$$

Consequently the combined transformation for velocity **v** is

$$\begin{aligned}\mathfrak{R}(\mathbf{v}/v, \mathbf{u})\Pi_L(\mathbf{u})\Lambda_L(\omega, \mathbf{u} = (1,0,0))\mathfrak{R}^{-1}(\mathbf{v}/v, \mathbf{u}) \\ = \mathfrak{R}(\mathbf{v}/v, \mathbf{u})\Pi_L(\mathbf{u})\mathfrak{R}^{-1}(\mathbf{v}/v, \mathbf{u})\Lambda_L(\omega, \mathbf{v}) \\ = \Pi_L(\mathbf{v}/v)\Lambda_L(\omega, \mathbf{v})\end{aligned} \quad \text{(D.7)}$$

Thus for a Lorentz transformation $\Lambda_L(\omega, \mathbf{v})$ for velocity **v** we see that we can define a subsidiary transformation $\Pi_L(\mathbf{v}/v)$ of the form

$$\Pi_L(\mathbf{v}/v) = \mathfrak{R}(\mathbf{v}/v, \mathbf{u})\Pi_L(\mathbf{u})\mathfrak{R}^{-1}(\mathbf{v}/v, \mathbf{u}) \quad \text{(D.8)}$$

The general form of $\mathfrak{R}(\mathbf{v}/v, \mathbf{u})$, is

$$\mathfrak{R}(\mathbf{v}/v, \mathbf{u}) = \begin{bmatrix} 1 & 0 & 0 & 0 \\ 0 & & & \\ 0 & & \mathfrak{R}_3(\mathbf{v}/v, \mathbf{u}) & \\ 0 & & & \end{bmatrix} \quad \text{(D.9)}$$

where $\mathfrak{R}_3(\mathbf{v}/v, \mathbf{u})$ is a 3×3 rotation matrix that can be expressed in terms of the generators of the 3-dimensional rotation group as

$$\mathfrak{R}_3(\mathbf{v}/v, \mathbf{u}) = \exp(i\boldsymbol{\theta}\cdot\mathbf{J}) \quad \text{(D.10)}$$

The rotation angles $\boldsymbol{\theta}$ are real numbers since we are rotating the real vector **u** to the real vector **v**/v. Given the form of eq. D.10 then we see that the form of $\Pi_L(\mathbf{v}//v)$ is

$$\Pi_L(\mathbf{v}/v) = \begin{bmatrix} -i & 0 & 0 & 0 \\ 0 & & & \\ 0 & \multicolumn{3}{c}{\mathfrak{R}_3(\mathbf{v}/v, \mathbf{u})\Pi_{L3}(\mathbf{u})\mathfrak{R}_3^{-1}(\mathbf{v}/v, \mathbf{u})} \\ 0 & & & \end{bmatrix} \quad (D.11)$$

where

$$\Pi_{L3}(\mathbf{u}) = \begin{bmatrix} -i & 0 & 0 \\ 0 & 1 & 0 \\ 0 & 0 & 1 \end{bmatrix} \quad (D.12)$$

If we consider the case of an infinitesimal rotation $\boldsymbol{\theta}$ to first order in $\boldsymbol{\theta}$

$$\mathfrak{R}_3(\mathbf{v}/v, \mathbf{u}) \simeq I + i\boldsymbol{\theta}\cdot\mathbf{J} \quad (D.13)$$

then

$$\Pi_{L3}(\mathbf{v}/v) = \mathfrak{R}_3(\mathbf{v}/v, \mathbf{u})\Pi_{L3}(\mathbf{u})\mathfrak{R}_3^{-1}(\mathbf{v}/v, \mathbf{u}) \simeq \Pi_{L3}(\mathbf{u}) + i\boldsymbol{\theta}\cdot\mathbf{J}\Pi_{L3}(\mathbf{u}) - i\Pi_{L3}(\mathbf{u})\boldsymbol{\theta}\cdot\mathbf{J}$$
$$\simeq \Pi_{L3}(\mathbf{u})[I + i\Pi_{L3}^{-1}(\mathbf{u})[\boldsymbol{\theta}\cdot\mathbf{J}, \Pi_{L3}(\mathbf{u})] \quad (D.14)$$

where $\Pi_{L3}^{-1}(\mathbf{u})$ is the inverse of $\Pi_{L3}(\mathbf{u})$ and [...] represents the commutator. Thus for arbitrary rotations eq. D.14 implies

$$\Pi_{L3}(\mathbf{v}/v) = \mathfrak{R}_3(\mathbf{v}/v, \mathbf{u})\Pi_{L3}(\mathbf{u})\mathfrak{R}_3^{-1}(\mathbf{v}/v, \mathbf{u}) = \Pi_{L3}(\mathbf{u})\exp\{i\Pi_{L3}^{-1}(\mathbf{u})[\boldsymbol{\theta}\cdot\mathbf{J}, \Pi_{L3}(\mathbf{u})]\} \quad (D.15)$$

We can find the general form of $\Pi_{L3}(\mathbf{v}/v)$ by considering the case of eq. D.6 in more detail. The exponentiated matrix expression in D.15 can be written

$$\Pi_{L3}^{-1}(\mathbf{u})[\boldsymbol{\theta}\cdot\mathbf{J}, \Pi_{L3}(\mathbf{u})] = \Pi_{L3}^{-1}(\mathbf{u})\boldsymbol{\theta}\cdot\mathbf{J}\Pi_{L3}(\mathbf{u}) - \boldsymbol{\theta}\cdot\mathbf{J} = \boldsymbol{\theta}\cdot\mathbf{Q} \quad (D.16)$$

where

$$\mathbf{Q} = \Pi_{L3}^{-1}(\mathbf{u})\mathbf{J}\Pi_{L3}(\mathbf{u}) - \mathbf{J} = \mathbf{Q}' - \mathbf{J} \quad (D.17)$$

The matrices Q_i can be evaluated using eq. D.12 and the matrix representations of rotation generators J_i: which are equivalent in form to the SU(2) generators T_i:

$$J_1 = \begin{bmatrix} 0 & 0 & 0 \\ 0 & 0 & -i \\ 0 & i & 0 \end{bmatrix} = T_1 \tag{D.18}$$

$$J_2 = \begin{bmatrix} 0 & 0 & i \\ 0 & 0 & 0 \\ -i & 0 & 0 \end{bmatrix} = T_2 \tag{D.19}$$

$$J_3 = \begin{bmatrix} 0 & -i & 0 \\ i & 0 & 0 \\ 0 & 0 & 0 \end{bmatrix} = T_3 \tag{D.20}$$

The rotation generators satisfy the commutation relations

$$[J_i, J_j] = i\epsilon_{ijk}J_k \tag{D.21}$$

as do the SU(2) generators:

$$[T_i, T_j] = i\epsilon_{ijk}T_k \tag{D.22}$$

We can calculate Q' and obtain

$$Q'_1 = \begin{bmatrix} 0 & 0 & 0 \\ 0 & 0 & -i \\ 0 & i & 0 \end{bmatrix} \tag{D.23}$$

$$Q'_2 = \begin{bmatrix} 0 & 0 & -1 \\ 0 & 0 & 0 \\ -1 & 0 & 0 \end{bmatrix} \tag{D.24}$$

$$Q'_3 = \begin{bmatrix} 0 & 1 & 0 \\ 1 & 0 & 0 \\ 0 & 0 & 0 \end{bmatrix} \tag{D.25}$$

We note that each Q'_i is hermitean and the Q'_i satisfy the commutation relations:

$$[Q'_i, Q'_j] = i\epsilon_{ijk}Q'_k \tag{D.26}$$

Consequently the set of Q'_i are also equivalent to SU(2) generators. As a result the exponential factor

$$\Pi_{L3}(\mathbf{v}/v) = \Pi_{L3}(\mathbf{u})\exp\{i\boldsymbol{\theta}\cdot(\mathbf{Q}' - \mathbf{J})\} \tag{D.27}$$

is equivalent to a combination of SU(2) rotations not only in this case but in general for superluminal transformations. The factor $\Pi_{L3}(\mathbf{u})$ is not an SU(2) matrix since its determinant is not 1. However

$$\Pi'_{L3}(\mathbf{u}) = -i\Pi_{L3}(\mathbf{u}) \tag{D.28}$$

is an SU(2) matrix since

$$\Pi'_{L3}{}^{-1}(\mathbf{u}) = \Pi'_{L3}{}^{\dagger}(\mathbf{u}) \tag{D.29}$$

$$\det \Pi'_{L3}(\mathbf{u}) = 1 \tag{D.30}$$

and

$$\Pi'_{L3}(\mathbf{v}/v) = \Pi'_{L3}(\mathbf{u})\exp\{i\boldsymbol{\theta}\cdot(\mathbf{Q}' - \mathbf{J})\} \tag{D.31}$$

is similarly an SU(2) rotation.

Thus the general form of superluminal, *real* velocity, transformation from a real set of coordinates to a real set of coordinates is[170]

$$\Pi_L(\mathbf{v}/v)\Lambda_L(\omega, \mathbf{v}) \tag{D.32}$$

---

[170] The choice of the unit vector **u** and the angle vector **θ** must be such that applying the transformation to a real set of coordinates yields a real set of coordinates.

where

$$\Pi_L(\mathbf{v}/v) = \begin{bmatrix} -i & 0 \quad 0 \quad 0 \\ 0 & \\ 0 & \Pi_{L3}(\mathbf{u})\exp\{i\,\boldsymbol{\theta}\cdot(\mathbf{Q}' - \mathbf{J})\} \\ 0 & \end{bmatrix} \qquad \text{(D.33)}$$

The Lorentz condition for real to real physical transformations generalizes to

$$\Lambda(\mathbf{v})^T\Pi_L(\mathbf{v}/v)^\dagger G\,\Pi_L(\mathbf{v})\Lambda(\mathbf{v}/v) = G \qquad \text{(D.34)}$$

Since superluminal transformations $\Lambda_L(\omega, \mathbf{v})$ transform real coordinates to complex coordinates in general, we can generalize the form of a real-to-real superluminal transformation to

$$e^{i\varphi}\Pi_L(\mathbf{v}'/v')\Lambda_L(\omega, \mathbf{v}) \qquad \text{(D.35)}$$

where φ is a constant phase and $\mathbf{v}'$ is an arbitrary velocity. This generalization will satisfy the generalized Lorentz condition

$$\Lambda(\mathbf{v})^T\Pi_L(\mathbf{v}'/v')^\dagger e^{-i\varphi}G\,e^{i\varphi}\Pi_L(\mathbf{v}'/v')\Lambda(\mathbf{v}) = G \qquad \text{(D.36)}$$

but the transformation will, in general, yield a complex set of coordinates when applied to a set of real coordinates.

These considerations imply:

1. Any observer in a coordinate system will treat a complex 4-dimensional coordinate system as if it were a real 4-dimensional coordinate system with complex-valued straight lines along each dimension (assuming rectangular coordinates).

2. The transformation $e^{i\varphi}\Pi'_{L3}(\mathbf{v}/v)$ is a SU(2)⊗U(1) transformation that takes complex 3-dimensional spatial coordinates to complex 3-dimensional spatial coordinates. In particular straight lines map to straight lines.

3. Physical observations in the observer's coordinate system are invariant under $SU(2)\otimes U(1)$ rotations of the spatial coordinates and the multiplication of the time component by an arbitrary phase.

4. The matrix

$$\Pi'_L(\mathbf{v}/v, \chi, \varphi) = \begin{bmatrix} e^{i\chi} & 0 & 0 & 0 \\ 0 & & & \\ 0 & & e^{i\varphi}\Pi'_{L3}(\mathbf{u})\exp\{i\,\boldsymbol{\theta}\cdot(\mathbf{Q}' - \mathbf{J})\} & \\ 0 & & & \end{bmatrix} \quad (D.37)$$

(where χ and φ are real numbers and $\mathbf{u}$ is a unit vector along any convenient coordinate axis) is a $SU(2)\otimes U(1)$ transformation that transforms complex 4-dimensional coordinates to complex 4-dimensional coordinates. Note, $\Pi_L(\mathbf{v}/v) = \Pi'_L(\mathbf{v}/v, 3\pi/2, \pi/2)$ is a special case of $\Pi'_L(\mathbf{v}/v, \chi, \varphi)$. Due to the manifest form of D.37 we see

$$\Pi'_L{}^{\mu}{}_{\alpha}{}^{*}\Pi'_L{}^{\mu}{}_{\beta} = [\Pi'_L{}^{\dagger}\Pi'_L]_{\alpha\beta} = I_{\alpha\beta} \quad (D.38)$$

(with an implied sum over μ) or, in matrix form,

$$\Pi'_L{}^{\dagger}\,\Pi'_L = I \quad (D.39)$$

and also[171]

$$\Pi'_L{}^{\dagger}G\Pi'_L = G \quad (D.40)$$

5. Complex coordinate values of the type generated by superluminal transformations with real-valued velocities are transformable to real coordinates. The complex coordinates are thus physically equivalent to corresponding real coordinate values in the sense that an observer in that frame would automatically use the real

---

[171] Eq.D.40 is close to the defining condition for a Lorentz group element but the presence of complex conjugation rather than a transpose means Π'_L is outside the real and complex Lorentz groups.

coordinates so obtained since rulers and clocks always measure real spatial coordinates and times. *Therefore physical theory is invariant under global SU(2)⊗U(1) coordinate transformations since complex coordinates, so generated, can be rotated back to real coordinates.*

6. The complex coordinates of any point obtained through a superluminal transformation can be transformed to a real set of coordinates by the above SU(2)⊗U(1) transformation. This SU(2)⊗U(1) invariance is the analogue of the SU(2)⊗U(1) symmetry of the ElectroWeak interactions.

Appendix E. Coordinate Reality Group Analogue for Dark ElectroWeak SU(2)⊗U(1)

In this appendix[172] we will consider superluminal transformations based on *complex-valued relative velocities*. The previous appendix considered the case of real-valued velocities. That case led to the Coordinate Reality Group analogue of ElectroWeak SU(2)⊗U(1).

We now consider Complex Lorentz transformations for complex-valued relative velocities. These transformations will require us to introduce another Coordinate Reality group with transformations that map complex coordinates to real coordinates. These transformations will be of significance because they lead to a hitherto unstated SU(2)⊗U(1) symmetry. We identify this SU(2)⊗U(1) symmetry as the analogue of the symmetry of the *Dark* ElectroWeak interactions of the Superstandard Model. Dark matter and interactions remain to be found experimentally but there may be some preliminary suggestive data from the CERN LHC.

We introduce these new Dark transformations by extending the previous simple example to Fig. E.1 in which one coordinate system is traveling at a complex-valued velocity u in the x direction with respect to the "laboratory" system. In these new transformations the relative velocity is complex-valued and has two components: a real-valued component in the x direction and an imaginary-valued component in the y direction. $\mathbf{u} = u_x\mathbf{i} + iu_y\mathbf{j}$. In the complex case $\beta = \tanh(\omega_L)$ is real-valued by eqs. H.20 – H.22 where $\omega_L = \omega + i\pi/2$ and ω is real.

The (left-handed[173]) Lorentz transformation is given by eq. H.23:

[172] Most of the material in this chapter appeared in Blaha (2011c) originally.
[173] The right-handed Lorentz transformation case is analogous.

$$\Lambda_L(\omega, \mathbf{u}) = \begin{bmatrix} i\gamma_s & -i\beta\gamma_s u_x & \beta\gamma_s u_y & 0 \\ -i\beta\gamma_s u_x & 1 + (i\gamma_s - 1)u_x^2 & i(i\gamma_s - 1)u_x u_y & 0 \\ \beta\gamma_s u_y & i(i\gamma_s - 1)u_x u_y & 1 - (i\gamma_s - 1)u_y^2 & 0 \\ 0 & 0 & 0 & 1 \end{bmatrix} \tag{E.1}$$

$$= \Lambda(\omega + i\pi/2, \mathbf{u})$$

implementing the coordinate transformation:

$$X' = \Lambda_L(\omega, \mathbf{u} = (u_x, iu_y, 0))X$$

or

$$\begin{aligned} t' &= i\gamma_s(t - \beta u_x - i\beta u_y) \\ x' &= -i\gamma_s\beta u_x t + i\gamma_s x + u_x x - u_x^2 x + i(i\gamma_s - 1)u_x u_y y \\ y' &= \gamma_s\beta u_y t + i(i\gamma_s - 1)u_x u_y x + [1 - (i\gamma_s - 1)u_y^2]y \\ z' &= z \end{aligned} \tag{E.2}$$

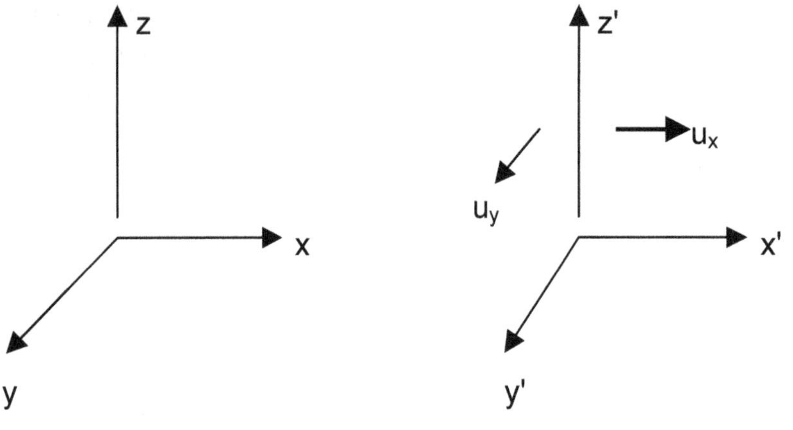

Figure E.1. Depiction of two coordinate systems. The “primed” coordinate system is moving with velocity u = u_xi + iu_yj with respect to the “unprimed” coordinate system. We choose parallel axes for convenience.

We now define a transformation that maps the real coordinates of the unprimed reference frame to real coordinates in the primed reference frame.

$$\Pi_L(\mathbf{u}, \mathbf{X}) = \begin{bmatrix} e^{ia} & e^{ib} & e^{ic} & 0 \\ e^{id} & e^{ie} & e^{if} & 0 \\ e^{ig} & e^{ih} & e^{ij} & 0 \\ 0 & 0 & 0 & 1 \end{bmatrix} \tag{E.3}$$

where **u** is the unit vector corresponding to the direction of the relative velocity.

It is important to note that $\Pi_L(\mathbf{u}, \mathbf{X})$ is position dependent and so the Coordinate Reality group is a local group. This is clear from eqs. E.1 and E.2. The Coordinate Reality Group in this case is SU(2)⊗U(1). It appears similar to the Reality group of Appendix D except for the crucial difference that the Coordinate Reality group here mixes time and spatial rows while the Coordinate Reality group of Appendix D only mixed spatial rows (See eqs. D.35 and D.37.).Again we have a local theory. Thus the combined Coordinate Reality group from this appendix and appendix D is SU(2)⊗U(1)⊗DSU(2)⊗DU(1) where we prepend 'D' to the Dark Reality group parts.

Elsewhere we defined complex boosts. We summarize the definition below:

$$\Lambda_L(\omega, \mathbf{u}) = \Lambda_L(\mathbf{v_c}) = \exp[i(\omega + i\pi/2)\mathbf{u}\cdot\mathbf{K}] \tag{E.4}$$

where ω remains

$$\omega = (\omega_r^{\ 2} - \omega_i^{\ 2})^{1/2}$$

and

$$\mathbf{u} = (\omega_r\mathbf{u_r} + i\omega_i\mathbf{u_i})/\omega$$
$$\mathbf{u}\cdot\mathbf{u} = \mathbf{u_r}\cdot\mathbf{u_r} = \mathbf{u_i}\cdot\mathbf{u_i} = 1$$
$$\mathbf{v_c} = \mathbf{u}\tanh(\omega + i\pi/2) = \mathbf{u}\,\mathrm{cotanh}(\omega)$$

In the example, that we are considering, we set $\mathbf{u_x} = \omega_r\mathbf{u_r}$ and $\mathbf{u_y} = \omega_i\mathbf{u_i}$.

Using $\Pi_L(\mathbf{u}, \mathbf{X})$ we obtain an overall transformation from real coordinates to real coordinates:

$$X'' = \Pi_L(\mathbf{u}, \mathbf{X})\Lambda_L(\omega, \mathbf{u} = (u_x, u_y, 0))X \quad \text{(E.5)}$$

with the coordinates of X and X" real-valued. An observer in the double primed reference frame would consider his/her time to be real when measured on a clock, and distances along the x and y axes to be real when measured with a ruler.

The velocity vectors: $u_x\mathbf{i}$ and $iu_y\mathbf{j}$ in our example define a plane in space. There are two types of rotations that are possible. 1) An angular rotation in the plane defined by the vectors. This is a U(1) transformation. 2) a spatial rotation of the plane that is an SU(2) rotation. Thus the joint rotations of **u** have an SU(2)⊗U(1) symmetry group. The Coordinate Reality group for 4-dimensions – has two SU(2)⊗U(1) factors that we denote SU(2)⊗U(1)⊗DSU(2)⊗DU(1). We see that the "newly found" group can be *assumed* to be the analogue of the Dark ElectroWeak symmetry group. We note that a further Coordinate Reality group part SU(3) will be introduced in the next appendix.

We consider Dark Matter and its interactions in greater detail elsewhere.

Appendix F. Coordinate Reality Group Analogue for Color SU(3)

F.1 Two Possible Approaches to Color SU(3)

There are two approaches to obtaining the Strong interaction and Color SU(3) symmetry:

Assume up-type and down-type quarks are in 3 representations of Color SU(3). This assumption sheds no light on a deeper origin of the Strong interaction and Color SU(3). It simply assumes the color SU(3) of the Strong interaction sector of the Standard Model. Thus our understanding is not deepened. A postulate corresponding to this assumption is:

Possible Postulate: Quarks are in the 3 representation of Color SU(3). The SU(3) symmetry is gauged with local Yang-Mills SU(3) fields called gluons that constitute the Strong interaction of the quark sector. Quarks are minimally coupled to the gluons in a gauge covariant fashion.

Comment: We will ***not*** use this postulate in our derivation.

1. In the preceding appendices the Internal Symmetry ElectroWeak interactions of the Unified Superstandard Model (modulo generations and their mixing) were shown to be an analogue to the Coordinate Reality Group associated with the Complex Lorentz group. The Reality group included SU(2)⊗U(1)⊗SU(2)⊗U(1).[174] Thus we found a significant geometrical analogue for the form of the ElectroWeak interactions of normal and Dark matter.

[174] We no longer prepend 'D' to the Dark SU(2)⊗U(1).

2. We now establish a similar geometrical analogue of the internal symmetry Strong interaction and Color SU(3). *If we extend the parameters to be real functions of the space-time coordinates (i.e. local SU(3) transformations), then we obtain an analogue for color SU(3). A key factor in this interpretation is the global covariance of complexon equations of motion under global SU(3).*[175]

Comment: We follow this approach in our derivation.

F.2 A Global SU(3) Symmetry of Complexon Quarks

We will now consider a global SU(3) covariance implicit in eqs.2.123 – 2.127. The defining property of the SU(3) group is that it preserves the invariance of inner products of complex 3-vectors of the form:

$$u^{*}\cdot v = u^{1*}v^{1} + u^{2*}v^{2} + u^{3*}v^{3} \tag{F.1}$$

If we examine the dynamical equation eq. 2.202 of Blaha (2018e) we see that the differential operator is covariant under a global SU(3) transformation U of the complex spatial 3-coordinates:

$$[i\gamma^{0}\partial/\partial t + i\mathbf{D}_{c}^{*}\cdot\gamma - m] = [i\gamma^{0}\partial/\partial t + i\mathbf{D}_{c}'^{*}\cdot\gamma' - m] \tag{F.2}$$

where

$$\mathbf{D}_{c}^{*} = \nabla_{c} = \nabla_{r} + i\nabla_{i}$$

and

$$\gamma^{a} = U^{ab}\gamma'^{b} \tag{F.3a}$$
$$D_{c}^{*a} = D_{c}'^{*b}U^{ab*} \tag{F.3b}$$

where $U^{\dagger} = U^{-1}$. We now exhibit the covariance of eq. 2.202. Since we can view the three spatial γ-matrices as SU(3) 3-vectors, we can express eq. F.3 as the result of a SU(3) rotation V of the γ-matrices (on the spinor indices)

175 This appendix was extracted from chapter 17 of Blaha (2011c) with some changes.

$$V\gamma^a V^{-1} = U^{ab}\gamma'^b \tag{F.4}$$

where V is a 4×4 reducible representation of SU(3), namely, $\underline{3}\oplus\underline{1}$. Since V commutes with γ^0 in the Pauli matrix representation of the γ matrices we see that V can have the form

$$V = \begin{bmatrix} A\exp(i\alpha_i\sigma_i) & 0 \\ 0 & B\exp(i\beta_i\sigma_i) \end{bmatrix}$$

where A, B, α_i and β_i are constants, and the zeroes represent 2×2 zero matrices. The inverse of V is $V^\dagger$. Thus eq. F.4 becomes

$$V\gamma^a V^{-1} = \begin{bmatrix} 0 & AB^*\exp(i\alpha_i\sigma_i)\sigma_a\exp(-i\beta_i\sigma_i) \\ -A^*B\exp(i\beta_i\sigma_i)\sigma_a\exp(-i\alpha_i\sigma_i) & 0(-i\beta_i\sigma_i) \end{bmatrix}$$

We now note the generators of the global SU(3) symmetry under discussion have a 4×4 matrix reducible representation ($\underline{3}\oplus\underline{1}$). The generators of this reducible representation are F_i and F_0 (a diagonal matrix diag(0,0,0,0,0,0,0,0,0,1) with F_i being the Gell-Mann SU(3) generators for i = 1, 2, …, 8.

Projection operators can be defined to project out the $\underline{3}$ representation piece P_3 and the $\underline{1}$ representations piece P_1. of the complexon spinor fields:

Thus the $\underline{3}$ complexon field is

$$\psi_{C3}(t, \mathbf{x_r}, \mathbf{x_i}) = P_3\psi_C(t, \mathbf{x_r}, \mathbf{x_i}) \tag{F.5}$$

while the $\underline{1}$ complexon field is

$$\psi_{C1}(t, \mathbf{x_r}, \mathbf{x_i}) = P_1\psi_C(t, \mathbf{x_r}, \mathbf{x_i}) \tag{F.6}$$

Since P_1 and P_3 do not commute with Lorentz transformations, a Lorentz transformation mixes ψ_{C1} and ψ_{C3}.[176] Since P_1 and P_3 do not commute with γ_5, left-handed and right-handed complexons would also be mixed by these projection operators. The matrix V has a $\underline{3}\oplus\underline{1}$ reducible representation.

In a manner similar to the covariance proof of the Dirac equation[177] we see that eq. F.2 is covariant under SU(3) transformations:

$$V[i\gamma^0\partial/\partial t + i\boldsymbol{\gamma}\cdot\mathbf{D}_c{}^* - m]V^{-1}V\psi_C(t, \mathbf{x}_r, \mathbf{x}_i) = 0$$

or

$$[i\gamma^{0\prime}\partial/\partial t' + i\mathbf{D}_c{}'^*\cdot\boldsymbol{\gamma}' - m]V\psi_C(t, \mathbf{x}_r, \mathbf{x}_i) = 0 \qquad \text{(F.7)}$$

(Note $\gamma^{0\prime} = V\gamma^0V^{-1}$ and $t' = t$.) The SU(3) transformed wave function $\psi_C'(t, \mathbf{x}')$ is

$$\psi_C'(t', \mathbf{x}') = V\psi_C(t, \mathbf{x}) = V\psi_C(t', U\mathbf{x}') \qquad \text{(F.8)}$$

Thus the complexon Dirac equation is covariant under global coordinate SU(3).

The subsidiary condition,

$$\nabla_r\cdot\nabla_i\,\psi_{Cu}(t, \mathbf{x}_r, \mathbf{x}_i) = 0 \qquad \text{(F.9)}$$

is also covariant under an SU(3) rotation:

$$\nabla_r'^*\cdot\nabla_i'\psi_C'(t, \mathbf{x}') = \nabla_r\cdot\nabla_i\,V\psi_C(t, \mathbf{x}) = V\nabla_r{}^*\cdot\nabla_i\,\psi_C(t, \mathbf{x}) = 0 \qquad \text{(F.10)}$$

---

[176] At this point it is worth noting that the construction of complexon fields, based on a boost from a particle rest state, guarantees that a reference frame exists in which any complexon particle has a single real time variable. Similarly a reference frame exists for a set of complexon particles (that is within a Lorentz of the center of momentum frame) with a single real time variable. The time variables of the individual complexon particles in the set are complex in general but are functions of the center of momentum real time variable. So there is only one real time variable for each complexon in the set although the time variable of an individual particle may be a complex function of the real-valued center of momentum time variable.

[177] For example see Bjorken (1964) pp. 18 – 20.

We now examine the transformation of the wave function eq. F.8 under the SU(3) transformation U. If we define

$$q^{*\mu} = (q^0, \mathbf{q}^*) = (p^0, \mathbf{p_r} + i\mathbf{p_i}) = (p^0, \mathbf{p}) = p^\mu \qquad \text{(F.11)}$$

then $\psi_C(t, \mathbf{x})$ will be seen to be covariant form under an SU(3) transformation:

$$\psi_C(t, x) = \sum_{\pm s} \int d^3q_r d^3q_i\, N_C(p^0)\delta(\mathbf{q}_r^* \cdot \mathbf{q}_i/m^2)[b_C(q^*,s)u_C(q^*,s)e^{-i(q^*.x\, +\, q.x^*)/2} +$$
$$+\; d_C^\dagger(q^*,s)v_C(q^*,s)e^{+i(q^*.x\, +\, q.x^*)/2}] \qquad \text{(F.12)}$$

Note both terms in each exponential are separately invariant under global SU(3). ($\mathbf{q}_r^* = \mathbf{q}_r$ since $\mathbf{q}_r$ is real.)

Eq. F.8 implies that the spinors appearing in eq. F.12 are covariant under SU(3) transformations

$$u_C'(q'^*,s') = Vu_C(q^*,s) \qquad \text{(F.13)}$$
$$v_C'(q'^*,s') = Vv_C(q^*,s) \qquad \text{(F.14)}$$

The fourier coefficients, if second quantized in a complex spatial coordinate generalization of the usual manner, also have covariant anti-commutation relations under an SU(3) transformation:

$$\{b_C(q,s), b_C^\dagger(q'^*,s')\} = \delta_{ss'}\delta^3(q_r - q'_{r'})\delta^3(q_i - q'_{i'}) \qquad \text{(F.15)}$$

Under an SU(3) transformation, z = Uq and z' = Uq', the right side of eq. F.15 transforms to

$$\delta^3(q_r - q'_{r'})\delta^3(q_i - q'_{i'}) \rightarrow \delta^3(z_r - z'_{r'})\delta^3(z_i - z'_{i'})/|\partial(q)/\partial(z)| = \delta^3(z_r - z'_{r'})\delta^3(z_i - z'_{i'}) \qquad \text{(F.16)}$$

where

$$|\partial(q)/\partial(z)| = |\partial(q_r^{\,1}, q_r^{\,2}, q_r^{\,3}, q_i^{\,1}, q_i^{\,2}, q_i^{\,3})/\partial(z_r^{\,1}, z_r^{\,2}, z_r^{\,3}, z_i^{\,1}, z_i^{\,2}, z_i^{\,3})| = 1 \qquad \text{(F.17)}$$

is the Jacobian of the transformation U. The fourier coefficients transform trivially under SU(3):

$$b_C(q^*, s) \rightarrow b_C(z^*, s) \qquad \text{(F.18)}$$

Since the integrand transforms as

$$\int d^3q_r d^3q_i \rightarrow \int d^3z_r d^3z_i\, |\partial(q)/\partial(z)| = \int d^3z_r d^3z_i \qquad \text{(F.19)}$$

we see that the wave function $\psi_C(t, \mathbf{x})$ transforms covariantly.

F.3 Local Color SU(3) and the Internal Symmetry Strong Interactions

In the previous section we showed that the equations of motion of free Dirac-like, complexon, up-type quarks are covariant under global SU(3) coordinate rotations.. The free, tachyon, complexon, down-type quark equations of motion are also easily seen to be covariant under this SU(3) subgroup. In this section we will show this covariance is the analogue of local Color SU(3) symmetry of quarks, and then we will introduce the Internal Symmetry Strong interaction analogue via minimal coupling to SU(3) Yang-Mills gluons in gauge covariant derivatives.

We now introduce a complexon field with a global SU(3) index a which takes values from 1 to 3 making the field a member of the 3 representation of global SU(3):

$$\psi_C^{\,a}(t, \mathbf{x}) \qquad \text{(F.20)}$$

Due to the SU(3) index the transformation property of $\psi_C^{\,a}(t, \mathbf{x})$ changes from eq. F.8 to

$$\psi_C{}''^{a}(t, \mathbf{x}') = U^{ab}V\psi_C^{\,b}(t, \mathbf{x}) = U^{ab}V\psi_C^{\,b}(t, U\mathbf{x}') \qquad \text{(F.21)}$$

where U^{ab} is an SU(3) rotation of 3 representation "vectors" such as $\psi_C{}^b$ and **x.** V is the corresponding rotation of the spinor indices of $\psi_C{}^b(t, \mathbf{x})$.

Note that the coordinate SU(3) rotation of the field factorizes into an SU(3) rotation of the three fields $\psi_C{}^b$ by U^{ab} and an SU(3) rotation of the four spinor components of each individual field $\psi_C{}^b$ by V.

This factorization enables us to consider a global SU(3) rotation of the $\psi_C{}^b$ fields while holding the coordinates fixed:

$$\psi_C{}'^a(t, \mathbf{x}) = U^{ab}\psi_C{}^b(t, \mathbf{x}) \tag{F.22}$$

The equations of motion are covariant under this global transformation

$$0 = U^{ab}[i\gamma^0\partial/\partial t + i\gamma\cdot\mathbf{D}_c{}^* - m]\psi_C{}^b(t, \mathbf{x}_r, \mathbf{x}_i)$$

$$= [i\gamma^0\partial/\partial t + i\gamma\cdot\mathbf{D}_c{}^* - m]\psi'_C{}^a(t, \mathbf{x}_r, \mathbf{x}_i) \tag{F.23}$$

We now note the form of eq. F.22 is the same as that of a *local* Yang-Mills rotation:

$$\psi_C{}'^a(t, \mathbf{x}) = \Theta^{ab}(t, \mathbf{x})\psi_C{}^b(t, \mathbf{x}) \tag{F.24}$$

where $\mathbf{x} = \mathbf{x}_r + i\mathbf{x}_i$. Therefore if we introduce a local SU(3) Yang-Mills field $A_{C\nu}(t, \mathbf{x}_r, \mathbf{x}_i)$ and define a covariant derivative we can convert eq. F.21 to the analogue case of Internal Symmetry local, color SU(3) if we do not perform the spinor rotation V.[178] The covariant derivative is

$$\mathcal{D}_\nu = D_\nu - igA_{C\nu} \tag{F.25}$$

---

[178] This approach, by analogy, enables us to avoid the dilemmas associated with mixing coordinate and internal symmetries as described by Coleman, S., Phys. Rev. **138** B1262 (1965) and others in the case of SU(6) in the 1960's. Note that the spinor rotation V is expressed in terms of numerical matrices while, in the second quantized formulation, the U^{ab} rotation is expressed in terms of second quantized fields as well as numeric matrices. Thus the factorization is reflected in the form of the transformation.

where

$$A_{C\nu} = A_C{}^a{}_\nu t^a \tag{F.26}$$

and where $D_\nu = D_{q\nu}$ is given by

$$D_0 = \partial/\partial x^0$$
$$D_k = \partial/\partial x_r{}^k + i\, \partial/\partial x_i{}^k \tag{F.27}$$

The SU(3) 3×3 matrix generators satisfy

$$[t^a, t^b] = if^{abc}t^c \tag{F.28}$$

We can represent $\Theta_{ab}(x)$ in the form:

$$\Theta_{ab}(x) = [\exp(-i\varphi_c(x)t^c)]_{ab} \tag{F.29}$$

where $\varphi_c(x)$ is a local parameter dependent on $x = (x^0, \mathbf{x} = \mathbf{x_r} + i\mathbf{x_i})$, and t^c is an SU(3) generator.

Applying a gauge transformation to the gauge covariant derivative of a complexon fermion field $\mathcal{D}_\nu\psi_C(x)$:

$$\Theta\mathcal{D}_\nu\psi_C(x) = \Theta D_\nu\psi_C(x) - ig\Theta A_{C\nu}\Theta^{-1}\Theta\psi_C(x) \tag{F.30}$$
$$= D_\nu\psi_C'(x) - igA_C'{}_\nu\psi_C'(x) = (\mathcal{D}_\nu\psi_C(x))'$$

where

$$\psi_C'(x) = \Theta(x)\psi_C(x) \tag{F.31}$$

we find

$$A_C'{}_\nu = (-i/g)(D_\nu\Theta(x))\Theta^{-1}(x) + \Theta(x)A_{C\nu}(x)\Theta^{-1}(x) \tag{F.32}$$

The reader will note that the form of eqs. F.25 – F.31 is identical to those associated with a conventional non-abelian gauge interaction with the replacement:

$$\partial/\partial x^\nu \rightarrow D_\nu \tag{F.33}$$

with D_ν given by eq. F.27. Note that $\varphi_c(x)$, the local parameter in eq. F.29 is dependent in general, on time, and the real and imaginary parts of the complex spatial 3-vector.

Introducing the SU(3) gauge covariant derivative transforms eq. F.23 to

$$0 = [i\gamma^\nu \mathfrak{D}_\nu - m]\psi_C{}^a(t, \mathbf{x_r}, \mathbf{x_i}) \quad (F.34)$$

The preceding argumentation supports the following postulate:

Postulate: Quarks are in a $\underline{3}$ representation of an internal symmetry global SU(3) group.

We note the case of tachyon complexon quarks differs only in small details from the above discussion of Dirac-type complexon quarks.

F.4 Internal Symmetry Interactions Resulting by Analogy from Complex Space-Time Projected to Real Physical Space-Time

This appendix, and the preceding appendices, have shown that the Complex Lorentz group and the Coordinate Reality Group generate analogues of the familiar interactions of The Standard Model: SU(3)⊗SU(2)⊗U(1) plus an additional set of SU(2)⊗U(1) interactions that we take to be the interactions of Dark Matter.

Appendix G. Summary of Complex Lorentz Boosts of Fermions

G.1 Boosts at Real-Valued Velocities Greater Than the Speed of Light

In chapter 5 of Blaha (2017b) we showed that Complex Lorentz boosts for real-valued velocities with magnitude greater than the speed of light transform a rest frame coordinate system to a new complex-valued coordinate system. In general an SU(2)⊗U(1) transformation is needed to transform the target complex-valued coordinate system to a real-valued coordinate system for this type of boosts. We can symbolize these transformations with

$$\text{Real-valued Coordinate System} \rightarrow R_{SU(2)\otimes U(1)}\Lambda(|\mathbf{v}| > \mathbf{c}) \rightarrow \text{Real-valued Coordinate System}$$

where $\Lambda(|\mathbf{v}| > \mathbf{c})$ is a Complex Lorentz transformation with real-valued relative velocity $|v| > c$, and $R_{SU(2)\otimes U(1)}$ is an SU(2)⊗U(1) transformation from complex-valued coordinates to real-valued coordinates.

The Complex Lorentz transformation $\Lambda(|\mathbf{v}| > \mathbf{c})$ when applied to a fermion at rest transforms it into a tachyon with real-valued energy. (See chapter 3 of Blaha (2017b).) The tachyons of this type are those of the neutral lepton species and of the Dark neutral lepton species.

G.2 Boosts at Complex-Valued Velocities

There are two types of boosts of this kind: those whose magnitude of velocity exceeds the speed of light yielding tachyons, and those whose magnitude of velocity is below the speed of light yielding a non-tachyon fermion. (See chapter 2, or chapters 6 and 7 of Blaha (2017b) for details. The forms of the tachyons of these varieties are described chapter 2.) The fermions generated by these boosts are identified as the up and down type normal and Dark quarks.

G.3 Mapping of the Boosted Fermions to Normal and Dark Fermion Species

Four types of fermion species, and the internal symmetries and interactions of particles, emerge from the Complex Lorentz group. We can summarize the map from our theory to the real world in the following way:

Normal Matter Fermions

Dirac fermions – Charged Leptons – Fields generated by Real Lorentz boosts – real v < c
Tachyon fermions – Neutrinos – Fields generated by Real Lorentz boosts – real v > c
Complexon fermions – Up-Type Quark Triplets – Fields generated by Lorentz boosts – complex v < c
Tachyon Complexon fermions – Down-Type Quark Triplets – From Lorentz boosts – complex v > c

Dark Matter Fermions

Dirac fermions – Dark Charged Leptons – Fields generated by Real Lorentz boosts – real v < c
Tachyon fermions – Dark Neutrinos – Fields generated by Real Lorentz boosts – real v > c
Complexon fermions – Dark Up-Type Quark Singlets – Fields generated by Lorentz boosts – complex v < c
Tachyon Complexon fermions – Dark Down-Type Quark Singlets – From Lorentz boosts – complex v > c

Normal Matter Gauge Bosons

SU(2)⊗U(1) - real space-time coordinates – not complexon coordinates
SU(3) - complex space-time coordinates – complexon coordinates

Dark Matter Gauge Bosons

SU(2)⊗U(1) - real space-time coordinates – not complexon coordinates

Appendix H. Particle Cores, The Four Fermion Species, and Their Second Quantization

Summary

Under transformations of the Complex Lorentz group[179] from a state of rest, a spin ½ fermion can have one of four forms which we call *fermion species*.[180] These forms (species) are Dirac fermions and tachyon fermions with real-valued momenta; and complexon (Dirac-like) fermions and complexon tachyon fermions with real energies and complex-valued spatial momenta (subject to the condition that the spin, real part of the spatial momentum, and the imaginary part of the spatial momentum are orthogonal to each other.) The energy of each boosted fundamental particle must be real-valued in the free field case since free fundamental particles do not decay.

The number of distinct species is not changed by the extension of the set of coordinate transformations to Complex General Coordinate Transformations.

We map the four fermion species to charged leptons, neutrinos, up-type quarks, and down-type quarks respectively.

This derivation of four types (species) of fermions *directly* from Complex Lorentz group boosts from a rest state to a state with real-valued energy (thus giving stable fundamental fermions in the absence of interactions) is the only derivation of four fermion species that does not make any assumptions about internal symmetries.

Later the derivation will be brought to a deeper level by the introduction of fermion functionals that can combine with coordinate fourier expansions via an inner

---

[179] Complex Lorentz group boosts are require to obtain four species. It is one of our axioms and a requirement of proofs in axiomatic quantum field theory. See Streater (2000).

[180] Most of the material in this appendix was presented in the First Edition, Blaha (2012a), (2015a) and (2017b) as well as earlier books.

product to generate the fermion quantum fields described in this appendix. In addition we will define functionals for bosons that embody boson spin within a logical construct.

H.0 The Logic Core of Fundamental Fermions and Bosons

The First Edition of this book opened the possibility that fermions (and bosons) might have a core that embodies logic in the form of spin as well as bare masses in the case of fermions. In this book we expand our discussion to better describe the functionals space within which the core functionals of each type of fundamental particle reside. We will define functionals of various spins: 0, ½, 1, and 2. We will see that the core of spin ½ fermion functionals (that we call *qubes*) have four varieties that we will describe in this appendix.[181]Fermion functionals have a bare mass denoted m_0.

Bosons have cores as well that are bosonic functionals with integer spin. We call a bosonic core a *quba*[182] in analogy with the fermion functionals name of qubes. Bosonic functionals are massless. Bosons acquire masses through interactions.

The rationales for logic cores for particles is discussed in detail in chapters 3 and 8 of Blaha (2018e). We find them necessary to establish internal particle symmetries and interactions. We also will show that the formalism based on a space of all particle functionals can lead to an explanation of the 'spooky' action at a distance of Quantum Entanglement that has been the subject of much discussion.

H.0.1 The Logic Building Block of Fermions – Qube Cores

If we consider all possible 'things' that might constitute a fundamental building block for a fundamental fermion theory they are all, at best, *ad hoc* and raise questions of their necessity and whether they are composed of yet a more fundamental substructure.

There is only one choice of building block that avoids these issues – a logic unit or qubit. A qubit is a fundamental entity that is a complex form of computer bit. A bit (and thus a qubit) is known to have an energy or equivalently a mass, and has no

[181] Each of the four varieties (which we call species) can be separated into left-handed and right-handed functionals.
[182] We use 'quba' simply because of its similarity to 'qube'. The leading 'b' signifies its bosonic use. We pronounce 'quba' as 'bub' with a silent 'e.' The word 'quba', itself, is the name of a Bantu language spoken by the Bubi people of Bioko Island in Equatorial Guinea.

constituents of a more primitive form.[183] We call a unit of logic that forms the core of a particle a *qube*.[184] It exists as the core of a particle. But, in itself, it has no *independent* material existence or space-time coordinates. A qube is a functional that acquires features such as coordinates, to become an elementary particle. We define a qube as a fermion field theory functional. (See chapters 3 and 8 of Blaha (2018e).) Later in this appendix we introduce physical features that will cloak qubes with properties and interactions making them into fundamental fermion quantum fields.

H.0.2 Mass of a Qube

Recent experiments have shown that a logical value of a qubit has an energy.associated with it. One bit of information has about 3 x 10^{-21} joules of energy[185] or a rest mass, m_0, or about 0.02 eV using $m_0 = E/c^2$. This result was confirmed by E. Lutz et al.[186] who showed that there is a minimum amount of heat produced per bit of erased data. This minimal heat is called the *Landauer[187] limit*. The equivalent mass we will call the *Landauer mass* and denote it as m_0. We will assume that a fundamental Landauer mass exists in our discussions although the precise value of the mass will not be used since we may expect all physical particle masses to be renormalized to different values when interactions are taken into account.

We will assume all fermions contain a qube within them. (As stated above bosons do not have qubes within them.) We call their core a quba. A qube is assumed to have mass m_0. *The masses of fermions are modified to their known values by interactions.*

It is intriguing that the mass of the electron neutrino has been measured in a variety of experiments and found to be within an order of magnitude or so larger than

---

[183] A qube is a physical manifestation of a logical value. The relation of a qube to a logical value is analogous to the relation of a penciled point placed on paper to the concept of a point as a primitive in geometry.

[184] In the First Edition we called qubes iotas. However, since the name iota was previously used as a particle name many years ago it seemed reasonable to use a different name. We chose the name 'qube' for self-evident reasons. *'Qube' is pronounced 'cube.'*

[185] E. Muneyuki et al, *Nature Physics*, DOI: 10.1038/NPHYS1821.

[186] E. Lutz et al, Nature **483** (7388): 187–190,10.1038/nature10872, (2012).

[187] R. Landauer, "Irreversibility and heat generation in the computing process", IBM Journal of Research and Development **5** (3): 183–191, (1961).

our estimate of the Landauer mass (as we would expect since particles acquire a 'cloud of virtual particles' due to interactions.) This 'cloud' can be expected to increase its mass above the Landauer mass. Since neutrinos only have the weak interaction it is not surprising that the increase due to interactions should not be large. The Mainz Neutrino Mass Experiment, for example, estimates the electron neutrino mass to be less than 2 eV. The new Karlsruhe Tritium Neutrino Experiment (September, 2019) found an upper limit of less than 1.1 eV.

A number of astronomical studies have also generated estimates of neutrino masses. In July 2010 the 3-D MegaZ DR7 galaxy survey found a limit for the combined mass of the three neutrino varieties to be less than 0.28 eV.[188] A smaller upper bound for the sum of neutrino masses, 0.23 eV, was found in March 2013 by the Planck collaboration,[189] In February 2014 a new estimate of the sum was found to be 0.320 ± 0.081 eV due to discrepancies between the Planck's measurements of the Cosmic Microwave Background, and other predictions, combined with the assumption that neutrinos are the cause of weaker gravitational lensing than implied by massless neutrinos.[190]

Thus the experimentally measured values of neutrino masses are consistent with the qube Landauer mass estimate of 0.02 eV given above. We thus assume that *a fermion particle consists of an qube with a certain mass,[191] that is renormalized, together with other features. These features will emerge later in the derivation of the complete theory.[192]*

---

[188] S. Thomas et al, "Upper Bound of 0.28 eV on Neutrino Masses from the Largest Photometric Redshift Survey", Physical Review Letters **105**: 031301 (2010).

[189] Planck Collaboration, arXiv:1303.5076 (2013).

[190] R. A. Battye et al, "Evidence for Massive Neutrinos from Cosmic Microwave Background and Lensing Observations", Phys. Rev. Lett. **112,** 051303 (2014).

[191] Leibniz first proposed the idea of logic 'particles' which he called monads. Our definition of a logic 'particle' does not include (or exclude) the presence of a spiritual part which was part of the definition of Leibniz's monads.

[192] A recent experiment claims to separate the spin part (which we identify as a logical value later) of a molecule from the rest of the molecule.

We view Reality as ultimately a representation (or painting) of logic values evolving through interactions in time and space.[193]

H.0.3 Qube Spin

The spin of a qube is assumed to be spin ½. Qubes are solely a building block of fermions.

H.0.4 Qubes as Fermion Field Functionals

At this point qubes have an insubstantial appearance with only the attributes of mass and spin. In chapters 3 and 8 of Blaha (2018e) we suggested that they can be mathematically represented as fermion field functionals[194] and used to develop the structure of the fermion spectrum and The Unified Superstandard Model.

We saw that the Standard Model interactions and features such as Quantum Entanglement *require* the use of a functional formalism for particle fields.

This new deeper formulation supports the theory presented in the First Edition (Blaha (2017f)). It adds a new level of depth that extends and clarifies the theory presented there.

We will begin here by defining a canonical functional approach to creating a simple Dirac fermion quantum field from a qube and a fourier quantum expression for the space-time part of a free fermion quantum field. We symbolize a qube for a fermion with no internal symmetry and spin ½ as f. We begin by defining a coordinate space Dirac fourier quantum expansion as

$$(s, x, t) = N(p)[b(p, s)u(p, s)e^{-ip\cdot x} + d^{\dagger}(p, s)v(p, s)e^{+ip\cdot x}]$$

where N(p) is a normalization factor, u and v are functions of spin and momentum, and b and $d^{\dagger}$ are creation/annihilation operators.

---

[193] Those who might suggest matter is substantial, and logic values are not, should remember that matter would be completely insubstantial if there were no forces in nature. Neutrinos which are close to insubstantial would be completely insubstantial if there were no weak interactions.

[194] Functionals are a mathematical primitive of our theory. They have been used extensively by Feynman and others in quantum theories.

A Dirac quantum wave function can be defined as an inner product of a qube functional and a coordinate space fourier quantum expansion. For example

$$\psi(x) = (f, (s, x, t)) = \sum_{\pm s} \int d^3p N(p)[b(p, s)u(p, s)e^{-ip\cdot x} + d^\dagger(p, s)v(p, s)e^{+ip\cdot x}] \quad \text{(H.1)}$$

where we use a functional inner product formalism in the manner of Riesz (1955)[195] and others. A functional inner product yields a numeric value. In the present case, it yields a numeric (possibly quantum) function. In general an inner product of a functional f with a variable function g is expressed as

$$G(x) = (f, g(x))$$

For each value of x, G(x) has one numeric value modulo quantum smearing.

In this appendix we will implicitly use the above inner product expression for a fermion field for each of the four general types of fermions presented here:[196] a Dirac type of fermion, a tachyonic type of fermion, a Dirac type of fermion with a complex 3-space momentum fourier expansion; and a tachyonic type of fermion with a complex 3-space momentum fourier expansion.

Thus there are four differing functionals initially – one for each of the four types of fourier expansions. Internal symmetries, introduced later, will lead to a multi-dimensional space of functionals. We discuss these points in detail in chapter 3 of Blaha (2018e). *In this appendix we use conventional quantum field theory expressions deferring the functional space inner product representation discussions to subsequent chapters of Blaha (2018e).*

H.0.5 Bosonic Elementary Particles – Quba Cores

In defining qubes above we have considered only the fermion case. For reasons presented in chapter 3 of Blaha (2018e) there is also a need for boson core functionals called qubas. We can define a corresponding boson functional quba for each type of

---

[195] For example see pp. 61-2 of Blaha (2018e) where linear functionals and their inner products are defined.

[196] These types of fermion fields can be further subdivided into left-handed and right-handed fields.

boson. We will designate a boson functional as b_s where s specifies the spin which may be 0, 1, or 2. Every boson contains a boson functional core within it. For consistency we called a boson functional a *quba*. A quba has the spin of the elementary boson within which it resides. It has zero mass since bosons are typically massless prior to symmetry breaking effects. The functional content embodied in each type of elementary particle is summarized in Table H.1.

PARTICLE TYPE	CORE	MASS	SPIN
Fermion	qube	m_0	½
Scalar Boson	quba	0	0
Vector Boson	quba	0	1
Graviton	quba	0	2

Table H.1 Core functionals within the various types of fundamental elementary particles.

Having established the core concept for fundamental femions we now determine the four basic species (types) of fermions with the implicit understanding that each fermion quantum field is the inner product of a core functional and a coordinate space fourier representation. The remainder of the appendix's derivations are based on the Complex Lorentz Group—as it was in the First Edition (Blaha (2017f)) and earlier books. [197]

H.1 Matrix Representation of Complex Lorentz Group L_C Boosts

The remainder of this appendix is based on the Complex Lorentz group which will be seen to have a primary role in defining the structure of the fermion spectrum.

We begin with Complex Lorentz Group (L_C) boosts because they will be crucial in the determination of the equations of motion of various types of spin ½ particles. An L_C boost can be expressed in the form

[197] The remainder of this appendix is the same as chapter 2 of the First Edition (Blaha (2017f)). A new section 2.10 has been added to depict the functional formulation of the quantum fields.

$$\Lambda_C(\mathbf{v_c}) = \exp[i\omega\hat{\mathbf{w}}\cdot\mathbf{K}] \quad \text{(H.2)}$$

where

$$\omega = (\omega_r^2 - \omega_i^2 + 2i\omega_r\omega_i\, \hat{\mathbf{u}}_r\cdot\hat{\mathbf{u}}_i)^{1/2} \quad \text{(H.3)}$$

and

$$\hat{\mathbf{w}} = (\omega_r\hat{\mathbf{u}}_r + i\omega_i\hat{\mathbf{u}}_i)/\omega \quad \text{(H.4)}$$

Since $\hat{\mathbf{u}}_r\cdot\hat{\mathbf{u}}_r = 1 = \hat{\mathbf{u}}_i\cdot\hat{\mathbf{u}}_i$

$$\hat{\mathbf{w}}\cdot\hat{\mathbf{w}} = 1 \quad \text{(H.5)}$$

and the complex relative velocity is

$$\mathbf{v_c} = \hat{\mathbf{w}}\tanh(\omega) \quad \text{(H.6)}$$

We now analytically continue to complex ω and complex unit vectors $\hat{\mathbf{w}}$. The resulting complex generalization will be the matrix form of proper L_C boosts:

$$\Lambda_C(\mathbf{v_c}) = \exp[i\omega\hat{\mathbf{w}}\cdot\mathbf{K}] \equiv \Lambda_C(\omega, \hat{\mathbf{w}})$$

$$= \begin{bmatrix} \cosh(\omega) & -\sinh(\omega)\hat{w}_x & -\sinh(\omega)\hat{w}_y & -\sinh(\omega)\hat{w}_z \\ -\sinh(\omega)\hat{w}_x & 1 + (\cosh(\omega) - 1)\hat{w}_x^2 & (\cosh(\omega) - 1)\hat{w}_x\hat{w}_y & (\cosh(\omega) - 1)\hat{w}_x\hat{w}_z \\ -\sinh(\omega)\hat{w}_y & (\cosh(\omega) - 1)\hat{w}_x\hat{w}_y & 1 + (\cosh(\omega) - 1)\hat{w}_y^2 & (\cosh(\omega) - 1)\hat{w}_y\hat{w} \\ -\sinh(\omega)\hat{w}_z & (\cosh(\omega) - 1)\hat{w}_x\hat{w}_z & (\cosh(\omega) - 1)\hat{w}_y\hat{w}_z & 1 + (\cosh(\omega) - 1)\hat{w}_z^2 \end{bmatrix}$$

(H.7)

Since analytic continuations are unique, the above form for $\Lambda_C(\mathbf{v_c})$ is well-defined and unique. It spans the complete set of proper L_C boosts.

H.2 Left-handed and Right-handed Parts of L_C

We now describe the Left-handed and Right-handed parts[198] of L_C boosts.

---

[198] The designations Left-handed and Right-handed are chosen to reflect the Left-handed and Right-handed fermion fields that will be used to construct The Standard Model later. See Blaha (2007b) for more detail.

H.2.1 Left-handed Part of L_C

If we let

$$\hat{\mathbf{u}}_i = \hat{\mathbf{u}}_r \equiv \hat{\mathbf{u}} \tag{H.8}$$

so that the vector $\hat{\mathbf{u}}_i$ is parallel to $\hat{\mathbf{u}}_r$, and

$$\omega_i = \pi/2 \tag{H.9}$$

then $\Lambda_C(\mathbf{v}_c)$ becomes a Left-handed L_C boost:

$$\Lambda_C(\mathbf{v}_c) = \Lambda_L(\omega_r, \mathbf{u}) \tag{H.10}$$

H.2.2 Right-handed part of L_C

If we let

$$\hat{\mathbf{u}}_i = -\hat{\mathbf{u}}_r \equiv -\hat{\mathbf{u}} \tag{H.11}$$

so that the vector $\hat{\mathbf{u}}_i$ is anti-parallel to $\hat{\mathbf{u}}_r$, and

$$\omega_i = -\pi/2 \tag{H.12}$$

then $\Lambda_C(\mathbf{v}_c)$ becomes a Right-handed L_C boost:

$$\Lambda_C(\mathbf{v}_c) = \Lambda_R(\omega_r, \mathbf{u}) \tag{H.13}$$

as described in Blaha (2007b).

H.3 Difference between the Parts of L_C Reduced to Parallelism of $\hat{\mathbf{u}}_r$ and $\hat{\mathbf{u}}_i$

Since the Left-handed L_C part leads to the Standard Model's left-handed features, it seems that the parallel case $\hat{\mathbf{u}}_i = \hat{\mathbf{u}}_r \equiv \hat{\mathbf{u}}$ is more favored by Nature.[199] To some extent this concept of parallel vectors $\hat{\mathbf{u}}_i$ and $\hat{\mathbf{u}}_r$, which leads to the Left-handed L_C, is more intuitively satisfying then the anti-parallel case that leads to the Right-handed L_C part. However, a deeper reason for Nature's choice remains to be found.

H.4 Free Spin ½ Particles – Leptons & Quarks

In this section we begin by developing dynamical equations for spin ½ particles based on the L_C parts. These spin ½ particles are conventional Dirac particles (Majorana particles are also allowed but not discussed), spin ½ tachyons, and "color" versions of both types totalling four species. We will identify leptons and quarks with these fields.

H.4.1 Introduction

Tachyons are particles that move faster than the speed of light. As we saw in earlier books tachyons exist inside Black Holes, and within current theories – particularly SuperString theories. There are also experimental indications that neutrinos are tachyons.

Attempts to create canonical tachyon quantum field theories began in the 1960's. These attempts were made within the framework of the Lorentz group and, consequently, were limited to spin 0 theories since there are no finite dimensional representations of the Lorentz group for negative m^2 except for the one-dimensional representation. None of these attempts, or attempts since then, succeeded in creating a canonically quantized spin 0 tachyon quantum field theory.[200]

In this section we will formulate a free spin ½ tachyon[201] Quantum Field Theory. We choose to develop a normal spin ½ theory first. Then we develop a free

---

[199] It is possible that parity violation might disappear at ultra-high energies. Then we would view the parity symmetric theory as broken to the left-handed Standard Model currently established by experiment with right-handed parts at higher energy.

[200] Except Blaha (2006).

[201] It fiffers significantly from tachyon theories such as those of G. Feinberg and E. C. Sudarshan.

spin ½ tachyon theory because, as we will see, spin ½ tachyon particles (quarks and leptons) play an extraordinary role in the Standard Model.

We will develop our spin ½ tachyon theory from the "ground up" by applying a Left-Handed L_C boost to the Dirac equation, and its Dirac spinor wave function, for a particle at rest. This procedure will give a tachyon spinor wave function, and the momentum space tachyon equation equivalent of the Dirac equation. Then we will obtain the coordinate space tachyon Dirac equation, define a lagrangian, and proceed to create a canonical quantum field theory for spin ½ tachyons.

The need for dynamical equations arises when we clothe each qube with coordinates to "make" a fermion. Having coordinates leads to describing the motion of particles. Dynamical equations specify the motion of particles. If they have a finite number of terms they can be derived from lagrangians. A lagrangian formalism yields a Hamiltonian (the energy) and the momentum.

H.4.2 First Step - Deriving the Conventional Dirac Equation

In this section we will review a method of obtaining the equation of motion of a particle using a free Dirac equation that is obtained by a Lorentz boost of a spinor wave function[202] of a particle at rest.

In the case of a Lorentz transformation the 4×4 matrix form of a Lorentz transformation of Dirac matrices is

$$S^{-1}(\Lambda(v))\gamma^{\nu}S(\Lambda(v)) = \Lambda^{\nu}{}_{\mu}(v)\gamma^{\mu} \qquad \text{(H.14)}$$

where $S(\Lambda(v))$ is

$$S(\Lambda(v)) = \exp(-i\omega\sigma_{0i}v_i/(2|\mathbf{v}|)) = \exp(-\omega\gamma^0\boldsymbol{\gamma}\cdot\mathbf{v}/(2|\mathbf{v}|))$$
$$= \cosh(\omega/2)I + \sinh(\omega/2)\gamma^0\boldsymbol{\gamma}\cdot\mathbf{p}/|\mathbf{p}| \qquad \text{(H.15)}$$

with $\omega = \text{arctanh}(|\mathbf{v}|)$, $\cosh(\omega/2) = [(E+m)/(2m)]^{1/2}$ and $\sinh(\omega/2) = |\mathbf{p}|[2m(E+m)]^{-1/2}$. Also

---

[202] The spinor wave function of a particle at rest is a 4-vector of the 4×4 matrix representation of 4-valued Asynchronous Logic.

$$S^{-1}(\Lambda(v)) = \gamma^0 S^\dagger(\Lambda(v))\gamma^0 = \exp(\omega\gamma^0\boldsymbol{\gamma}\cdot\mathbf{v}/(2|\mathbf{v}|))$$
$$= \cosh(\omega/2)I - \sinh(\omega/2)\gamma^0\boldsymbol{\gamma}\cdot\mathbf{p}/|\mathbf{p}| \qquad \text{(H.16)}$$

In constructing fermion dynamical equations *we shall assume that they are linear in derivatives* (although a quadratic form is possible.) We will use the sixteen 4×4 matrices that span the set of transformations of the four values of Asynchronous Logic. Since by theorem[203] all 4 × 4 γ matrices are equivalent up to a unitary transformation we can rotate any constant matrix into a multiple of γ^0 without loss of generality.

We begin by defining a generic positive energy plane wave solution of the Dirac equation for a normal fermion particle at rest with rest mass m, *which we take to be the qube bare mass in the absence of interactions,*[204] as

$$\psi(x) = e^{-imt}w(0) \qquad \text{(H.17)}$$

with w(0) a four component logic spinor column vector. *For a free particle at rest, the rest energy* $m = m_0$, *the qube mass.* The wave function satisfies the momentum space Dirac equation for a fermion at rest:

$$(m\gamma^0 - m)e^{-imt}w(0) = 0 \qquad \text{(H.18)}$$

Subsequently we will use a similar procedure to construct the free tachyonic Dirac equation.

If we now apply S(Λ(v)) we find

$$0 = S(\Lambda(v))(m\gamma^0 - m)e^{-imt}w(0) = [mS(\Lambda(v))\gamma^0 S^{-1}(\Lambda(v)) - m]S(\Lambda(v))w(0)$$

A straightforward evaluation shows

---

[203] R. H. Good, Rev. Mod. Phys., **27**, 187 (1955).

[204] As stated earlier the derivation proceeds in steps from the Complex Lorentz Group to free fermions and thence to interacting fermions. **We use the bare qube mass throughout the free fermion discussions in this appendix.** $\mathbf{m = m_0}$.

$$mS(\Lambda(v))\gamma^0 S^{-1}(\Lambda(v)) = g_{\mu\nu}p^\mu\gamma^\nu = \not{p} \tag{H.19}$$

where $p^0 = (p^2 + m^2)^{1/2}$, $\mathbf{p} = \gamma m\mathbf{v}$, and $p = |\mathbf{p}|$. In addition

$$S(\Lambda(v))w(0) = w(p) \tag{H.20}$$

is a positive energy Dirac spinor. Therefore the Dirac equation for a fermion in motion in momentum space has the form:

$$(\not{p} - m)e^{-ip\cdot x}w(p) = 0 \tag{H.21}$$

where the exponential factor, mt, is also boosted to p·x. Eq. H.21 implies the well-known free, coordinate space Dirac equation:

$$(i\gamma^\mu\partial/\partial x^\mu - m)\psi(x) = 0 \tag{H.22}$$

H.4.3 Derivation of the Tachyon Dirac Equation

The Left-handed boost has the form:

$$\Lambda_L(\omega, \mathbf{u}) = \Lambda(\omega + i\pi/2, \mathbf{u}) = \exp[i\omega_L\hat{\mathbf{u}}\cdot\mathbf{K}] \tag{H.23}$$

where $\omega_L = \omega + i\pi/2$ and

$$\cosh(\omega_L) = i\sinh(\omega) = -\gamma = i\gamma_s \tag{H.24}$$
$$\sinh(\omega_L) = i\cosh(\omega) = -\beta\gamma = i\beta\gamma_s$$

with, $\beta = v > 1$, $\gamma_s = (\beta^2 - 1)^{-1/2}$, and $\boldsymbol{\omega} \geq \mathbf{0}$. Thus

$$\sinh(\omega) = \gamma_s \tag{H.25}$$
$$\cosh(\omega) = \beta\gamma_s$$

The corresponding spinor transformation is:

$$S_L(\Lambda_L(\omega, \mathbf{u})) = \exp(-i\omega_L\sigma_{0i}v_i/(2|\mathbf{v}|)) = \exp(-\omega_L\gamma^0\boldsymbol{\gamma}\cdot\mathbf{v}/(2|\mathbf{v}|))$$

$$= \cosh(\omega_L/2)I + \sinh(\omega_L/2)\gamma^0\boldsymbol{\gamma}\cdot\mathbf{p}/|\mathbf{p}| \quad \text{(H.26)}$$

The inverse transformation is

$$S_L^{-1}(\Lambda_L(\omega, \mathbf{u})) = \gamma^2\gamma^0 K^{-1} S_L^{\dagger} K \gamma^0\gamma^2 = \gamma^2\gamma^0 S_L^{T}\gamma^0\gamma^2 = \exp(\omega_L\gamma^0\boldsymbol{\gamma}\cdot\mathbf{v}/(2|\mathbf{v}|))$$
$$= \cosh(\omega_L/2)I - \sinh(\omega_L/2)\gamma^0\boldsymbol{\gamma}\cdot\mathbf{p}/|\mathbf{p}| \quad \text{(H.27)}$$

where the superscript T denotes the transpose and K is the complex conjugation operator (that also appears in the time-reversal operator). Note that S_L is not unitary just as the equivalent spinor Lorentz transformation $S(\Lambda(v))$ is not unitary.
We can now apply a left-handed superluminal transformation to the generic positive energy plane wave solution of the Dirac equation for a particle of mass m at rest. The result is

$$0 = S_L(\Lambda_L(\omega, \mathbf{u}))(m\gamma^0 - m)e^{-imt}w(0)$$
$$= [mS_L\gamma^0 S_L^{-1} - m]e^{-imt}S_L w(0)$$

where $S_L = S_L(\Lambda_L(\omega, \mathbf{u}))$. After some algebra

$$mS_L\gamma^0 S_L^{-1} = m[\cosh(\omega_L)\gamma^0 - \sinh(\omega_L)\boldsymbol{\gamma}\cdot\mathbf{p}/|\mathbf{p}|]$$
$$= i\gamma^0 E - i\boldsymbol{\gamma}\cdot\mathbf{p} = i\not{p} \quad \text{(H.28)}$$

using the tachyon energy and momentum expressions

$$\mathbf{p} = m\mathbf{v}\gamma_s \qquad E = m\gamma_s \quad \text{(H.29)}$$

Also

$$S_L w(0) = w_T(p) \quad \text{(H.30)}$$

is a tachyon spinor. See section H.12 for a discussion of tachyon spinors.

The momentum space tachyonic Dirac equation is

$$(i\not{p} - m)e^{ip\cdot x}w_T(p) = 0 \quad \text{(H.31)}$$

where p·x = Et – **p·x** after performing a corresponding left-handed superluminal coordinate transformation in the exponential factor. Thus a positive energy wave is transformed into a negative energy wave by the superluminal transformation.

If we apply $i\not{\partial}$ to we find the tachyon mass condition is satisfied

$$-E^2 + \mathbf{p}^2 = m^2 \qquad \text{(H.32)}$$

Transforming back to coordinate space we obtain the *tachyon Dirac equation*:

$$(\gamma^\mu \partial/\partial x^\mu - m)\psi_T(x) = 0 \qquad \text{(H.33)}$$

The "missing" factor of i in the first term of eq. H.33 requires the lagrangian to be different from the conventional Dirac lagrangian in order for the lagrangian to be real. The simplest, physically acceptable, free spin ½ tachyon lagrangian density is:

$$\mathcal{L}_T = \psi_T{}^S(\gamma^\mu \partial/\partial x^\mu - m)\psi_T(x) \qquad \text{(H.34)}$$

where

$$\psi_T{}^S = \psi_T{}^\dagger i\gamma^0\gamma^5 \qquad \text{(H.35)}$$

The corresponding action is

$$I = \int d^4x \mathcal{L}_T \qquad \text{(H.36)}$$

Appendix 3-B of Blaha (2007b) proves I is real. The Hamiltonian density is

$$\mathcal{H} = \pi_T\dot{\psi}_T - \mathcal{L} = i\psi_T{}^\dagger\gamma^5(\boldsymbol{\alpha}\cdot\nabla + \beta m)\psi_T = -i\psi_T{}^\dagger\gamma^5\dot{\psi}_T \qquad \text{(H.37)}$$

using the tachyon Dirac equation to obtain the last equality. The reader will note that the tachyon hamiltonian is hermitean by explicit calculation up to an irrelevant total spatial divergence.

H.4.3.1 Probability Conservation Law

The tachyon Dirac equation implies a probability conservation law:

$$\partial\rho_5/\partial t = \nabla\cdot\mathbf{j}_5 \tag{H.38}$$

where

$$\rho_5 = \psi_T^{\dagger}\gamma^5\psi_T \qquad\qquad \mathbf{j}_5 = \psi_T^{\dagger}\gamma^5\boldsymbol{\alpha}\psi_T \tag{H.39}$$

We are thus led to define the conserved axial charge Q_5

$$Q_5 = \int d^3x\, \psi_T^{\dagger}\gamma^5\psi_T \tag{H.40}$$

H.4.3.2 Energy-Momentum Tensor

The tachyon energy-momentum tensor is

$$\mathfrak{T}_{T\mu\nu} = -\, g_{\mu\nu}\, \mathcal{L}_T + \partial\mathcal{L}_T/\partial(\partial\psi_T/\partial x_\mu)\ \partial\psi_T/\partial x^\nu \tag{H.41}$$

$$= i\psi_T^{\dagger}\gamma^0\gamma^5\gamma_\mu\partial\psi_T/\partial x^\nu \tag{H.42}$$

and thus the conserved energy and momentum are

$$P^0 = H = \int d^3x\, \mathfrak{T}_T^{\ 00} = i\int d^3x\psi_T^{\dagger}\gamma^5(\boldsymbol{\alpha}\cdot\nabla + \beta m)\psi_T \tag{H.43}$$

and

$$P^i = \int d^3x\, \mathfrak{T}_T^{\ 0i} = -\, i\int d^3x\, \psi_T^{\dagger}\gamma^5\partial\psi_T/\partial x_i \tag{H.44}$$

Both the energy and momentum differ significantly from the corresponding quantities for conventional Dirac fields.

H.4.4 Tachyon Canonical Quantization

Having defined a suitable tachyon lagrangian we can now proceed to its canonical quantization. The conjugate momentum can be calculated from the above lagrangian density:

$$\pi_{Ta} = \partial\mathcal{L}_T/\partial\dot{\psi}_{Ta} \equiv \partial\mathcal{L}_T/\partial(\partial\psi_{Ta}/\partial t) = -i(\psi_T^{\dagger}\gamma^5)_a \tag{H.45}$$

The resulting non-zero, canonical anti-commutation relations are

$$\{\pi_{T_a}(x), \psi_{Tb}(x')\} = i\,\delta_{ab}\,\delta^3(x - x')$$

or

$$\{\psi_{T_a}^{\dagger}(x), \psi_{Tb}(x')\} = -[\gamma^5]_{ab}\,\delta^3(x - x') \qquad \text{(H.46)}$$

At this point we might attempt to complete the canonical quantization procedure in the conventional manner by fourier expanding the quantum field and specifying anti-commutation relations for the fourier component amplitudes. However the incompleteness of the set of plane waves, which are limited by the restriction $|p| \geq m$, causes the anti-commutator of the fields not to yield a $\delta^3(x - x')$. Thus the conventional approach fails to yield the required anti-commutation relations.[205]

Other approaches: 1) decompose the tachyon field into left-handed and right-handed parts and then second quantize each part; and 2) second quantize in light-front coordinates ($x^{\pm} = (x^0 \pm x^3)/\surd 2$). These approaches also both fail.[206]

The only approach that does succeed[207] *is to decompose the tachyon field into left-handed and right-handed parts and then second quantize in light-front coordinates. We follow that procedure in the following subsections.*

H.4.4.1 Separation into Left-Handed and Right-Handed Fields

We will use a transformed set of Dirac matrices to develop our left-handed and right-handed tachyon formulations:

$$\gamma^0 = \begin{bmatrix} 0 & -I \\ -I & 0 \end{bmatrix} \qquad \gamma^i = \begin{bmatrix} 0 & \sigma_i \\ -\sigma_i & 0 \end{bmatrix} \qquad \gamma^5 = \begin{bmatrix} I & 0 \\ 0 & -I \end{bmatrix}$$

(H.47)

which are obtained from the usual Dirac matrices by applying the unitary transformation $U = 2^{-1/2}(I + \gamma^5\gamma^0)$. *I is the 4×4 identity matrix in eq. H.47.* The γ^5 chirality

[205] See G. Feinberg, Phys. Rev. **159**, 1089 (1967) for example.

[206] See Blaha (2006) where these possibilities were considered and found to fail.

[207] Blaha (2006) discusses this case in detail.

operator's eigenvalues define handedness: +1 corresponds to right-handed; and –1 corresponds to left-handed:

$$\gamma^5\psi_L = -\psi_L \qquad\qquad \gamma^5\psi_R = \psi_R \tag{H.48}$$

Consequently, we can define left-handed and right-handed tachyon fields with the projection operators:

$$\begin{aligned} C^{\pm} &= ½(I \pm \gamma^5) \\ C^+ + C^- &= I \\ C^{\pm\,2} &= C^{\pm} \\ C^+C^- &= 0 \end{aligned} \tag{H.49}$$

with the result

$$\begin{aligned} \psi_{TL} &= C^-\psi_T \\ \psi_{TR} &= C^+\psi_T \end{aligned} \tag{H.50}$$

We can calculate the commutation relations of the left-handed and right-handed tachyon fields from eq. H.46 by pre-multiplying and post-multiplying by $½(1 - \gamma^5)$ and $½(1 + \gamma^5)$. The results are:

$$\{\psi_{TLa}^{\dagger}(x), \psi_{TLb}(x')\} = ½(1 - \gamma^5)_{ab}\, \delta^3(x - x') \tag{H.51}$$

$$\{\psi_{TRa}^{\dagger}(x), \psi_{TRb}(x')\} = -½(1 + \gamma^5)_{ab}\, \delta^3(x - x') \tag{H.52}$$

$$\{\psi_{TLa}^{\dagger}(x), \psi_{TRb}(x')\} = \{\psi_{TRa}^{\dagger}(x), \psi_{TLb}(x')\} = 0 \tag{H.53}$$

The lagrangian density above decomposes into left-handed and right-handed parts:

$$\mathcal{L}_T = \psi_{TL}^{\dagger}\gamma^0 i\gamma^{\mu}\partial_{\mu}\psi_{TL} - \psi_{TR}^{\dagger}\gamma^0 i\gamma^{\mu}\partial_{\mu}\psi_{TR} - im[\psi_{TR}^{\dagger}\gamma^0\psi_{TL} - \psi_{TL}^{\dagger}\gamma^0\psi_{TR}] \tag{H.54}$$

H.4.4.2 Further Separation into + and – Light-Front Fields

There have been many studies of light-front (infinite momentum frame) physics in the past forty years.[208] Light-front coordinates *cannot* be obtained by a Lorentz transformation, or by a superluminal transformation, from a standard set of coordinate system variables even in a limiting sense. Instead they are a defined set of variables that have been used to develop quantum field theories that have been shown to be equivalent to quantum field theories based on conventional coordinates. In particular, light-front quantum field theories have been shown to yield fully Lorentz covariant S matrix elements that are the same as S matrix elements calculated in the conventional way.

Light-front variables can be defined by:

$$x^{\pm} = (x^0 \pm x^3)/\sqrt{2} \tag{H.55}$$

$$\partial/\partial x^{\pm} \equiv \partial^{\mp} \equiv (\partial/\partial x^0 \pm \partial/\partial x^3)/\sqrt{2}$$

with the "transverse" coordinate variables, x^1 and x^2, unchanged.

The inner product of two 4-vectors has the form

$$x{\cdot}y = x^+y^- + y^+x^- - x^1y^1 - x^2y^2 \tag{H.56}$$

and the light-front definition of Dirac matrices is:

$$\gamma^{\pm} = (\gamma^0 \pm \gamma^3)/\sqrt{2} \tag{H.57}$$

with transverse matrices γ^1 and γ^2 defined as usual. Note the useful identity:

---

[208] L. Susskind, Phys. Rev. **165**, 1535 (1968); K. Bardakci and M. B. Halpern Phys. Rev. **176**, 1686 (1968), S. Weinberg, Phys. Rev. **150**, 1313 (1966); J. Kogut and D. Soper, Phys. Rev. **D1**, 2901 (1970); J. D. Bjorken, J. Kogut, and D. Soper, Phys. Rev. **D3**, 1382 (1971); R. A. Neville and F. Rohrlich, Nuov. Cim. **A1**, 625 (1971); F. Rohrlich, Acta Phys Austr. Suppl. **8**, 277 (1971); S-J Chang, R. Root, and T-M Yan, Phys. Rev. **D7**, 1133 (1973); S-J Chang, and T-M Yan, Phys. Rev. **D7**, 1147 (1973); T-M Yan, Phys. Rev. **D7**, 1761 (1973); T-M Yan, Phys. Rev. **D7**, 1780 (1973); C. Thorn, Phys. Rev. **D19**, 639 (1979); and references therein.

$$\gamma^{\pm 2} = 0$$

We define "+" and "–" tachyon fields with the projection operators:

$$R^{\pm} = ½(I \pm \gamma^0\gamma^3) \tag{H.58}$$

They are:

$$\text{Left-handed, } \pm \text{ light-front fields:} \qquad \psi_{TL}{}^{\pm} = R^{\pm}C^{-}\psi_T$$

$$\text{Right-handed, } \pm \text{ light-front fields:} \qquad \psi_{TR}{}^{\pm} = R^{\pm}C^{+}\psi_T \tag{H.59}$$

Now if we transform to light-front variables and fields as above we obtain the light-front free tachyon lagrangian:

$$\begin{aligned}\mathcal{L}_T = {} & 2^{½}\psi_{TL}{}^{+\dagger}i\partial^{-}\psi_{TL}{}^{+} + 2^{½}\psi_{TL}{}^{-\dagger}i\partial^{+}\psi_{TL}{}^{-} - \psi_{TL}{}^{+\dagger}\gamma^0 i\gamma^j\partial^j\psi_{TL}{}^{-} - \psi_{TL}{}^{-\dagger}\gamma^0 i\gamma^j\partial^j\psi_{TL}{}^{+} - \\ & - 2^{½}\psi_{TR}{}^{+\dagger}i\partial^{-}\psi_{TR}{}^{+} - 2^{½}\psi_{TR}{}^{-\dagger}i\partial^{+}\psi_{TR}{}^{-} + \psi_{TR}{}^{+\dagger}\gamma^0 i\gamma^j\partial^j\psi_{TR}{}^{-} + \psi_{TR}{}^{-\dagger}\gamma^0 i\gamma^j\partial^j\psi_{TR}{}^{+} - \\ & - im[\psi_{TR}{}^{+\dagger}\gamma^0\psi_{TL}{}^{-} - \psi_{TL}{}^{+\dagger}\gamma^0\psi_{TR}{}^{-} + \psi_{TR}{}^{-\dagger}\gamma^0\psi_{TL}{}^{+} - \psi_{TL}{}^{-\dagger}\gamma^0\psi_{TR}{}^{+}]\end{aligned} \tag{H.60}$$

with implied sums over j = 1,2. In contrast to the light-front tachyon lagrangian we note the corresponding light-front "normal" Dirac fermion lagrangian is

$$\begin{aligned}\mathcal{L}_{Dirac} = {} & 2^{½}\psi_L{}^{+\dagger}i\partial^{-}\psi_L{}^{+} + 2^{½}\psi_L{}^{-\dagger}i\partial^{+}\psi_L{}^{-} - \psi_L{}^{+\dagger}\gamma^0 i\gamma^j\partial^j\psi_L{}^{-} - \psi_L{}^{-\dagger}\gamma^0 i\gamma^j\partial^j\psi_L{}^{+} - \\ & - 2^{½}\psi_R{}^{+\dagger}i\partial^{-}\psi_R{}^{+} + 2^{½}\psi_R{}^{-\dagger}i\partial^{+}\psi_R{}^{-} - \psi_R{}^{+\dagger}\gamma^0 i\gamma^j\partial^j\psi_R{}^{-} - \psi_R{}^{-\dagger}\gamma^0 i\gamma^j\partial^j\psi_R{}^{+} - \\ & - im[\psi_R{}^{+\dagger}\gamma^0\psi_L{}^{-} + \psi_L{}^{+\dagger}\gamma^0\psi_R{}^{-} + \psi_R{}^{-\dagger}\gamma^0\psi_L{}^{+} + \psi_L{}^{-\dagger}\gamma^0\psi_R{}^{+}]\end{aligned} \tag{H.61}$$

The difference in signs between these lagrangians will turn out to be a crucial factor in the derivation of features of the Standard Model later.

Returning to the tachyon lagrangian eq. H.60 we obtain equations of motion through the standard variational techniques:

$$2^{\frac{1}{2}}i\partial^{-}\psi_{TL}{}^{+} - \gamma^{0}i\gamma^{j}\partial^{j}\psi_{TL}{}^{-} + im\gamma^{0}\psi_{TR}{}^{-} = 0$$
$$2^{\frac{1}{2}}i\partial^{-}\psi_{TR}{}^{+} - \gamma^{0}i\gamma^{j}\partial^{j}\psi_{TR}{}^{-} + im\gamma^{0}\psi_{TL}{}^{-} = 0$$
$$2^{\frac{1}{2}}i\partial^{+}\psi_{TL}{}^{-} - \gamma^{0}i\gamma^{j}\partial^{j}\psi_{TL}{}^{+} + im\gamma^{0}\psi_{TR}{}^{+} = 0$$
$$2^{\frac{1}{2}}i\partial^{+}\psi_{TR}{}^{-} - \gamma^{0}i\gamma^{j}\partial^{j}\psi_{TR}{}^{+} + im\gamma^{0}\psi_{TL}{}^{+} = 0 \qquad \text{(H.62)}$$

Eqs. H.62 show that $\psi_{TL}{}^{-}$ and $\psi_{TR}{}^{-}$ are dependent fields that are functions of $\psi_{TL}{}^{+}$ and $\psi_{TR}{}^{+}$ on the light-front where x^{+} equals a constant. They can be expressed in an integral form as well. (The independent fields $\psi_{TL}{}^{+}$ and $\psi_{TR}{}^{+}$ play a fundamental role in tachyon theory and are used to define "in" and "out" tachyon states in perturbation theory.)

The conjugate momenta are

$$\pi_{TL}{}^{+} = \partial\mathcal{L}/\partial(\partial^{-}\psi_{TL}{}^{+}) = 2^{\frac{1}{2}}i\psi_{TL}{}^{+\dagger} \qquad \text{(H.63)}$$
$$\pi_{TL}{}^{-} = \partial\mathcal{L}/\partial(\partial^{-}\psi_{TL}{}^{-}) = 0$$
$$\pi_{TR}{}^{+} = \partial\mathcal{L}/\partial(\partial^{-}\psi_{TR}{}^{+}) = -2^{\frac{1}{2}}i\psi_{TR}{}^{+\dagger} \qquad \text{(H.64)}$$
$$\pi_{TR}{}^{-} = \partial\mathcal{L}/\partial(\partial^{-}\psi_{TR}{}^{-}) = 0$$

Quantization on surfaces of constant x^{+} (light-front surfaces) has been shown to support satisfactory formulations of Quantum Electrodynamics and other quantum field theories. Thus x^{+} plays the role of the "time" variable in light-front quantized theories. So we will define canonical equal x^{+} anti-commutation relations for spin ½ tachyons.

The resulting canonical equal-light-front ($x^{+} = y^{+}$) anti-commutation relations of the independent fields are:

$$\{\psi_{TL}{}^{+\dagger}{}_{a}(x), \psi_{TL}{}^{+}{}_{b}(y)\} = 2^{-1}[C^{-}R^{+}]_{ab}\,\delta(x^{-} - y^{-})\delta^{2}(x - y) \qquad \text{(H.65)}$$
$$\{\psi_{TR}{}^{+\dagger}{}_{a}(x), \psi_{TR}{}^{+}{}_{b}(y)\} = -2^{-1}[C^{+}R^{+}]_{ab}\,\delta(x^{-} - y^{-})\delta^{2}(x - y) \qquad \text{(H.66)}$$
$$\{\psi_{TL}{}^{+}{}_{a}{}^{\dagger}(x), \psi_{TR}{}^{+}{}_{b}(y)\} = \{\psi_{TR}{}^{+}{}_{a}{}^{\dagger}(x), \psi_{TL}{}^{+}{}_{b}(y)\} = 0 \qquad \text{(H.67)}$$
$$\{\psi_{TL}{}^{+}{}_{a}(x), \psi_{TR}{}^{+}{}_{b}(y)\} = \{\psi_{TR}{}^{+}{}_{a}{}^{\dagger}(x), \psi_{TL}{}^{+\dagger}{}_{b}(y)\} = 0 \qquad \text{(H.68)}$$

where the factors of 2^{-1} are the result of the $2^{½}$ factor in eqs. H.63 and H.64, and the factor of $2^{-½}$ in the definition of x^- above.

If we compare eqs. H.65 and H.66 with the corresponding anti-commutation relations of *conventional* <u>*Dirac*</u> quantum fields:

$$\{\psi_{L}{}^{+\dagger}{}_{a}(x), \psi_{L}{}^{+}{}_{b}(y)\} = 2^{-1}[C^{-}R^{+}]_{ab}\, \delta(x^{-} - y^{-})\delta^{2}(x - y) \qquad \text{(H.69)}$$

$$\{\psi_{R}{}^{+\dagger}{}_{a}(x), \psi_{R}{}^{+}{}_{b}(y)\} = 2^{-1}[C^{+}R^{+}]_{ab}\, \delta(x^{-} - y^{-})\delta^{2}(x - y) \qquad \text{(H.70)}$$

we see that the right-handed tachyon anti-commutation relation has a minus sign relative to the corresponding right-handed conventional anti-commutation relation. The right-handed tachyon anti-commutation relation with its minus sign will require compensating minus signs in its creation and annihilation Fourier component operators' anti-commutation relations.

The sign differences between the lagrangian terms in eqs. H.63 and H.64 ultimately lead to parity violating features in the Standard Model lagrangian and thus resolve the long-standing question:

Why parity violation? Answer: Nature preferentially chooses the Left-handed part of the complex Lorentz group.. This choice is not a consequence of Ockham's Razor. But it does conform to Leibniz's Minimax Principle – a minor differentiation based on parity results in "maximal" physical consequences.

H.4.4.3 Left-Handed Tachyons

The free, "+" light-front, left-handed tachyon wave function Fourier expansion is:

$$\psi_{TL}{}^{+}(x) = \sum_{\pm s} \int d^{2}p\,dp^{+} N_{TL}{}^{+}(p)\theta(p^{+})[b_{TL}{}^{+}(p, s)u_{TL}{}^{+}(p, s)e^{-ip\cdot x} + d_{TL}{}^{+\dagger}(p, s)v_{TL}{}^{+}(p, s)e^{+ip\cdot x}] \qquad \text{(H.71)}$$

and its hermitean conjugate is

$$\psi_{TL}{}^{+\dagger}(x) = \sum_{\pm s} \int d^{2}p\,dp^{+} N_{TL}{}^{+}(p)\theta(p^{+})\, [b_{TL}{}^{+\dagger}(p, s)u_{TL}{}^{+\dagger}(p,s)e^{+ip\cdot x} + d_{TL}{}^{+}(p, s)v_{TL}{}^{+\dagger}(p, s)e^{-ip\cdot x}] \qquad \text{(H.72)}$$

where † indicates hermitean conjugate, where

$$N_{TL}^{\ +}(p) = [2m|\mathbf{p}|/((2\pi)^3(p^+(p^+ - p^-) + p_\perp^{\ 2}))]^{1/2} \qquad \text{(H.73)}$$

where the anti-commutation relations of the Fourier coefficient operators are

$$\begin{aligned}
&\{b_{TL}^{\ +}(q,s),\ b_{TL}^{\ +\dagger}(p,s')\} = \delta_{ss'}\delta^2(\mathbf{q} - \mathbf{p})\delta(q^+ - p^+)\\
&\{d_{TL}^{\ +}(q,s),\ d_{TL}^{\ +\dagger}(p,s')\} = \delta_{ss'}\delta^2(\mathbf{q} - \mathbf{p})\delta(q^+ - p^+)\\
&\{b_{TL}^{\ +}(q,s),\ b_{TL}^{\ +}(p,s')\} = \{d_{TL}^{\ +}(q,s),\ d_{TL}^{\ +}(p,s')\} = 0 \qquad \text{(H.74)}\\
&\{b_{TL}^{\ +\dagger}(q,s),\ b_{TL}^{\ +\dagger}(p,s')\} = \{d_{TL}^{\ +\dagger}(q,s),\ d_{TL}^{\ +\dagger}(p,s')\} = 0\\
&\{b_{TL}^{\ +}(q,s),\ d_{TL}^{\ +\dagger}(p,s')\} = \{d_{TL}^{\ +}(q,s),\ b_{TL}^{\ +\dagger}(p,s')\} = 0\\
&\{b_{TL}^{\ +\dagger}(q,s),\ d_{TL}^{\ +\dagger}(p,s')\} = \{d_{TL}^{\ +}(q,s),\ b_{TL}^{\ +}(p,s')\} = 0
\end{aligned}$$

and where the spinors are

$$\begin{aligned}
&u_{TL}^{\ +}(p,\ s) = C^- R^+ S_L(\Lambda_L(\mathbf{p}))w^1(0)\\
&u_{TL}^{\ +}(p,\ -s) = C^- R^+ S_L(\Lambda_L(\mathbf{p}))w^2(0)\\
&v_{TL}^{\ +}(p,\ s) = C^- R^+ S_L(\Lambda_L(\mathbf{p}))w^3(0)\\
&v_{TL}^{\ +}(p,\ -s) = C^- R^+ S_L(\Lambda_L(\mathbf{p}))w^4(0) \qquad \text{(H.75)}\\
&u_{TL}^{\ +\dagger}(p,\ s) = w^{1T}(0)S_L^{\ \dagger}(\Lambda_L(\mathbf{p}))R^+C^-\\
&u_{TL}^{\ +\dagger}(p,\ -s) = w^{2T}(0)S_L^{\ \dagger}(\Lambda_L(\mathbf{p}))R^+C^-\\
&v_{TL}^{\ +\dagger}(p,\ s) = w^{3T}(0)S_L^{\ \dagger}(\Lambda_L(\mathbf{p}))R^+C^-\\
&v_{TL}^{\ +\dagger}(p,\ -s) = w^{4T}(0)S_L^{\ \dagger}(\Lambda_L(\mathbf{p}))R^+C^-
\end{aligned}$$

where the superscript "T" indicates the transpose. (These spinors are described in section H.12.)

The canonical left-handed, light-front anti-commutation relation results in:

$$\{\psi_{TL\ a}^{\ +}(x),\ \psi_{TL\ b}^{\ +\dagger}(y)\} = \sum_{\pm s,s'} \int d^2p dp^+ \int d^2p' dp'^+\ N_{TL}^{\ +}(p)N_{TL}^{\ +}(p')\theta(p^+)\theta(p'^+)\cdot$$

$$\cdot[\{b_{TL}^{+\dagger}(p',s'),b_{TL}^{+}(p,s)\}u_{TL}{}^{+}{}_{a}(p,s)u_{TL}{}^{+\dagger}{}_{b}(p',s')e^{+ip'\cdot y - ip\cdot x} + \\ + \{d_{TL}^{+}(p',s'),d_{TL}^{+\dagger}(p,s)\}v_{TL}{}^{+}{}_{a}(p,s)v_{TL}{}^{+\dagger}{}_{b}(p',s')e^{-ip'\cdot y + ip\cdot x}]$$

$$= \sum_{\pm s}\int d^2p dp^+ N_{TL}{}^{+2}(p)\theta(p^+)[u_{TL}{}^{+}{}_{a}(p,s)u_{TL}{}^{+\dagger}{}_{b}(p,s)e^{+ip\cdot(y-x)} + \\ + v_{TL}{}^{+}{}_{a}(p,s)v_{TL}{}^{+\dagger}{}_{b}(p,s)e^{-ip\cdot(y-x)}]$$

$$= -i\int d^2p dp^+\theta(p^+)N_{TL}{}^{+2}(p)(2m|\mathbf{p}|)^{-1}\{[C^-R^+(i\not{p} - m)\boldsymbol{\gamma}\cdot\mathbf{p}R^+C^-]_{ab}e^{+ip\cdot(y-x)} + \\ + [C^-R^+(i\not{p} + m)\boldsymbol{\gamma}\cdot\mathbf{p}R^+C^-]_{ab}e^{-ip\cdot(y-x)}\}$$

$$= -i\int d^2p_\perp\int_0^\infty dp^+N_{TL}{}^{+2}(p)\{[C^-R^+(ip^+(p^+ - p^-) + ip_\perp{}^2 - mp_\perp\cdot\gamma_\perp)C^-]_{ab}e^{+ip^+(y^- - x^-) - ip_\perp\cdot(y_\perp - x_\perp)} - \\ - [C^-R^+(-ip^+(p^+ - p^-) - ip_\perp{}^2 - mp_\perp\cdot\gamma_\perp)C^-]_{ab}e^{-ip^+(y^- - x^-) + ip_\perp\cdot(y_\perp - x_\perp)}\}/(2m|\mathbf{p}|)$$

$$= \int d^2p_\perp\int_{-\infty}^{\infty} dp^+N_{TL}{}^{+2}(p)[C^-R^+(p^+(p^+ - p^-) + p_\perp{}^2)]_{ab}\, e^{+ip^+(y^- - x^-) - ip_\perp\cdot(y_\perp - x_\perp)}/(2m|\mathbf{p}|)$$

upon letting $p^+ \to -p^+$ and $\mathbf{p}_\perp \to -\mathbf{p}_\perp$ in the second term after using $N_{TL}{}^{+2}(p)(p^+(p^+ - p^-) + p_\perp{}^2) = 1$. The result

$$= \tfrac{1}{2}\int d^2p_\perp\int_{-\infty}^{\infty} dp^+(2\pi)^{-3}[C^-R^+]_{ab}e^{+ip^+(y^- - x^-) - ip_\perp\cdot(y_\perp - x_\perp)}$$

$$= 2^{-1}[C^-R^+]_{ab}\,\delta(y^- - x^-)\delta^2(\mathbf{y} - \mathbf{x}) \tag{H.76}$$

Therefore we have left-handed, light-front quantized tachyons with canonical commutation relations and localized tachyons. As a result we have a canonical Tachyon Quantum Field Theory unlike previous efforts.

H.4.4.4 Right-Handed Tachyons

The case of right-handed tachyons is similar to the left-handed case with only two differences: a minus sign in the creation and annihilation operator anti-commutation relations, and the use of right-handed projection operators. The right-handed tachyon wave function light-front Fourier expansion is:

$$\psi_{TR}{}^{+}(x) = \sum_{\pm s} \int d^2p dp^+ N_{TR}{}^{+}(p)\theta(p^+)[b_{TR}{}^{+}(p, s)u_{TR}{}^{+}(p, s)e^{-ip\cdot x} + d_{TR}{}^{+\dagger}(p, s)v_{TR}{}^{+}(p, s)e^{+ip\cdot x}] \tag{H.77}$$

and its hermitean conjugate is

$$\psi_{TR}{}^{+\dagger}(x) = \sum_{\pm s} \int d^2p dp^+ N_{TR}{}^{+}(p)\theta(p^+) [b_{TR}{}^{+\dagger}(p, s)u_{TR}{}^{+\dagger}(p, s)e^{+ip\cdot x} + d_{TR}{}^{+}(p, s)v_{TR}{}^{+\dagger}(p, s)e^{-ip\cdot x}] \tag{H.78}$$

where $N_{TR}{}^{+}(p) = N_{TL}{}^{+}(p)$, where the anti-commutation relations of the Fourier coefficient operators are

$$\begin{aligned}
\{b_{TR}{}^{+}(q,s), b_{TR}{}^{+\dagger}(p,s')\} &= -\delta_{ss'}\delta^2(\mathbf{q} - \mathbf{p})\delta(q^+ - p^+) \\
\{d_{TR}{}^{+}(q,s), d_{TR}{}^{+\dagger}(p,s')\} &= -\delta_{ss'}\delta^2(\mathbf{q} - \mathbf{p})\delta(q^+ - p^+) \\
\{b_{TR}{}^{+}(q,s), b_{TR}{}^{+}(p,s')\} &= \{d_{TR}{}^{+}(q,s), d_{TR}{}^{+}(p,s')\} = 0 \\
\{b_{TR}{}^{+\dagger}(q,s), b_{TR}{}^{+\dagger}(p,s')\} &= \{d_{TR}{}^{+\dagger}(q,s), d_{TR}{}^{+\dagger}(p,s')\} = 0 \\
\{b_{TR}{}^{+}(q,s), d_{TR}{}^{+\dagger}(p,s')\} &= \{d_{TR}{}^{+}(q,s), b_{TR}{}^{+\dagger}(p,s')\} = 0 \\
\{b_{TR}{}^{+\dagger}(q,s), d_{TR}{}^{+\dagger}(p,s')\} &= \{d_{TR}{}^{+}(q,s), b_{TR}{}^{+}(p,s')\} = 0
\end{aligned} \tag{H.79}$$

and where the spinors are

$$u_{TR}{}^{+}(p, s) = C^+R^+u_T(p,s) \tag{H.80}$$

$$v_{TR}{}^{+}(p, s) = C^+R^+v_T(p,s) \tag{H.81}$$

by section H.12 (eq. H.7).

The right-handed anti-commutation relation with the minus sign follows in particular because of the minus signs found earlier.

H.4.5 Interpretation of Tachyon Creation and Annihilation Operators

To properly discuss the physical interpretation of tachyon creation and annihilation operators we must first determine the Hamiltonian and momentum operators in terms of creation and annihilation operators.

The energy-momentum tensor density is the symmetrized version of

$$\mathfrak{T}^{\mu\nu} = \sum_i \partial\mathfrak{L}/\partial(\partial\chi_i/\partial x_\mu)\, \partial\chi_i/\partial x_\nu - g^{\mu\nu}\mathfrak{L} \tag{H.82}$$

where the sum over i is over the fields. The light-front hamiltonian is

$$H \equiv P^- = T^{+-} = \int dx^- d^2x\, \mathfrak{T}^{+-} \tag{H.83}$$

and the "momenta" are

$$P^+ = T^{++} = \int dx^- d^2x\, \mathfrak{T}^{++} \tag{H.84}$$

$$P^i = T^{+i} = \int dx^- d^2x\, \mathfrak{T}^{+i} \tag{H.85}$$

for i = 1,2.

The light-front, left-handed and right-handed tachyon lagrangian $\mathfrak{L}_T$ and its equations of motion imply

$$H = i2^{-1/2}\int dx^- d^2x\, [\psi_{TL}{}^{+\dagger}\partial^-\psi_{TL}{}^{+} - \partial^-\psi_{TL}{}^{+\dagger}\psi_{TL}{}^{+} + \psi_{TL}{}^{-\dagger}\partial^+\psi_{TL}{}^{-} - \partial^+\psi_{TL}{}^{-\dagger}\psi_{TL}{}^{-} - \\ - \psi_{TR}{}^{+\dagger}\partial^-\psi_{TR}{}^{+} + \partial^-\psi_{TR}{}^{+\dagger}\psi_{TR}{}^{+} - \psi_{TR}{}^{-\dagger}\partial^+\psi_{TR}{}^{-} + \partial^+\psi_{TR}{}^{-\dagger}\psi_{TR}{}^{-} + \text{mass terms}] \tag{H.86}$$

After substituting for the various fields we find the *independent fields* (which create the in and out particle states) have the hamiltonian terms:

$$H = \sum_{\pm s} \int d^2p dp^+ \, p^- [b_{TL}^{+\dagger}(p,s) b_{TL}^{+}(p,s) - d_{TL}^{+}(p,s) d_{TL}^{+\dagger}(p,s) - b_{TR}^{+\dagger}(p,s) b_{TR}^{+}(p,s) + d_{TR}^{+}(p,s) d_{TR}^{+\dagger}(p,s)] \quad \text{(H.87)}$$

$$= \sum_{\pm s} \int d^2p dp^+ \, p^- [b_{TL}^{+\dagger}(p,s) b_{TL}^{+}(p,s) + d_{TL}^{+\dagger}(p,s) d_{TL}^{+}(p,s) - b_{TR}^{+\dagger}(p,s) b_{TR}^{+}(p,s) - d_{TR}^{+\dagger}(p,s) d_{TR}^{+}(p,s)] \quad \text{(H.88)}$$

up to the usual infinite constants due to left-handed operator rearrangement and right-handed operator rearrangement that are discarded. Eq. H.88 is the basis for our particle interpretation of tachyon creation and annihilation operators based on Dirac's hole theory. Dirac hole theory as applied in light-front coordinates assumes all negative p^- ("energy") states are filled.

H.4.5.1 Left-Handed Tachyon Creation and Annihilation Operators

1. We identify $b_{TL}^{+\dagger}(p,s)$ and $d_{TL}^{+}(p,s)$ as creation operators for left-handed tachyons. $b_{TL}^{+\dagger}(p,s)$ creates a positive p^- ("energy") state and $d_{TL}^{+}(p,s)$ creates a negative p^- ("energy") state.

2. $b_{TL}^{+}(p,s)$ and $d_{TL}^{+\dagger}(p,s)$ are the corresponding annihilation operators for left-handed tachyons. $b_{TL}^{+}(p,s)$ annihilates a positive p^- ("energy") state and $d_{TL}^{+\dagger}(p,s)$ annihilates a negative p^- ("energy") state.

3. We assume Dirac hole theory holds for the left-handed tachyon vacuum with all negative energy states filled. There is no tachyon energy gap as there is for Dirac fermions. There is also the problem that the left-handed tachyon vacuum is not invariant under ordinary Lorentz transformations or Superluminal transformations. *However if we confine ourselves to light-front coordinates for computations no ambiguity can result and the Lorentz covariant quantities that we calculate, such as the S matrix, are well-defined.*

4. Using tachyon hole theory we identify $b_{TL}{}^{+}(p,s)$ and $d_{TL}{}^{+\dagger}(p,s)$ as annihilation operators for left-handed tachyons. $b_{TL}{}^{+}(p,s)$ annihilates a positive p^{-} ("energy") state and $d_{TL}{}^{+\dagger}(p,s)$ annihilates a negative p^{-} ("energy") state – thus creating a hole in the tachyon sea that we view as the creation of a positive p^{-} ("energy"), left-handed antitachyon. $d_{TL}{}^{+}(p,s)$ annihilates a positive p^{-} ("energy"), left-handed antitachyon.

H.4.5.2 Right-Handed Tachyon Creation and Annihilation Operators

The anti-commutation relations of right-handed tachyon creation and annhilation operators and the right-handed Hamiltonian terms have the "wrong" sign compared to corresponding Dirac operators and left-handed tachyon operators. This situation is completely analogous to the situation of time-like photons in the covariant formulation of quantum Electrodynamics.[209] In the case of time-like photons it was possible to introduce an indefinite metric (Gupta-Bleuler formulation), and then to use the subsidiary condition $\partial A^{\nu}/\partial x^{\nu} = 0$ to reduce the dynamics of QED to the transverse components. Thus the time-like photons were intermediate artifacts needed to have a manifestly covariant formulation while QED observables depended solely on the transverse components of the electromagnetic field.

In the present case of free tachyons, and in leptonic ElectroWeak Theory there is no evident "subsidiary condition" to eliminate the right-handed tachyon fields. But since the only manner in which the right-handed leptonic tachyon fields[210] interact is through mass terms, which can be easily 'integrated out", right-handed leptonic tachyon fields are removed from the observable part of the leptonic ElectroWeak Theory by their "lack of interaction" with left-handed fields.

In the case of quark ElectroWeak Theory right-handed tachyon quark fields have charge (–1/3) and thus experience an electromagnetic interaction as well as a Z interaction. However, since quarks are totally confined, right-handed tachyon quarks will not be able to continuously emit photons or Z's due to energy conservation and

---

[209] Bogoliubov (1959) pp. 130-136.

[210] The tachyon fields are provisionally assumed to be neutrino fields in the leptonic sector, and d, s and b quarks in the quark sector.

their confinement to bound states of fixed positive energy. Earlier, when we consider complex Lorentz group boosts, we will suggest that quarks may not consist of Dirac particles or tachyons of the type considered up to this point in this appendix. Rather they may be variants on Dirac particles and tachyons satisfying different dynamical equations. However, the preceding comments on quarks would still apply.

Thus right-handed tachyons are analogous to time-like photons – necessary theoretically but prevented from causing a negative energy disaster by the forms of their interactions. We discuss this subject in more detail in the following appendices.

H.4.6 Tachyon Feynman Propagator

In this section we develop the light-front propagator for tachyons. We begin with a subsection describing the light-front propagators of Dirac fields.

H.4.6.1 Dirac Field Light-Front Propagators

The light-front Feynman propagator for the ψ^+ field of a Dirac fermion is

$$iS^+_F(x,y)\gamma^0 = \theta(x^+ - y^+)<0|\psi^+(x)\psi^{+\dagger}(y)|0> - \theta(y^+ - x^+)<0|\psi^{+\dagger}(y)\psi^+(x)|0> \qquad (H.89)$$

and does not contain a non-covariant piece due to the projection operators:

$$\begin{aligned} iS^+_F(x,y) &= \int d^2p dp^+ \theta(p^+)[1/(2(2\pi)^3 p^+)]\{\theta(x^+ - y^+)[R^+(\not p + m)R^-]\, e^{-ip\cdot(x-y)} + \\ &\qquad + \theta(y^+ - x^+)[R^+(-\not p + m)R^-]e^{+ip\cdot(x-y)}\} \\ &= R^+ iS_F(x,y)R^- \end{aligned} \qquad (H.90)$$

where $S_F(x,y)$ is the usual Feynman propagator.

The light-front Feynman propagator for a *left-handed* Dirac field ψ^+ is

$$\begin{aligned} iS^+_{LF}(x,y) &= \int d^2p dp^+ \theta(p^+)[1/(2(2\pi)^3 p^+)]\{\theta(x^+ - y^+)[C^- R^+(\not p + m)R^- C^-]e^{-ip\cdot(x-y)} + \\ &\qquad + \theta(y^+ - x^+)[C^- R^+(-\not p + m)R^- C^-]e^{+ip\cdot(x-y)}\} \\ &= C^- R^+ iS_F(x,y)R^- C^- \end{aligned} \qquad (H.91)$$

H.4.6.2 Tachyon Field Light Front Propagators

Turning now to tachyons, the light-front Feynman propagator for the left-handed ψ_{TL}^{+} *tachyon* field is (using the previous Fourier expansion of the left-handed tachyon field):

$$\begin{aligned} iS^{+}_{TLF}(x,y) &= \theta(x^{+} - y^{+})\langle 0|\psi_{TL}^{+}(x)\psi_{TL}^{+\dagger}(y)\gamma^{0}|0\rangle - \theta(y^{+} - x^{+})\langle 0|\psi_{TL}^{+\dagger}(y)\gamma^{0}\psi_{TL}^{+}(x)|0\rangle \\ &= -i\int d^{2}p\,dp^{+}\theta(p^{+})N_{TL}^{+2}(2m|\mathbf{p}|)^{-1}C^{-}R^{+}\{\theta(x^{+} - y^{+})[(i\not{p} - m)\boldsymbol{\gamma}\cdot\mathbf{p}]e^{-ip\cdot(x-y)} + \\ &\qquad + \theta(y^{+} - x^{+})[(i\not{p} + m)\boldsymbol{\gamma}\cdot\mathbf{p}]e^{+ip\cdot(x-y)}\}R^{+}C^{-}\gamma^{0} \end{aligned}$$

If we define the on-shell momentum variable

$$p_{0}^{-} = (p_{0}^{1}p_{0}^{1} + p_{0}^{2}p_{0}^{2} - m^{2})/(2p_{0}^{+}),\ p_{0}^{+} = p^{+},\ p_{0}^{j} = p^{j}\ (\text{for } j = 1, 2),\ p_{\perp 0}^{2} = p_{0}^{j}p_{0}^{j}$$

and

$$\not{p}_{0} = p_{0}\cdot\gamma$$

then the above equation can be rewritten as

$$\begin{aligned} S^{+}_{TLF}(x,y) &= -C^{-}R^{+}\int d^{4}p[32\pi^{4}(p_{0}^{+}(p_{0}^{+} - p_{0}^{-}) + p_{0\perp}^{2})]^{-1}e^{-ip\cdot(x-y)}\{\theta(p^{+})(i\not{p}_{0} - m)\boldsymbol{\gamma}\cdot\mathbf{p}_{0}]/[p^{-} - \\ &\qquad - p_{0}^{-} + i\varepsilon] + \theta(-p^{+})(i\not{p}_{0} + m)\boldsymbol{\gamma}\cdot\mathbf{p}_{0}]/[p^{-} + p_{0}^{-} - i\varepsilon]\}R^{+}C^{-}\gamma^{0} \\ &= -\tfrac{1}{2}\, i\int d^{4}p(2\pi)^{-4}[C^{-}R^{+}(i\not{p} - m)\boldsymbol{\gamma}\cdot\mathbf{p}R^{+}C^{-}\gamma^{0}]e^{-ip\cdot(x-y)}[(p^{2} + m^{2} + i\varepsilon)(p^{+}(p^{+} - p^{-}) + p_{\perp}^{2}))]^{-1} \end{aligned}$$

and using $C^{-}R^{+}(i\not{p} - m)\boldsymbol{\gamma}\cdot\mathbf{p}R^{+}C^{-} = i\,C^{-}R^{+}(p^{+}(p^{+} - p^{-}) + p_{\perp}^{2})$ we find

$$iS^{+}_{TLF}(x,y) = \tfrac{1}{2}C^{-}R^{+}\gamma^{0}\int d^{4}p(2\pi)^{-4}\,p^{+}e^{-ip\cdot(x-y)}/(p^{2} + m^{2} + i\varepsilon) \qquad \text{(H.92)}$$

Similarly the light-front Feynman propagator for the right-handed ψ_{TR}^{+} tachyon field is

$$iS^{+}_{TRF}(x,y) = \theta(x^{+} - y^{+})\langle 0|\psi_{TR}^{+}(x)\psi_{TR}^{+\dagger}(y)\gamma^{0}|0\rangle - \theta(y^{+} - x^{+})\langle 0|\psi_{TR}^{+\dagger}(y)\gamma^{0}\psi_{TR}^{+}(x)|0\rangle$$

$$= -\frac{1}{2}C^{+}R^{+}\gamma^{0}\int d^{4}p(2\pi)^{-4}\, p^{+}e^{-ip\cdot(x-y)}/(p^{2} + m^{2} + i\varepsilon) \qquad \text{(H.93)}$$

where the relative minus sign between eqs. H.92 and H.93 is due to the relative minus signs of the Fouier component operator anti-commutation relations.

Thus we find *tachyon* pole terms in the tachyon propagators as one would expect.

H.5 Complex Space and 3-Momentum & Real-Valued Energy Fermions (Quarks)

In this section we will use L_C boosts to develop a wider set of dynamical equations for free spin ½ fermions with real-valued energy and complex-valued 3-momentum.[211] We defined L_C boosts with

$$\Lambda_C(\mathbf{v_c}) = \exp[i\omega\hat{\mathbf{w}}\cdot\mathbf{K}] \qquad \text{(H.94)}$$
$$\omega = (\omega_r^{\,2} - \omega_i^{\,2} + 2i\omega_r\omega_i\, \hat{\mathbf{u}}_r\cdot\hat{\mathbf{u}}_i)^{1/2} \qquad \text{(H.95)}$$
$$\hat{\mathbf{w}} = (\omega_r\hat{\mathbf{u}}_r + i\omega_i\hat{\mathbf{u}}_i)/\omega \qquad \text{(H.96)}$$
$$\hat{\mathbf{w}}\cdot\hat{\mathbf{w}} = \hat{\mathbf{u}}_r\cdot\hat{\mathbf{u}}_r = \hat{\mathbf{u}}_i\cdot\hat{\mathbf{u}}_i = 1 \qquad \text{(H.97)}$$
$$\mathbf{v_c} = \hat{\mathbf{w}}\tanh(\omega) \qquad \text{(H.98)}$$

H.5.1 L_C Spinor "Normal" Lorentz Boosts & More Spin ½ Particle Types

Spinor boost transformations were used in previous sections to develop the dynamical equations for Dirac fields and tachyon fields. In this section we will use L_C spinor boosts to generate additional fermion field dynamical equations.

The form of the L_C spinor boost transformation corresponding to the coordinate transformation is:

$$S_C(\omega, \mathbf{v_c}) = \exp(-i\omega\sigma_{0k}\hat{w}_k/2) = \exp(-\omega\gamma^0\boldsymbol{\gamma}\cdot\hat{\mathbf{w}}/2)$$
$$= \cosh(\omega/2)I + \sinh(\omega/2)\gamma^0\boldsymbol{\gamma}\cdot\hat{\mathbf{w}} \qquad \text{(H.99)}$$

---

[211] The complexon theory that we develop and use for quark dynamics in the Standard Model is not required. Our Standard Model could use Dirac fermion dynamics for the up-type quarks and tachyon dynamics for down-type quarks. We choose to use complexon dynamics for all quark types because they have an internal SU(3)-like structure suggestive of color SU(3). More importantly, their spin dynamics is different and thus may resolve the differences between theory and experiment – particularly for the deep inelastic parton spin-dependent structure functions.

The inverse transformation is

$$S_C^{-1}(\omega, \mathbf{v_c}) = \gamma^2\gamma^0 K^{-1} S_C^{\dagger} K \gamma^0\gamma^2 = \gamma^2\gamma^0 S_C^{\,T} \gamma^0\gamma^2 = \exp(\omega\gamma^0\boldsymbol{\gamma}\cdot\hat{\mathbf{w}}/2)$$
$$= \cosh(\omega/2)I - \sinh(\omega/2)\gamma^0\boldsymbol{\gamma}\cdot\hat{\mathbf{w}} \qquad \text{(H.100)}$$

where the superscript T denotes the transpose and K is the complex conjugation operator (that also appears in the time-reversal operator). Note that S_C is not unitary just as in previous cases considered in this appendix.

We now redo the development of spin ½ dynamical equations of motion of earlier sections for this more general case of complex ω and $\hat{\mathbf{w}}$. Again we apply a boost to a Dirac equation for a positive energy plane wave particle of mass m at rest:

$$0 = S_C(\omega, \mathbf{v_c}))(m\gamma^0 - m)e^{-imt}w(0)$$
$$= [mS_C\gamma^0 S_C^{-1} - m]e^{-imt} S_C w(0) \qquad \text{(H.101)}$$

where $S_C = S_C(\omega, \hat{\mathbf{w}})$. After some algebra

$$mS_C\gamma^0 S_C^{-1} = m[\cosh(\omega)\gamma^0 - \sinh(\omega)\boldsymbol{\gamma}\cdot\hat{\mathbf{w}}] \qquad \text{(H.102)}$$

H.5.1.1 Case 1: Parallel Real and Imaginary Relative Vectors

If the real and imaginary relative vectors parts of $\hat{\mathbf{w}}$, namely $\hat{\mathbf{u}}_r$ and $\hat{\mathbf{u}}_i$, are parallel, then $\hat{\mathbf{u}}_r\cdot\hat{\mathbf{u}}_i = 1$ and

$$\omega = \omega_r + i\omega_i \qquad \text{(H.103)}$$

Eq. H.102 can be re-expressed as

$$mS_C\gamma^0 S_C^{-1} = m[\cosh(\omega_r)\cos(\omega_i) + i\sinh(\omega_r)\sin(\omega_i)]\gamma^0 - m[\sinh(\omega_r)\cos(\omega_i) +$$
$$+ i\cosh(\omega_r)\sin(\omega_i)]\boldsymbol{\gamma}\cdot\hat{\mathbf{u}}_r \qquad \text{(H.104)}$$

or equivalently

$$mS_C\gamma^0 S_C^{-1} = \cos(\omega_i)\gamma\cdot p_r + i\sin(\omega_i)\gamma\cdot p_i \qquad \text{(H.105)}$$

where

$$p_r^0 = m\cosh(\omega_r) \qquad p_i^0 = m\sinh(\omega_r) \qquad \text{(H.106)}$$

and

$$\mathbf{p_r} = m\hat{\mathbf{u}}_r\sinh(\omega_r) \qquad \mathbf{p_i} = m\hat{\mathbf{u}}_r\cosh(\omega_r) \qquad \text{(H.107)}$$

If $\omega_i = 0$, then we recover the momentum space Dirac equation. If $\omega_i = \pi/2$, then we obtain the left-handed momentum space tachyon equation. Since the range of ω_i is $[0, \infty>$ (due to the cut along the real ω-plane axis) eq. H.105 corresponds to the results of the Left-Handed Lorentz boost part discussed earlier.

H.5.1.2 Case 2: Anti-Parallel Real and Imaginary Relative Vectors

If the real and imaginary relative vectors parts of $\hat{\mathbf{w}}$, $\hat{\mathbf{u}}_r$ and $\hat{\mathbf{u}}_i$, are anti-parallel $\hat{\mathbf{u}}_r = -\hat{\mathbf{u}}_i$, then $\hat{\mathbf{u}}_r\cdot\hat{\mathbf{u}}_i = -1$ and

$$\omega = \omega_r - i\omega_i \qquad \text{(H.108)}$$

We can then express eq. H.105 as

$$mS_C\gamma^0S_C^{-1} = m[\cosh(\omega_r)\cos(\omega_i) - i\sinh(\omega_r)\sin(\omega_i)]\gamma^0 - m[\sinh(\omega_r)\cos(\omega_i) - \\ - i\cosh(\omega_r)\sin(\omega_i)]\boldsymbol{\gamma}\cdot\hat{\mathbf{u}}_r \qquad \text{(H.109)}$$

or

$$mS_C\gamma^0S_C^{-1} = \cos(\omega_i)\gamma\cdot p_r - i\sin(\omega_i)\gamma\cdot p_i \qquad \text{(H.110)}$$

where

$$p_r^0 = m\cosh(\omega_r) \qquad p_i^0 = m\sinh(\omega_r) \qquad \text{(H.111)}$$

and

$$\mathbf{p_r} = m\hat{\mathbf{u}}_r\sinh(\omega_r) \qquad \mathbf{p_i} = m\hat{\mathbf{u}}_r\cosh(\omega_r) \qquad \text{(H.112)}$$

If $\omega_i = 0$, then we again recover the momentum space Dirac equation, If $\omega_i = \pi/2$, then we obtain the right-handed momentum space tachyon equation. (The range of ω_i is again $[0, \infty>$.)

Note: Since the matrix elements in the boost depend on $\gamma = (1 - \beta^2)^{-1/2}$ with a singularities at $\beta = \pm 1$, which in turn corresponds to $\omega = \pm\infty$, there is a branch cut along the ω axis in the complex ω-plane. Therefore we point out again the product of three Left-handed transformations is not equivalent to a Right-handed transformation.

H.5.1.3 Case 3: Complexons: A New Type of Particle with Perpendicular Real and Imaginary 3-Momenta

If the real and imaginary relative vectors parts of $\hat{\mathbf{w}}$, namely $\hat{\mathbf{u}}_r$ and $\hat{\mathbf{u}}_i$, are perpendicular, $\hat{\mathbf{u}}_r \cdot \hat{\mathbf{u}}_i = 0$, then

$$\omega = (\omega_r^2 - \omega_i^2)^{1/2} \tag{H.113}$$

Thus ω is either pure real ($\omega_r \geq \omega_i$) or pure imaginary ($\omega_r < \omega_i$).

The momentum space equation generated by the corresponding L_C spinor boost is

$$\{m \cosh(\omega)\gamma^0 - m \sinh(\omega)\boldsymbol{\gamma}\cdot(\omega_r\hat{\mathbf{u}}_r + i\omega_i\hat{\mathbf{u}}_i)/\omega - m\}e^{-ip\cdot x}w_c(p) = 0 \tag{H.114}$$

Defining the momentum 4-vector

$$p = (p^0, \mathbf{p}) \tag{H.115}$$

where

$$p^0 = m \cosh(\omega) \qquad \mathbf{p} = \mathbf{p}_r + i\mathbf{p}_i \tag{H.116}$$

$$\mathbf{p}_r = m\omega_r\hat{\mathbf{u}}_r \sinh(\omega)/\omega \qquad \mathbf{p}_i = m\omega_i\hat{\mathbf{u}}_i \sinh(\omega)/\omega \tag{H.117}$$

and

$$\mathbf{p}_r\cdot\mathbf{p}_i = 0 \tag{H.118}$$

then we obtain a positive energy Dirac-like equation with complex 3-momentum

$$[p\cdot\gamma - m]e^{-ip\cdot x}w_c(p) = 0$$

or, explicitly, (H.119)

$$[p^0\gamma^0 - (\mathbf{p}_r + i\mathbf{p}_i)\cdot\boldsymbol{\gamma} - m]e^{-ip\cdot x}w_c(p) = 0$$

with a complex 3-momentum $\mathbf{p}$ and the 4-momentum mass shell condition:

$$p^2 = p^{0\,2} - \mathbf{p}_r\cdot\mathbf{p}_r + \mathbf{p}_i\cdot\mathbf{p}_i = m^2 \tag{H.120}$$

Note

$$|\mathbf{v}| = |\mathbf{p}|/p^0 = [(\mathbf{p}_r + i\mathbf{p}_i)\cdot(\mathbf{p}_r + i\mathbf{p}_i)]^{1/2}/p^0 = \tanh(\omega) \tag{H.121}$$

and thus the Lorentz factor

$$\gamma = \cosh(\omega) \qquad \text{(H.122)}$$

Eq. H.119 is the momentum space equivalent of the wave equation

$$[i\gamma^0\partial/\partial t + i\boldsymbol{\gamma}\cdot(\nabla_r + i\nabla_i) - m]\psi_C(t, \mathbf{x}_r, \mathbf{x}_i) = 0 \qquad \text{(H.123)}$$

where

$$x_c = (t, \mathbf{x}_r - i\mathbf{x}_i) \qquad \text{(H.123a)}$$

and where the grad operators ∇_r and ∇_i are with respect to $\mathbf{x}_r$ and $\mathbf{x}_i$ respectively. Since $\hat{\mathbf{u}}_r\cdot\hat{\mathbf{u}}_i = 0$, we see that there is a subsidiary condition on the wave function

$$\nabla_r\cdot\nabla_i\ \psi_C(t, \mathbf{x}_r, \mathbf{x}_i) = 0 \qquad \text{(H.124)}$$

We will call the particles satisfying eqs.H.123 and H.124 *complexons*. In addition eq. H.118 implies the anti-commutation relation

$$\{\boldsymbol{\gamma}\cdot\mathbf{p}_r, \boldsymbol{\gamma}\cdot\mathbf{p}_i\} = 0 \qquad \text{(H.125)}$$

which in turn implies

$$\boldsymbol{\gamma}\cdot\nabla_r\boldsymbol{\gamma}\cdot\nabla_i\psi_C(t, \mathbf{x}_r, \mathbf{x}_i) = \boldsymbol{\gamma}\cdot\nabla_i\boldsymbol{\gamma}\cdot\nabla_r\psi_C(t, \mathbf{x}_r, \mathbf{x}_i) = 0 \qquad \text{(H.126)}$$

We note that eq. H.125 is covariant under the real Lorentz group and eq. H.126 can be easily put into covariant form since the difference of these 4-vectors squared is a real Lorentz group invariant: $[\gamma^0\partial/\partial t + \boldsymbol{\gamma}\cdot(\nabla_r + i\nabla_i)]^2 - [\gamma^0\partial/\partial t + i\boldsymbol{\gamma}\cdot(\nabla_r - i\nabla_i)]^2 = 4\nabla_r\cdot\nabla_i$.

Before considering a lagrangian formulation and the Fourier operator representation of $\psi_C(t, \mathbf{x}_r, \mathbf{x}_i)$ we will define the spinors and associated real and imaginary spin operators.

The spinor generated from a spin up Dirac spinor at rest by a complex boost is

$$w_c(p) = S_C(p)w(0) = [\cosh(\omega/2)I + \sinh(\omega/2)\gamma^0\boldsymbol{\gamma}\cdot\hat{\mathbf{w}}]w(0) \qquad \text{(H.127)}$$

Following a procedure similar to section H.12 (which the reader may wish to examine first) we define four spinors for Dirac particles at rest:

$$w^k(0) = \begin{bmatrix} \delta_{1k} \\ \delta_{2k} \\ \delta_{3k} \\ \delta_{4k} \end{bmatrix} \qquad \text{(H.282)}$$

where Kronecker deltas appear in the brackets. Then by applying eq. H.127 to the spinors defined by eq. H.282 we find the L_C spinors

$$S_C w^k(0) = w_{Cr}{}^k(p) + i w_{Ci}{}^k(p) \qquad \text{(H.128)}$$

where

$$\begin{aligned} S_{Cr} &= \cosh(\omega/2)I + (\omega_r/\omega)\sinh(\omega/2)\gamma^0\boldsymbol{\gamma}\cdot\hat{\mathbf{u}}_r \\ &= [(m + E)/(2m)]^{1/2}I + [m(m + E)]^{-1/2}\gamma^0\boldsymbol{\gamma}\cdot\mathbf{p}_r = aI + b\gamma^0\boldsymbol{\gamma}\cdot\mathbf{p}_r \end{aligned} \qquad \text{(H.129)}$$

Thus the "real" spinors $w_{Cr}{}^k(p)$ are the columns of

$$S_{Cr} = \begin{bmatrix} \underline{w_{Cr}{}^1(p)} & \underline{w_{Cr}{}^2(p)} & \underline{w_{Cr}{}^3(p)} & \underline{w_{Cr}{}^4(p)} \\ a & 0 & bp_{r\,z} & bp_{r-} \\ 0 & a & bp_{r+} & -bp_{r\,z} \\ bp_{r\,z} & bp_{r-} & a & 0 \\ bp_{r+} & -bp_{r\,z} & 0 & a \end{bmatrix} \qquad \text{(H.130)}$$

where $p_{r\pm} = p_{r\,x} \pm ip_{r\,y}$. The "imaginary" spinors are the columns of

$$S_{Ci} = (\omega_i/\omega)\sinh(\omega/2)\gamma^0\boldsymbol{\gamma}\cdot\hat{\mathbf{u}}_i = [m(m + E)]^{-1/2}\gamma^0\boldsymbol{\gamma}\cdot\mathbf{p}_i = b\gamma^0\boldsymbol{\gamma}\cdot\mathbf{p}_i \qquad \text{(H.131)}$$

$$S_{Ci} = \begin{array}{cccc} \underline{w_{Ci}^1(p)} & \underline{w_{Ci}^2(p)} & \underline{w_{Ci}^3(p)} & \underline{w_{Ci}^4(p)} \end{array}$$
$$S_{Ci} = \begin{bmatrix} 0 & 0 & bp_{i\,z} & bp_{i-} \\ 0 & 0 & bp_{i+} & -bp_{i\,z} \\ bp_{i\,z} & bp_{i-} & 0 & 0 \\ bp_{i+} & -bp_{i\,z} & 0 & 0 \end{bmatrix} \quad \text{(H.132)}$$

where $p_{i\pm} = p_{i\,x} \pm ip_{i\,y}$.

Eqs. H.127 through H.132 imply that the wave function solution of eq. H.123, subject to the subsidiary condition eq. H.124, is[212, 213]

$$\psi_C(x_r, x_i) = \sum_{\pm s} \int d^3p_r d^3p_i\, N_C(p)\delta(\mathbf{p_r}\cdot\mathbf{p_i}/m^2)[b_C(p,s)u_C(p, s)e^{-i(p\cdot x + p^*\cdot x^*)/2} + $$
$$+ d_C^{\dagger}(p,s)v_C(p, s)e^{+i(p\cdot x + p^*\cdot x^*)/2}] \quad \text{(H.133)}$$

where $\mathbf{p} = \mathbf{p_r} + i\mathbf{p_i}$ (eq. 3.95), $\mathbf{x} = \mathbf{x_r} - i\mathbf{x_i}$, $p\cdot x = p^0x^0 - \mathbf{p}\cdot\mathbf{x}$, and where we use

$$(p\cdot x + p^*\cdot x^*)/2 = p^0x^0 - \mathbf{p_r}\cdot\mathbf{x_r} - \mathbf{p_i}\cdot\mathbf{x_i} \quad \text{(H.134)}$$

in the exponentials in order to avoid divergences that would appear in the calculation of the equal-time commutator, the Feynman propagator and other quantities of interest after second quantization. Note that

$$(\nabla_\mathbf{r} + i\nabla_\mathbf{i})e^{-i(p\cdot x + p^*\cdot x^*)/2} = i(\mathbf{p_r} + i\mathbf{p_i})e^{-i(p\cdot x + p^*\cdot x^*)/2} \quad \text{(H.135)}$$

---

[212] Note that when $|\mathbf{p_i}| \geq |\mathbf{p_r}|$ (for imaginary $\omega = (\omega_r^2 - \omega_i^2)^{1/2}$) the 3-momentum becomes imaginary $\mathbf{p}\cdot\mathbf{p} < 0$. However, since we will be identifying confined quarks with this type of particle – much modified by a confining color quark interaction – the issue of an imaginary 3-momentum in the hypothetical free quark case becomes moot. We note the energy gap between positive and negative energy states disappears so E = 0 is possible. Thus real Lorentz transformations can mix positive and negative energy states. The solution is to do all calculations in the light-front frame as we do for tachyons. Then the mixing issue is resolved. In the present case we second quantize on the "time-front" for illustrative purposes.

[213] We scale $\mathbf{p_r}\cdot\mathbf{p_i}$ with m^2 in the delta function for convenience. All fermions have at least a minimal mass – the mass of the qube.

and

$$(\nabla_{\mathbf{r}} + i\nabla_{\mathbf{i}})e^{-ip^*\cdot x^*} = 0 \tag{H.136}$$

for all p.
The wave function's conjugate (the hermitean conjugate modified by letting $\mathbf{x_i} \rightarrow -\mathbf{x_i}$ in addition to hermitean conjugation) is

$$\psi_C^{\dagger}(x) = \psi_C^{\dagger}(x_r, -x_i) = \sum_{\pm s} \int d^3p_r d^3p_i\, \delta(\mathbf{p_r}\cdot\mathbf{p_i}/m^2) N_C(p^*)\cdot$$
$$\cdot[b_C^{\dagger}(p^*,s)u_C^{\dagger}(p^*,s)e^{+i(p\cdot x^* + p^*\cdot x)/2} + d_C(p^*,s)v_C^{\dagger}(p^*,s)e^{-i(p\cdot x^* + p^*\cdot x)/2}] \tag{H.137}$$

where $\mathbf{p} = \mathbf{p_r} + i\mathbf{p_i}$, $\mathbf{x} = \mathbf{x_r} - i\mathbf{x_i}$, $p\cdot x = p^0x^0 - \mathbf{p}\cdot\mathbf{x}$, and † indicates hermitean hermitean conjugation.

The spinors are

$$\begin{aligned}
u_C(p, s) &= S_C(p)w^1(0) \\
u_C(p, -s) &= S_C(p)w^2(0) \\
v_C(p, s) &= S_C(p)w^3(0) \\
v_C(p, -s) &= S_C(p)w^4(0) \\
u_C^{\dagger}(p^*, s) &= w^{1T}(0)S_C^{\dagger}(p^*) = w^{1T}(0)S_C(p) \\
u_C^{\dagger}(p^*, -s) &= w^{2T}(0)S_C^{\dagger}(p^*) = w^{2T}(0)S_C(p) \\
v_C^{\dagger}(p^*, s) &= w^{3T}(0)S_C^{\dagger}(p^*) = w^{3T}(0)S_C(p) \\
v_C^{\dagger}(p^*, -s) &= w^{4T}(0)S_C^{\dagger}(p^*) = w^{4T}(0)S_C(p)
\end{aligned} \tag{H.138}$$

with the superscript "T" indicating the transpose. Note that

$$S_C^{\dagger}(p^*) = [S_C(p^*)]^{\dagger} = S_C(p) \tag{H.139}$$

The normalization factor $N_C(p)$ is

$$N_C(p) = [2m/((2\pi)^6p^0)]^{1/2} \tag{H.140}$$

Since $\mathbf{p_r} = \mathbf{p_i} = 0$ in the particle rest frame prior to the complex group boost, the boosted particle spin 4-vector s^μ satisfies

$$s^\mu p_r^\mu = s^\mu p_i^\mu = 0 \tag{H.141}$$

Note that s^μ is itself complex[214] and, if the spin points in the z-direction prior to the complex boost, then the boosted s^μ has the form

$$s^\mu = (-\sinh(\omega)\hat{w}_z\,,\,(0,0,1) + (\cosh(\omega) - 1)\hat{w}_z\hat{\mathbf{w}}) \tag{H.142}$$

with $\hat{\mathbf{w}}$ defined earlier: $\hat{\mathbf{w}} = (\omega_r\hat{\mathbf{u}}_r + i\omega_i\hat{\mathbf{u}}_i)/\omega = \mathbf{p}/(m\sinh(\omega))$.

H.5.1.4 A Global SU(3) Symmetry Revealed

Before proceeding to consider the second quantization of this case, we will consider a global SU(3) symmetry implicit in the previous equations. The defining property of the group SU(3) is that it preserves the invariance of inner products of complex 3-vectors of the form:

$$u^*\cdot v = u^{1*}v^1 + u^{2*}v^2 + u^{3*}v^3 \tag{H.143}$$

If we examine the dynamical equation eq. H.123 we see that the differential operator is invariant under an SU(3) transformation U (using $\nabla_c = (\nabla_c{}^*)^* = \mathbf{D}_c{}^*$)

$$[i\gamma^0\partial/\partial t + i\mathbf{D}_c{}^*\cdot\gamma - m] = [i\gamma^0\partial/\partial t + i\mathbf{D}_c{}'^*\cdot\gamma' - m] \tag{H.144}$$

where

$$\mathbf{D}_c{}^* = \nabla_c = \nabla_r + i\nabla_i$$

and

$$\gamma'^a = U^{ab}\gamma'^b$$
$$D_c{}'^{*a} = D_c{}'^{*b}U^{\dagger ab}$$

[214] This feature of partons, which is not present in ordinary Dirac particles, might be the source of the discrepancies between theory and experiment in deep inelstic parton spin physics which is based on conventional real parton spins.

where U is a global SU(3) transformation and $U^\dagger = U^{-1}$. By theorem[215] all 4×4 γ matrices such as $\boldsymbol{\gamma}'$ are equivalent up to a unitary transformation V. Thus $V^\dagger\boldsymbol{\gamma}'V = \boldsymbol{\gamma}$ and eq. H.144 is equivalent to

$$[i\gamma^0\partial/\partial t + i\mathbf{D}_c{}^*\cdot\boldsymbol{\gamma} - m] = [i\gamma^0\partial/\partial t + i\mathbf{D}_c{}'^*\cdot\boldsymbol{\gamma} - m] \qquad \text{(H.145)}$$

$$= [i\gamma^0\partial/\partial t + i\boldsymbol{\nabla}_c{}'\cdot\boldsymbol{\gamma} - m]$$

where $\boldsymbol{\nabla}_c{}'_a = U^{ab}\boldsymbol{\nabla}_{cb}$, This demonstrates that eq. H.123 is invariant under an SU(3) transformation if

$$\psi_C(t, \mathbf{x}_c) = \psi_C(t, U\mathbf{x}_c) = \psi_C{}'(t, \mathbf{x}_c{}') \qquad \text{(H.146)}$$

where $\psi_C(t, \mathbf{x}_c) \equiv \psi_C(t, \mathbf{x}_r, \mathbf{x}_i)$.

The subsidiary condition eq. H.124 can be seen to transform as

$$\boldsymbol{\nabla}_r\cdot\boldsymbol{\nabla}_i\, \psi_C(t, \mathbf{x}_c) = \boldsymbol{\nabla}_r{}^*\cdot\boldsymbol{\nabla}_i\, \psi_C(t, \mathbf{x}_c) = \boldsymbol{\nabla}_r{}'^*\cdot\boldsymbol{\nabla}_i{}'\psi_C{}'(t, \mathbf{x}_c{}') = 0 \qquad \text{(H.147)}$$

under an SU(3) rotation. The invariance of the orthogonality condition is preserved.

The wave function (eq. H.123) transforms in the following way under the SU(3) transformation U. If we define

$$q^{*\mu} = (q^0, \mathbf{q}^*) = (p^0, \mathbf{p}_r + i\mathbf{p}_i) = (p^0, \mathbf{p}) = p^\mu \qquad \text{(H.148)}$$

then eq. H.133 can be rewritten in an invariant form under a SU(3) transformation:

$$\psi_C(x) = \sum_{\pm s} \int d^3q_r d^3q_i\, N_C(p^0)\delta(\mathbf{q}_r{}^*\cdot\mathbf{q}_i/m^2)[b_C(q^*,s)u_C(q^*,s)e^{-i(q^*.x + q.x^*)/2} +$$
$$+ d_C{}^\dagger(q^*,s)v_C(q^*,s)e^{+i(q^*.x + q.x^*)/2}] \qquad \text{(H.149)}$$

---

[215] R. H. Good, Rev. Mod. Phys., **27**, 187 (1955).

where $x = x_c$ subject to an examination of the transformation properties of the fourier coefficients and spinors. Note both terms in each exponential are separately invariant under global SU(3). (Note also $\mathbf{q}_r^* = \mathbf{q}_r$ since $\mathbf{q}_r$ is real.)

From the form of S_C above it is clear that an argument similar to that for the dynamical equations shows S_C is invariant under an SU(3) transformation and thus their spinors are also invariant under SU(3) transformations. The fourier coefficients, if second quantized in a direct generalization of the usual manner, have covariant anti-commutation relations under an SU(3) transformation. For example

$$\{b_C(q,s), b_C^{\dagger}(q'^*,s')\} = \delta_{ss'}\delta^3(q_r - q'_{r'})\delta^3(q_i - q'_{i'}) \tag{H.150}$$

Under an SU(3) transformation, z = Uq and z' = Uq', the right side of eq. H.150 transforms to

$$\delta^3(q_r - q'_{r'})\delta^3(q_i - q'_{i'}) \to \delta^3(z_r - z'_{r'})\delta^3(z_i - z'_{i'})/|\partial(q)/\partial(z)| = \delta^3(z_r - z'_{r'})\delta^3(z_i - z'_{i'}) \tag{H.151}$$

where

$$|\partial(q)/\partial(z)| = |\partial(q_r^1, q_r^2, q_r^3, q_i^1, q_i^2, q_i^3)/\partial(z_r^1, z_r^2, z_r^3, z_i^1, z_i^2, z_i^3)| = 1 \tag{H.152}$$

is the Jacobian of the transformation U. Thus the fourier coefficients transform trivially under SU(3). For example,

$$b_C(q^*,s) \to b_C(z^*,s \tag{H.153}$$

Since the integrand transforms as

$$\int d^3q_r d^3q_i \to \int d^3z_r d^3z_i\, |\partial(q)/\partial(z)| = \int d^3z_r d^3z_i \tag{H.154}$$

the wave function $\psi_C(t, \mathbf{x})$ transforms as an SU(3) scalar up to an inessential unitary transformation V of γ matrices: $\psi_C(t, \mathbf{x}) \rightarrow V\psi_C(t, \mathbf{x})$.[216]

H.5.1.5 Global SU(3) Spin ½ Complexon Fields

Having uncovered an SU(3) symmetry in the scalar field equations of Case 3A the generalization of the scalar field equations to the 3 representation of SU(3) is direct:

$$\psi_C{}^a(x) = \sum_{\pm s} \int d^3p_r d^3p_i\, N_C(p)\delta(\mathbf{p}_r\cdot\mathbf{p}_i/m^2)[b_C(p,a,s)u_C{}^a(p, s)e^{-i(p\cdot x + p^*\cdot x^*)/2} + $$
$$+ d_C^\dagger(p,a,s)v_C{}^a(p, s)e^{+i(p\cdot x + p^*\cdot x^*)/2}] \qquad \text{(H.155)}$$

where $x = x_c$ for a = 1,2, 3 with $u_C{}^a(p, s)$ and $v_C{}^a(p, s)$ being the product a spinor of type eq. H.138 and a 3 element column vector c^a with b^{th} element

$$b^a(b) = \delta^{ab} \qquad \text{(H.156)}$$

Under a global SU(3) transformation U the 3 complexon wave functions transform as

$$\psi_C{}'^a(x) = U^{ab}\psi_C{}^b(x) \qquad \text{(H.157)}$$

In a subsequent discussion we will extend the global SU(3) symmetry described in these subsections to be color local SU(3) upon the introduction of the Yang-Mills color gluon interaction.

H.5.1.6 Lagrangian Formulation and Second Quantization of Complexons

In this subsection we will outline the canonical quantization of SU(3) singlet complexons with the quantum field equation

---

[216] The spinors $u_C(q^*,s)$ and $v_C(q^*,s)$ are unchanged up to a unitary transformation of the γ matrices ($V^\dagger\gamma' V = \gamma$). Thus the term $(U\mathbf{w})^*\cdot\gamma' = \mathbf{w}^*\cdot V\gamma V^\dagger \equiv \mathbf{w}^*\cdot\gamma$ in the expressions for the $u_C(q^*,s)$ and $v_C(q^*,s)$ spinors.

$$[i\gamma^0\partial/\partial t + i\boldsymbol{\gamma}\cdot(\nabla_r + i\nabla_i) - m]\psi_C(t, \mathbf{x_r}, \mathbf{x_i}) = 0 \tag{H.158}$$

and subsidiary condition

$$\nabla_r\cdot\nabla_i\, \psi_C(t, \mathbf{x_r}, \mathbf{x_i}) = 0 \tag{H.159}$$

We begin with the Lagrangian density

$$\mathcal{L} = \bar{\psi}_C(i\gamma^\mu D_\mu - m)\psi_C(x) \tag{H.160}$$

where $\bar{\psi}_C = \psi_C^{\dagger}\gamma^0$:

$$\psi_C^{\dagger} = [\psi_C(\mathbf{x_r}, \mathbf{x_i})]^{\dagger}\,|_{\mathbf{x_i} = -\mathbf{x_i}} \tag{H.161}$$

$$D_0 = \partial/\partial x^0$$
$$D_k = \partial/\partial x^k + i\,\partial/\partial x_i^k \tag{H.162}$$

with $x^k = x_r^k$ for k = 1, 2, 3. The invariant action (under real Lorentz transformations) is

$$I = \int d^7x\mathcal{L} \tag{H.163}$$

It is easy to show that the action is real

$$I^* = I \tag{H.164}$$

in a manner similar to the case considered in section H.12 due to the form of $\psi_C^{\dagger}$ in eq. H.161. (One has to change the integration over $\mathbf{x_i}$ to $-\mathbf{x_i}$ after taking the complex conjugate of I and performing manipulations similar to those in section H.12.)

The conjugate momentum is

$$\pi_{Ca} = \partial\mathcal{L}/\partial\dot{\psi}_{Ca} \equiv \partial\mathcal{L}/\partial(\partial\psi_{Ca}/\partial x^0) = i\psi_{C\,a}^{\dagger} \tag{H.165}$$

where a is a spinor index. It yields the non-zero anti-commutation relation

$$\{\psi_{C\,a}^{\dagger}(x), \psi_{Cb}(y)\} = \delta_{ab}\,\delta^3(x_r - y_r)\delta^3(x_i - y_i) \tag{H.166}$$

where x and y are complex. However we will see that the constraint eq. H.159 is required. So the correct anti-commutator turns out to be

$$\{\psi_{C}{}^{\dagger}{}_{a}(x), \psi_{Cb}(y)\} = -\delta_{ab}\delta'(\nabla_r\cdot\nabla_i/m^2)[\delta^3(x_r - y_r)\delta^3(x_i - y_i)] \tag{H.167}$$

where all ∇_r and ∇_i are ∇ derivatives with respect to x, and where $\delta'(\nabla_r\cdot\nabla_i)$ is the derivative of a delta function with the argument being differential operators such as those in eq. H.159. The minus sign is due to the presence of a *derivative* of a delta-function and is not an issue.

The hamiltonian density is

$$\mathcal{H} = \pi_C\dot{\psi}_C - \mathcal{L} = \psi_C{}^{\dagger}(-i\boldsymbol{\alpha}\cdot\mathbf{D} + \beta m)\psi_C \tag{H.168}$$

and the (unsymmetrized) energy-momentum tensor is

$$\mathcal{T}_{\mu\nu} = -\, g_{\mu\nu}\,\mathcal{L} + \partial\mathcal{L}/\partial(D^{\mu}\psi_C)D_{\nu}\psi_C \tag{H.169}$$

The conserved energy and momentum are

$$P^0 = H = \int d^3x_r d^3x_i\; \mathcal{T}^{00} = \int d^3x_r d^3x_i\; \mathcal{H} \tag{H.170}$$

and

$$P^i = \int d^3x_r d^3x_i\; \mathcal{T}^{0i} \tag{H.171}$$

We now proceed to establish the canonical anti-commutation relations. First, the second quantization of the complexon field uses the above fourier coefficient anti-commutation relations (suitably rewritten):

$$\begin{aligned}
&\{b_C(p,s), b_C{}^{\dagger}(p'^*,s')\} = \delta_{ss'}\delta^3(\mathbf{p}_r - \mathbf{p'}_{r'})\delta^3(\mathbf{p}_i + \mathbf{p'}_{i'})\\
&\{d_C(p,s), d_C{}^{\dagger}(p'^*,s')\} = \delta_{ss'}\delta^3(\mathbf{p}_r - \mathbf{p'}_{r'})\delta^3(\mathbf{p}_i + \mathbf{p'}_{i'})\\
&\{b_C(p,s), b_C(p'^*,s')\} = \{d_C(p,s), d_C(p'^*,s')\} = 0\\
&\{b_C{}^{\dagger}(p,s), b_C{}^{\dagger}(p'^*,s')\} = \{d_C{}^{\dagger}(p,s), d_C{}^{\dagger}(p'^*,s')\} = 0\\
&\{b_C(p,s), d_C{}^{\dagger}(p'^*,s')\} = \{d_C(p,s), b_C{}^{\dagger}(p'^*,s')\} = 0
\end{aligned} \tag{H.172}$$

$$\{b_C^\dagger(p,s), d_C^\dagger(p'^*,s')\} = \{d_C(p,s), b_C(p'^*,s')\} = 0$$

The delta-function arguments $\delta^3(\mathbf{p}_i + \mathbf{p'}_{i'})$ above have a positive sign in order to obtain $\delta^3(\mathbf{x}_i - \mathbf{y}_i)$ in the field anti-commutator eq. H.167.

The spinors, eq. H.138, satisfy

$$\sum_{\pm s} u_\alpha(p, s)\bar{u}_\beta(p^*, s) = (2m)^{-1}(\not{p} + m)_{\alpha\beta} \tag{H.173}$$

$$\sum_{\pm s} v_\alpha(p, s)\bar{v}_\beta(p^*, s) = (2m)^{-1}(\not{p} - m)_{\alpha\beta}$$

remembering

$$\bar{u}_C(p^*,s) = w^{1T}(0)S_C(p)\gamma^0 = w^{1T}(0)[\cosh(\omega/2)I + \sinh(\omega/2)\gamma^0\boldsymbol{\gamma}\cdot\hat{\mathbf{w}}]\gamma^0 \tag{H.174}$$

by eqs. H.137 since $\hat{\mathbf{w}}^{**} = \hat{\mathbf{w}}$.

We will now evaluate the equal-time anti-commutation relation using eqs. H.136 and H.137:

$$\{\psi_{C\,a}^\dagger(x), \psi_{Cb}(y)\} = \sum_{\pm s,\, s'} \int d^3p_r d^3p_i\, d^3p'_r d^3p'_i\, \delta(\mathbf{p}_r\cdot\mathbf{p}_i/m^2)\delta(\mathbf{p'}_r\cdot\mathbf{p'}_i/m^2)\, N_C(p')N_C(p)\cdot$$
$$\cdot[\{b_C^\dagger(p^*,s)u_{Ca}^\dagger(p^*,s)e^{+i(p.x^* + p^*.x)/2}, b_C(p',s')u_{Cb}(p', s')e^{-i(p'.y + p'^*.y^*)/2}\} +$$
$$+ \{d_C(p^*,s)v_{Ca}^\dagger(p^*,s)e^{-i(p.x^* + p^*.x)/2}, d_C^\dagger(p',s')v_{Cb}(p', s')e^{+i(p'.y + p'^*.y^*)/2}\}]$$

$$= \int d^3p_r d^3p_i\, N_C^2(p)[\delta(\mathbf{p}_r\cdot\mathbf{p}_i/m^2)]^2[((\not{p} + m)\gamma^0)_{ba}e^{+i(p.x^* + p^*.x)/2 - i(p^*.y + p.y^*)/2} +$$
$$+((\not{p} - m)\gamma^0)_{ba}e^{-i(p.x^* + p^*.x)/2 + i(p^*.y + p.y^*)/2}]/(2m)$$

Next we use eq. H.140 and the identity

$$[\delta(x - y)]^2 = -\tfrac{1}{2}\,\delta'(x - y) \equiv -\tfrac{1}{2}\, d\delta(x - y)/dx \tag{H.175}$$

which can be derived from the step function identity $\theta(x - y) = [\theta(x - y)]^2$ to obtain

$$\{\psi_{C}{}^{\dagger}{}_{a}(x),\psi_{Cb}(y)\} = -\tfrac{1}{2}\int d^3p_r d^3p_i N_C{}^2(p)\delta'(\mathbf{p}_r\cdot\mathbf{p}_i/m^2)[((\not{p}+m)\gamma^0)_{ba}e^{-ipr\cdot(xr-yr)+ipi\cdot(xi-yi)} + $$
$$+ ((\not{p}-m)\gamma^0)_{ba}e^{+ipr\cdot(xr-yr)-ipi\cdot(xi-yi)}]/(2m)$$

$$= -\tfrac{1}{2}\delta_{ba}\int d^3p_r d^3p_i N_C{}^2(p)\delta'(\mathbf{p}_r\cdot\mathbf{p}_i/m^2)p^0 e^{-ipr\cdot(xr-yr)+ipi\cdot(xi-yi)}/m$$

$$= -\delta_{ab}\,\delta'(\nabla_r\cdot\nabla_i/m^2)[\delta^3(x_r-y_r)\delta^3(x_i-y_i)] \qquad (H.176)$$

The grad operators, ∇_r and ∇_i, are derivatives are with respect to real and imaginary x in the Dirac delta functions. The factor[217] $\delta'(\nabla_r\cdot\nabla_i)$ expresses the orthogonality constraint in coordinate space on the momenta. It is analogous to the transversality constraint on the electromagnetic vector potential commutator:

$$[\pi_A{}^j(x), A_k(y)] = -i\,\delta^{tr}{}_{jk}(x-y) \qquad (H.177)$$

$$\delta^{tr}{}_{jk}(x-y) = (\delta_{jk} - \partial_j\partial_k/\nabla^2)\,\delta^3(x-y) \qquad (H.178)$$

where $\partial_k = \partial/\partial x_k$.

H.5.1.7 Complexon Feynman Propagator

The complexon Feynman propagator for ψ_C is[218]

$$iS_C(x, y) = \theta(x^0-y^0)<0|\psi_C(x)\psi_C{}^{\dagger}(y)\gamma^0|0> - \theta(y^0-x^0)<0|\psi_C{}^{\dagger}(y)\gamma^0\psi_C(x)|0> \qquad (H.179)$$
$$= \int d^3p_r d^3p_i N_C{}^2(p)[\delta(\mathbf{p}_r\cdot\mathbf{p}_i/m^2)]^2\{\theta(x^0-y^0)(\not{p}+m)e^{-i(p^*\cdot(x-y)+p\cdot(x^*-y^*))/2} -$$
$$-\theta(y^0-x^0)(\not{p}-m)e^{+i(p^*\cdot(x-y)+p\cdot(x^*-y^*))/2}\}/(2m)$$

---

217 A derivative of a delta function containing grad operators.

218 The reader, upon seeing the additional integrations $\int d^3p_i$ might suspect that they would ultimately lead to divergence issues in perturbation theory calculations. However the $\delta'(\mathbf{p}_r\cdot\mathbf{p}_i/m^2)$ term compensates in part for the additional integrations by four powers of momentum since $\delta'(\mathbf{p}_r\cdot\mathbf{p}_i/m^2) = (|\mathbf{p}_r||\mathbf{p}_i|/m^2)^{-2}\delta'(\cos\theta_{ri})$ where θ_{ri} is the angle between the momenta. As a result only 2 fermion and 3 fermion loop integrations would potentially have difficulties if one uses the conventional approach to perturbation theory. If one uses the approach of Blaha (2003) and (2005a) then there are no divergences.

$$= -(4\pi)^{-1}\int dp^0 d^3p_r d^3p_i (2\pi)^{-6}\delta'(\mathbf{p_r}\cdot\mathbf{p_i}/m^2)(\not{p}+m)e^{-i(p^*\cdot(x-y)+p\cdot(x^*-y^*))/2}/(p^2-m^2+i\varepsilon)$$

$$= -\tfrac{1}{2}\int dp^0 d^3p_r d^3p_i\, \delta'(\mathbf{p_r}\cdot\mathbf{p_i}/m^2)(\not{p}+m)(2\pi)^{-7}\exp[-ip^0(x^0-y^0)+$$
$$+\, i\mathbf{p_r}\cdot(\mathbf{x_r}-\mathbf{y_r}) - i\mathbf{p_i}\cdot(\mathbf{x_i}-\mathbf{y_i})]/(p^2-m^2+i\varepsilon) \qquad \text{(H.180)}$$

The integral can be written in the form:

$$I = \int dp^0 d^3p_r d^3p_i \delta'(\mathbf{p_r}\cdot\mathbf{p_i}/m^2)(\not{p}+m)\exp[-ip^0(x^0-y^0)+i\mathbf{p_r}\cdot(\mathbf{x_r}-\mathbf{y_r})-i\mathbf{p_i}\cdot(\mathbf{x_i}-\mathbf{y_i})]/(p^2-m^2+i\varepsilon)$$
$$= \int d^4p_r dM^2\delta'(\nabla_r\cdot\nabla_i/m^2)(p^0\gamma^0-(\mathbf{p_r}-\nabla_i)\cdot\boldsymbol{\gamma}+m)\exp[-ip^0(x^0-y^0)+i\mathbf{p_r}\cdot(\mathbf{x_r}-\mathbf{y_r})]\cdot$$
$$\cdot J(\mathbf{x_i}-\mathbf{y_i},M^2)/(p_r^2-M^2+i\varepsilon) \qquad \text{(H.181)}$$

where $p_r^2 = p^{0\,2} - \mathbf{p_r}\cdot\mathbf{p_r}$ and

$$J(\mathbf{x_i}-\mathbf{y_i}, M^2) = (2\pi)^{-3}\int d^3p_i\,\delta(M^2+\mathbf{p_i}^2-m^2)\exp[-i\mathbf{p_i}\cdot(\mathbf{x_i}-\mathbf{y_i})] \qquad \text{(H.182)}$$
$$= (2\pi)^{-2}|\mathbf{x_i}-\mathbf{y_i}|^{-1}\theta(m^2-M^2)\sin((m^2-M^2)^{1/2}|\mathbf{x_i}-\mathbf{y_i}|)$$

The complexon Feynman propagator can be rearranged into the form of a spectral integral:

$$iS_C(x, y) = -\int dM\,(i\gamma^0\partial/\partial x^0 - i(\nabla_r - i\nabla_i)\cdot\boldsymbol{\gamma}+m)\delta'(\nabla_r\cdot\nabla_i/m^2)J(\mathbf{x_i}-\mathbf{y_i}, M^2)\Delta_F(x-y, M)$$
$$\text{(H.183)}$$

where

$$\Delta_F(x-y, M) = (2\pi)^{-4}\int d^4p_r\,\exp[-ip^0(x^0-y^0)+i\mathbf{p_r}\cdot(\mathbf{x_r}-\mathbf{y_r})]/(p_r^2-M^2+i\varepsilon)$$
$$\text{(H.184)}$$

H.5.1.8 Case 4: Left-handed Tachyon Complexons

In this case $\hat{\mathbf{u}}_r \cdot \hat{\mathbf{u}}_i = 0$ again. However we add an imaginary term to ω to obtain a manifest Left-handed L_C boost[219]

$$\Lambda_{CL}(\mathbf{v}_c) = \exp[i(\omega + i\pi/2)\hat{\mathbf{w}}\cdot\mathbf{K}] \qquad \text{(H.185)}$$

where ω remains

$$\omega = (\omega_r^2 - \omega_i^2)^{\frac{1}{2}} \qquad \text{(H.186)}$$

and

$$\hat{\mathbf{w}} = (\omega_r \hat{\mathbf{u}}_r + i\omega_i \hat{\mathbf{u}}_i)/\omega \qquad \text{(H.187)}$$

$$\hat{\mathbf{w}}\cdot\hat{\mathbf{w}} = \hat{\mathbf{u}}_r\cdot\hat{\mathbf{u}}_r = \hat{\mathbf{u}}_i\cdot\hat{\mathbf{u}}_i = 1 \qquad \text{(H.188)}$$

$$\mathbf{v}_c = \hat{\mathbf{w}} \tanh(\omega + i\pi/2) = \hat{\mathbf{w}} \operatorname{cotanh}(\omega) \qquad \text{(H.189)}$$

Letting $\omega_L = \omega + i\pi/2$ we find, as before,

$$\cosh(\omega_L) = i \sinh(\omega) = -\gamma = i\,\gamma_s \qquad \text{(H.190)}$$
$$\sinh(\omega_L) = i \cosh(\omega) = -\beta\gamma = i\beta\gamma_s$$

with, $\beta = v_c = |\mathbf{v}_c| > 1$, $\gamma_s = (\beta^2 - 1)^{-\frac{1}{2}}$, and

$$\sinh(\omega) = \gamma_s \qquad \text{(H.191)}$$
$$\cosh(\omega) = \beta\gamma_s$$

Thus we denote $\Lambda_{CL}(\mathbf{v}_c)$ by

$$\Lambda_{CL}(\mathbf{v}_c) \equiv \Lambda_{CL}(\omega, \hat{\mathbf{w}}) \qquad \text{(H.192)}$$

The corresponding spinor boost transformation is:

$$S_{CL}(\Lambda_{CL}(\omega, \hat{\mathbf{w}})) = \exp(-i\omega_L\sigma_{0i}\hat{w}_i/2) = \exp(-\omega_L\gamma^0\boldsymbol{\gamma}\cdot\hat{\mathbf{w}}/2)$$
$$= \cosh(\omega_L/2)I + \sinh(\omega_L/2)\gamma^0\boldsymbol{\gamma}\cdot\hat{\mathbf{w}} \qquad \text{(H.193)}$$

The momentum space equation generated by $S_{CL}(\Lambda_{CL}(\omega, \hat{\mathbf{w}}))$ is

---

[219] The reader can readily verify the form is consistent that generated by an L_C boost transformation.

$$\{m \cosh(\omega_L)\gamma^0 - m \sinh(\omega_L)\boldsymbol{\gamma}\cdot(\omega_r\hat{\mathbf{u}}_r + i\omega_i\hat{\mathbf{u}}_i)/\omega - m\}e^{+ip\cdot x}w_{cL}(p) = 0 \quad (H.194)$$

or

$$\{im \sinh(\omega)\gamma^0 - im \cosh(\omega)\boldsymbol{\gamma}\cdot(\omega_r\hat{\mathbf{u}}_r + i\omega_i\hat{\mathbf{u}}_i)/\omega - m\}e^{+ip\cdot x}w_{cL}(p) = 0 \quad (H.195)$$

where $p\cdot x = Et - \mathbf{p}\cdot\mathbf{x}$ after performing a corresponding left-handed superluminal coordinate transformation in the exponential factor. Thus the positive energy wave is transformed into a negative energy wave by the transformation.

The momentum 4-vector is defined by

$$p = (p^0, \mathbf{p}) \quad (H.196)$$

where

$$p^0 = m \sinh(\omega) \qquad \mathbf{p} = \mathbf{p}_r + i\mathbf{p}_i \quad (H.197)$$

with

$$\mathbf{p}_r = m\omega_r\hat{\mathbf{u}}_r \cosh(\omega)/\omega \quad \mathbf{p}_i = m\omega_i\hat{\mathbf{u}}_i \cosh(\omega)/\omega \quad (H.198)$$

and

$$\mathbf{p}_r\cdot\mathbf{p}_i = 0 \quad (H.199)$$

then eq. H.195 becomes the complexon tachyon equation

$$[ip\cdot\gamma - m]e^{+ip\cdot x}w_{cL}(p) = 0 \quad (H.200)$$

with a complex 3-momentum $\mathbf{p}$ and the tachyon 4-momentum mass shell condition:[220]

$$p^2 = p^{0\,2} - \mathbf{p}_r^2 + \mathbf{p}_i^2 = -m^2 \quad (H.201)$$

Eq. H.200 is the momentum space equivalent of the wave equation

$$[\gamma^0\partial/\partial t + \boldsymbol{\gamma}\cdot(\nabla_r + i\nabla_i) - m]\psi_{CL}(t, \mathbf{x}_r, \mathbf{x}_i) = 0 \quad (H.202)$$

or

---

[220] Note that the presence of the $\mathbf{p}_i^2$ term does not change the tachyon requirement that $\mathbf{p}_r^2 \geq m^2$ as seen in the previous cases.

$$[\gamma\cdot\nabla - m]\psi_{CL}(t, \mathbf{x}_r, \mathbf{x}_i) = 0 \qquad \text{(H.203)}$$

with the subsidiary condition on the wave function

$$\nabla_r\cdot\nabla_i\, \psi_{CL}(t, \mathbf{x}_r, \mathbf{x}_i) = 0 \qquad \text{(H.204)}$$

also holds. We note that eq. H.202 is covariant under the real Lorentz group and eq. H.204 can be easily put into (real Lorentz group) covariant form.

Before considering a lagrangian formulation and the Fourier operator representation of $\psi_{CL}(t, \mathbf{x}_r, \mathbf{x}_i)$ we will define the tachyon spinors, and its associated real and imaginary spin operators.

The spinor generated from a spin up Dirac spinor at rest by the L_C spinor boost eq. H.193 is

$$w_{cL}(p) = S_{CL}w(0) = [\cosh(\omega_L/2)I + \sinh(\omega_L/2)\gamma^0\boldsymbol{\gamma}\cdot\hat{\mathbf{w}}]w(0) \qquad \text{(H.205)}$$

Following a procedure similar to section H.12 (which the reader may wish to examine first) we define four spinors for Dirac particles at rest with eq. H.282. Then by applying a boost to these rest spinors we find the L_C tachyon spinors:

$$S_{CL}w^k(0) = w_{cL}{}^k(p) \qquad \text{(H.206)}$$

and from these tachyon spinors we generalize to tachyon spinors $u_{CL}(p, s)$ and $v_{CL}(p, s)$ in a manner similar tothat of the previous case.

Eqs. H.200 through H.204 imply that the wave function solution of eq. H.200, subject to the subsidiary condition eq. H.204, has the form

$$\psi_{CL}(x) = \sum_{\substack{\pm s \\ \mathbf{p}_r{}^2 \geq m^2}} \int d^3p_r d^3p_i\, N_{CL}(p)\delta(\mathbf{p}_r\cdot\mathbf{p}_i/m^2)[b_{CL}(p,s)u_{CL}(p, s)e^{-i(p\cdot x + p^*\cdot x^*)/2} + $$
$$+ d_{CL}{}^\dagger(p,s)v_{CL}(p, s)e^{+i(p\cdot x + p^*\cdot x^*)/2}] \qquad \text{(H.207)}$$

where $\mathbf{p} = \mathbf{p_r} + i\mathbf{p_i}$, $\mathbf{x} = \mathbf{x_r} - i\mathbf{x_i}$, $p\cdot x = p^0x^0 - \mathbf{p}\cdot\mathbf{x}$, and $b_{CL}(p, s)$ and $d_{CL}(p,s)$ are tachyon fourier coefficients.

H.5.1.9 Global SU(3) Symmetry

We can show that there is also a global SU(3) symmetry present here as shown in the previous case. The demonstration is similar to that of eqs. H.143 – H.156.

H.5.1.10 Light-Front Quantization of Tachyonic Complexons

Because of the momentum constraint $\mathbf{p_r}^2 \geq m^2$ the set of solutions of the form of eq. H.207 is incomplete and the result of second quantization would not be an equal time anti-commutator expression consisting of derivatives of delta functions (eq. H.176) but rather an analogue to previous unsuccessful attempts to create a second quantized tachyon theory.[221]

Therefore we will use light-front coordinates, and left and right handed field operators (as previously) to obtain a successful second quantization of this new type of tachyon.

The "missing" factor of i in the first term of eq. H.203 requires the lagrangian to be different from the conventional Dirac lagrangian in order for the lagrangian to be real. The simplest, physically acceptable, free spin ½ tachyon lagrangian density for ψ_{CL} is:

$$\mathcal{L}_{CL} = \psi_{CL}{}^{C}(x)(\gamma\cdot\nabla - m)\psi_{CL}(x) \tag{H.208}$$

where

$$\psi_{CL}{}^{C}(x) = [\psi_{CL}(x)]^{\dagger}|_{\mathbf{x_i} = -\mathbf{x_i}}\, i\gamma^0\gamma^5 \tag{H.209}$$

is similar to eq. H.161. In words, eq. H.209 states: take the hermitean conjugate of $\psi_{CL}(x)$; change $\mathbf{x_i}$ to $-\mathbf{x_i}$; and then post-multiply by the indicated factors.

The free complexon invariant action (under real Lorentz transformations) is

[221] Such as G. Feinberg, Phys. Rev. **159**, 1089 (1967).

$$I = \int d^7x \mathcal{L}_{CL} \quad \text{(H.210)}$$

The action can be shown to be real

$$I^* = I \quad \text{(H.211)}$$

in a manner similar to the case considered in section H.12. The tachyonic complexon's energy-momentum tensor is

$$\begin{aligned} \mathfrak{T}_{CL\mu\nu} &= -\, g_{\mu\nu}\, \mathcal{L}_{CL} + \partial\mathcal{L}_{CL}/\partial(D^{\mu}\psi_{CL})\, D_{\nu}\psi_{CL} \\ &= i\psi_{CL}{}^{C}\gamma^0\gamma^5\gamma_{\mu}D_{\nu}\psi_{CL} \end{aligned} \quad \text{(H.212)}$$

where

$$\begin{aligned} D_0 &= \partial/\partial x^0 \\ D_k &= \partial/\partial x_r{}^k + i\, \partial/\partial x_i{}^k \end{aligned} \quad \text{(H.213)}$$

and thus the conserved energy and momentum are

$$P^0 = H = \int d^3x_r d^3x_i\, \mathfrak{T}_{CL}{}^{00} = i\int d^3x_r d^3x_i \psi_{CL}{}^{C}\gamma^5(\boldsymbol{\alpha}\cdot\mathbf{D} + \beta m)\psi_{CL} \quad \text{(H.214)}$$

$$P^k = \int d^3x_r d^3x_i\, \mathfrak{T}_{CL}{}^{0k} = -\, i\int d^3x_r d^3x_i\, \psi_{CL}{}^{C}\gamma^5 D^k\psi_{CL} \quad \text{(H.215)}$$

Having defined a suitable tachyon lagrangian we can now proceed to its canonical quantization. The conjugate momentum can be calculated from the lagrangian density eq. H.212:

$$\pi_{CLa} = \partial\mathcal{L}_{CL}/\partial\dot{\psi}_{CLa} \equiv \partial\mathcal{L}_{CL}/\partial(\partial\psi_{CLa}/\partial t) = -i([\psi_{CL}(x)]^{\dagger}|_{\mathbf{x}_i = -\mathbf{x}_i}\gamma^5)_a \quad \text{(H.216)}$$

The resulting non-zero, canonical anti-commutation relations are

$$\{\pi_{CL_a}(x), \psi_{CLb}(y)\} = i\, \delta_{ab}\, \delta^3(x_r - y_r)\delta^3(x_i - y_i)$$

based on locality in both real and imaginary coordinates:

$$\{\psi_{CL}{}^{\dagger}{}_{a}(x)|_{\mathbf{x_i} = -\mathbf{x_i}}, \psi_{Tb}(y)\} = -[\gamma^5]_{ab}\,\delta^3(x_r - y_r)\delta^3(x_i - y_i) \qquad \text{(H.217)}$$

At this point we might attempt to complete the canonical quantization procedure in the conventional manner by Fourier expanding the field and specifying anti-commutation relations for the fourier component amplitudes. However the incompleteness of the set of plane waves, which are limited by the restriction $\mathbf{p_r}^2 \geq m^2$, causes the equal time anti-commutator of the fields *not* to yield a δ-functions.

Therefore we turn to the previous successful approach to tachyon quantization[222] and decompose the tachyonic complexon field into left-handed and right-handed parts and then second quantize in light-front coordinates.

H.5.2 Separation into Left-Handed and Right-Handed Fields

As before we will use a transformed set of Dirac matrices to develop our left-handed and right-handed tachyon formulations. The γ^5 chirality operator's eigenvalues define handedness: +1 corresponds to right-handed; and −1 corresponds to left-handed:

$$\gamma^5\psi_{CLL} = -\psi_{CLL} \qquad\qquad \gamma^5\psi_{CLR} = \psi_{CLR} \qquad \text{(H.218)}$$

We define left-handed and right-handed tachyon fields with the projection operators:

$$\begin{aligned} C^{\pm} &= \tfrac{1}{2}(I \pm \gamma^5) \\ C^{+} + C^{-} &= I \\ C^{\pm 2} &= C^{\pm} \\ C^{+}C^{-} &= 0 \end{aligned} \qquad \text{(H.219)}$$

with the result

$$\begin{aligned} \psi_{CLL} &= C^{-}\psi_{CL} \\ \psi_{CLR} &= C^{+}\psi_{CL} \end{aligned} \qquad \text{(H.220)}$$

[222] Blaha (2006) discusses this case in detail.

We can calculate the commutation relations of the left-handed and right-handed tachyonic complexon fields from eq. H.217 by pre-multiplying and post-multiplying by ½(1 – γ^5) and ½(1 + γ^5). The results are:

$$\{\psi_{CLLa}{}^{\dagger}(x)|_{\mathbf{x_i} = -\mathbf{x_i}}, \psi_{CLLb}(y)\} = C^{-}{}_{ab}\,\delta^6(x - y) \qquad \text{(H.221)}$$

$$\{\psi_{CLRa}{}^{\dagger}(x)|_{\mathbf{x_i} = -\mathbf{x_i}}, \psi_{CLRb}(y)\} = -C^{+}{}_{ab}\,\delta^6(x - y) \qquad \text{(H.222)}$$

$$\{\psi_{CLLa}{}^{\dagger}(x)|_{\mathbf{x_i} = -\mathbf{x_i}}, \psi_{CLRb}(y)\} = \{\psi_{CLRa}{}^{\dagger}(x)|_{\mathbf{x_i} = -\mathbf{x_i}}, \psi_{CLLb}(x')\} = 0 \qquad \text{(H.223)}$$

where

$$\delta^6(x - y) = \delta^3(x_r - y_r)\delta^3(x_i - y_i) \qquad \text{(H.224)}$$

The lagrangian density of eq. H.208 decomposes into left-handed and right-handed parts: (The change $\mathbf{x_i}$ to $-\mathbf{x_i}$ will be understood in $\psi_{CLL}{}^{\dagger}(x)$ and $\psi_{CLR}{}^{\dagger}(x)$ in the following.)

$$\mathcal{L}_{CL} = \psi_{CLL}{}^{\dagger}\gamma^0 i\gamma^{\mu}\partial_{\mu}\psi_{CLL} - \psi_{CLR}{}^{\dagger}\gamma^0 i\gamma^{\mu}\partial_{\mu}\psi_{CLR} - im[\psi_{CLR}{}^{\dagger}\gamma^0\psi_{CLL} - \psi_{CLL}{}^{\dagger}\gamma^0\psi_{CLR}] \qquad \text{(H.225)}$$

H.5.3 Further Separation into + and – Light-Front Complexon Fields

As previously, we now use light-front coordinates and quantization to obtain a successful second quantization of this form of tachyon field. Light-front variables, in the present case where we have to contend with complex 3-vectors, are defined by real coordinates and derivatives:

$$x^{\pm} = (x^0 \pm x_r{}^3)/\sqrt{2} \qquad \text{(H.226)}$$
$$\partial/\partial x^{\pm} \equiv \partial^{\mp} \equiv (\partial/\partial x^0 \pm \partial/\partial x_r{}^3)/\sqrt{2}$$

with the "transverse" real coordinate variables, $x_r{}^1$ and $x_r{}^2$, and imaginary coordinate variables $x_i{}^1$, $x_i{}^2$, and $x_i{}^3$.

The inner product of two 4-vectors has the form

$$x \cdot y = x^+ y^- + y^+ x^- + i[y_i^3(x^+ - x^-) + x_i^3(y^+ - y^-)]/\sqrt{2} + x_i^3 y_i^3 - (\mathbf{x}_{\mathbf{r}\perp} - i\mathbf{x}_{\mathbf{i}\perp})\cdot(\mathbf{y}_{\mathbf{r}\perp} - i\mathbf{y}_{\mathbf{i}\perp}) \tag{H.227}$$

with

$$\begin{aligned} \mathbf{x}_{\mathbf{r}\perp} &= (x_r^1, x_r^2) \qquad & \mathbf{x}_{\mathbf{i}\perp} &= (x_i^1, x_i^2) \\ \mathbf{y}_{\mathbf{r}\perp} &= (y_r^1, y_r^2) \qquad & \mathbf{y}_{\mathbf{i}\perp} &= (y_i^1, y_i^2) \end{aligned} \tag{H.228}$$

where $x = (x^0, \mathbf{x} = \mathbf{x_r} - i\mathbf{x_i})$ and $y = (y^0, \mathbf{y} = \mathbf{y_r} - i\mathbf{y_i})$. Momenta are always defined as $p = (p^0, \mathbf{p} = \mathbf{p_r} + i\mathbf{p_i})$.

The light-front definition of Dirac matrices is:

$$\gamma^\pm = (\gamma^0 \pm \gamma^3)/\sqrt{2} \tag{H.229}$$

with transverse matrices γ^1 and γ^2 defined as usual. Note:

$$\gamma^{\pm\,2} = 0$$

We define "+" and "–" tachyon fields with the projection operators:

$$R^\pm = \tfrac{1}{2}(I \pm \gamma^0\gamma^3) \tag{H.230}$$

$$\begin{aligned} &\text{Left-handed, } \pm \text{ light-front fields:} \qquad & \psi_{CLL}{}^\pm &= R^\pm C^- \psi_{CL} \\ &\text{Right-handed, } \pm \text{ light-front fields:} \qquad & \psi_{CLR}{}^\pm &= R^\pm C^+ \psi_{CL} \end{aligned} \tag{H.231}$$

Transforming to light-front variables and fields as above we obtain the light-front free tachyon lagrangian:

$$\begin{aligned} \mathcal{L}_{CL} = {} & 2^{1/2}\psi_{CLL}{}^{+\dagger} i\partial^- \psi_{CLL}{}^+ + 2^{1/2}\psi_{CLL}{}^{-\dagger} i\partial^+ \psi_{CLL}{}^- - \psi_{CLL}{}^{+\dagger}\gamma^0[i\boldsymbol{\gamma}_\perp\cdot\boldsymbol{\nabla}_{\mathbf{r}\perp} - \boldsymbol{\gamma}\cdot\boldsymbol{\nabla}_\mathbf{i}]\psi_{CLL}{}^- - \\ & - \psi_{CLL}{}^{-\dagger}\gamma^0[i\boldsymbol{\gamma}_\perp\cdot\boldsymbol{\nabla}_{\mathbf{r}\perp} - \boldsymbol{\gamma}\cdot\boldsymbol{\nabla}_\mathbf{i}]\psi_{CLL}{}^+ - 2^{1/2}\psi_{CLR}{}^{+\dagger} i\partial^- \psi_{CLR}{}^+ - 2^{1/2}\psi_{CLR}{}^{-\dagger} i\partial^+ \psi_{CLR}{}^- + \end{aligned}$$

$$+ \psi_{CLR}^{+\dagger}\gamma^0[i\gamma_\perp\cdot\nabla_{r_\perp} - \gamma\cdot\nabla_i]\psi_{CLR}^{-} + \psi_{CLR}^{-\dagger}\gamma^0[i\gamma_\perp\cdot\nabla_{r_\perp} - \gamma\cdot\nabla_i]\psi_{CLR}^{+} -$$
$$- im[\psi_{CLR}^{+\dagger}\gamma^0\psi_{CLL}^{-} - \psi_{CLL}^{+\dagger}\gamma^0\psi_{CLR}^{-} + \psi_{CLR}^{-\dagger}\gamma^0\psi_{CLL}^{+} - \psi_{CLL}^{-\dagger}\gamma^0\psi_{CLR}^{+}]$$
(H.232)

(Note the similarity to the previous tachyon case.) Again the difference in signs between the left-handed and right-handed terms will be a crucial factor in the derivation of the left-handed features of the Standard Model.

Eq. H.232 generates the equations of motion:

$$2^{1/2}i\partial^{-}\psi_{CLL}^{+} - \gamma^0[i\gamma_\perp\cdot\nabla_{r_\perp} - \gamma\cdot\nabla_i]\psi_{CLL}^{-} + im\gamma^0\psi_{CLR}^{-} = 0 \qquad \text{(H.233)}$$
$$2^{1/2}i\partial^{-}\psi_{CLR}^{+} - \gamma^0[i\gamma_\perp\cdot\nabla_{r_\perp} - \gamma\cdot\nabla_i]\psi_{CLR}^{-} + im\gamma^0\psi_{CLL}^{-} = 0$$
$$2^{1/2}i\partial^{+}\psi_{CLL}^{-} - \gamma^0[i\gamma_\perp\cdot\nabla_{r_\perp} - \gamma\cdot\nabla_i]\psi_{CLL}^{+} + im\gamma^0\psi_{CLR}^{+} = 0$$
$$2^{1/2}i\partial^{+}\psi_{CLR}^{-} - \gamma^0[i\gamma_\perp\cdot\nabla_{r_\perp} - \gamma\cdot\nabla_i]\psi_{CLR}^{+} + im\gamma^0\psi_{CLL}^{+} = 0$$

Eqs. H.233 show that ψ_{CLL}^{-} and ψ_{CLR}^{-} are dependent fields that are functions of ψ_{CLL}^{+} and ψ_{CLR}^{+} on the light-front where x^+ equals a constant. They can be expressed in an integral form as well. (The independent fields ψ_{CLL}^{+} and ψ_{CLR}^{+} play a fundamental role in tachyonic complexon theory and are used to define "in" and "out" tachyon states in perturbation theory.)

The conjugate momenta implied by eq. H.232 are

$$\pi_{CLL}^{+} = \partial\mathcal{L}/\partial(\partial^{-}\psi_{CLL}^{+}) = 2^{1/2}i\psi_{CLL}^{+\dagger} \qquad \text{(H.234)}$$
$$\pi_{CLL}^{-} = \partial\mathcal{L}/\partial(\partial^{-}\psi_{CLL}^{-}) = 0$$
$$\pi_{CLR}^{+} = \partial\mathcal{L}/\partial(\partial^{-}\psi_{CLR}^{+}) = -2^{1/2}i\psi_{CLR}^{+\dagger} \qquad \text{(H.235)}$$
$$\pi_{CLR}^{-} = \partial\mathcal{L}/\partial(\partial^{-}\psi_{CLR}^{-}) = 0$$

x^+ plays the role of the "time" variable in light-front quantized theories. So we define canonical equal x^+ anti-commutation relations for spin ½ tachyonic complexons also.

The canonical equal-light-front ($x^+ = y^+$) anti-commutation relations of the independent fields would normally be:

$$\{\psi_{CLL}{}^{+\dagger}{}_a(x), \psi_{CLL}{}^{+}{}_b(y)\} = 2^{-1}[C^-R^+]_{ab}\delta(x^- - y^-)\delta^2(x_r - y_r)\delta^3(x_I - y_i) \qquad (H.236)$$

$$\{\psi_{CLR}{}^{+\dagger}{}_a(x), \psi_{CLR}{}^{+}{}_b(y)\} = -2^{-1}[C^+R^+]_{ab}\,\delta(x^- - y^-)\delta^2(x_r - y_r)\delta^3(x_I - y_i) \qquad (H.237)$$

$$\{\psi_{CLL}{}^{+}{}_a{}^{\dagger}(x), \psi_{CLR}{}^{+}{}_b(y)\} = \{\psi_{CLR}{}^{+}{}_a{}^{\dagger}(x), \psi_{CLL}{}^{+}{}_b(y)\} = 0 \qquad (H.238)$$

$$\{\psi_{CLL}{}^{+}{}_a(x), \psi_{CLR}{}^{+}{}_b(y)\} = \{\psi_{CLR}{}^{+}{}_a{}^{\dagger}(x), \psi_{CLL}{}^{+\dagger}{}_b(y)\} = 0 \qquad (H.239)$$

But as in the previous case they will be modified.

Again we see that the right-handed tachyon anti-commutation relation (eq. H.237) has a minus sign relative to the corresponding conventional right-handed anti-commutation relation.

The sign differences between the left-handed and right-handed lagrangian terms ultimately lead to parity violating features in the Standard Model lagrangian.

H.5.3.1 Left-Handed Tachyonic Complexons

The free, "+" light-front, left-handed tachyonic complexon Fourier expansion is:

$$\psi_{CLL}{}^{+}(x_r, x_i) = \sum_{\pm s} \int d^2p_r dp^+ d^3p_i\, N_{CLL}{}^{+}(p)\theta(p^+)\delta((p_i{}^3(p^+ - p^-)/\surd 2 + \mathbf{p}_{r\perp}\cdot\mathbf{p}_{i\perp})/m^2)\cdot$$

$$\cdot[b_{CLL}{}^{+}(p, s)u_{CLL}{}^{+}(p, s)e^{-i(p\cdot x + p^*\cdot x^*)/2} + d_{CLL}{}^{+\dagger}(p, s)v_{CLL}{}^{+}(p, s)e^{+i(p\cdot x + p^*\cdot x^*)/2}] \qquad (H.240)$$

Its hermitean conjugate is

$$\psi_{CLL}{}^{+\dagger}(x_r, x_i) = \sum_{\pm s} \int d^2p_r dp^+ d^3p_i\, N_{CLL}{}^{+}(p)\theta(p^+)\delta((p_i{}^3(p^+ - p^-)/\surd 2 + \mathbf{p}_{r\perp}\cdot\mathbf{p}_{i\perp})/m^2)\cdot$$

$$\cdot[b_{CLL}{}^{\dagger}(p^*,s)u_{CLL}{}^{\dagger}(p^*,s)e^{+i(p^*.x + p.x^*)/2} + d_{CLL}(p^*,s)v_{CLL}{}^{\dagger}(p^*,s)e^{-i(p^*.x + p.x^*)/2}] \quad \text{(H.241)}$$

where $\mathbf{p} = \mathbf{p_r} + i\mathbf{p_i}$, $\mathbf{x} = \mathbf{x_r} - i\mathbf{x_i}$, $p{\cdot}x = p^0x^0 - \mathbf{p{\cdot}x}$, and † indicates hermitean conjugate. The spinors are

$$\begin{aligned}
u_{CLL}{}^{+}(p, s) &= C^- R^+ S_{CL}w^1(0) \\
u_{CLL}{}^{+}(p, -s) &= C^- R^+ S_{CL}w^2(0) \\
v_{CLL}{}^{+}(p, s) &= C^- R^+ S_{CL}w^3(0) \\
v_{CLL}{}^{+}(p, -s) &= C^- R^+ S_{CL}w^4(0) \\
u_{CLL}{}^{+\dagger}(p^*, s) &= w^{1T}(0)S_{CL}R^+C^- \\
u_{CLL}{}^{+\dagger}(p^*, -s) &= w^{2T}(0)S_{CL}R^+C^- \\
v_{CLL}{}^{+\dagger}(p^*, s) &= w^{3T}(0)S_{CL}R^+C^- \\
v_{CLL}{}^{+\dagger}(p^*, -s) &= w^{4T}(0)S_{CL}R^+C^-
\end{aligned} \quad \text{(H.242)}$$

where the superscript "T" indicates the transpose (These spinors are described in section H.12.) and

$$N_{CLL}{}^{+}(p) = (2\pi)^{-3}(2m/p^+)^{½} \quad \text{(H.243)}$$

The anti-commutation relations of the Fourier coefficient operators are

$$\begin{aligned}
\{b_{CLL}(p,s), b_{CLL}{}^{\dagger}(p'^*,s')\} &= 2^{-½}\delta_{ss'}\delta(p^+ - p'^+)\delta^2(\mathbf{p_r} - \mathbf{p'_{r'}})\delta^3(\mathbf{p_i} + \mathbf{p'_{i'}}) \\
\{d_{CLL}(p,s), d_{CLL}{}^{\dagger}(p'^*,s')\} &= 2^{-½}\delta_{ss'}\delta(p^+ - p'^+)\delta^2(\mathbf{p_r} - \mathbf{p'_{r'}})\delta^3(\mathbf{p_i} + \mathbf{p'_{i'}}) \\
\{b_{CLL}(p,s), b_{CLL}(p'^*,s')\} &= \{d_{CLL}(p,s), d_{CLL}(p'^*,s')\} = 0 \\
\{b_{CLL}{}^{\dagger}(p,s), b_{CLL}{}^{\dagger}(p'^*,s')\} &= \{d_{CLL}{}^{\dagger}(p,s), d_{CLL}{}^{\dagger}(p'^*,s')\} = 0 \\
\{b_{CLL}(p,s), d_{CLL}{}^{\dagger}(p'^*,s')\} &= \{d_{CLL}(p,s), b_{CLL}{}^{\dagger}(p'^*,s')\} = 0 \\
\{b_{CLL}{}^{\dagger}(p,s), d_{CLL}{}^{\dagger}(p'^*,s')\} &= \{d_{CLL}(p,s), b_{CLL}(p'^*,s')\} = 0
\end{aligned} \quad \text{(H.244)}$$

The delta-function arguments $\delta^3(\mathbf{p_i} + \mathbf{p'_{i'}})$ above have a positive sign in order to obtain $\delta^3(\mathbf{x_i} - \mathbf{y_i})$ in the field anti-commutators.

The spinors, eq. H.242, satisfy

$$\sum_{\pm s} u_{CLL}{}^{+}{}_{\alpha}(p, s)\bar{u}_{CLL}{}^{+}{}_{\beta}(p^*, s) = (2m)^{-1}[C^{-}R^{+}(i\not{p} + m)R^{-}C^{+}]_{\alpha\beta}$$
$$\sum_{\pm s} v_{CLL}{}^{+}{}_{\alpha}(p, s)\bar{v}_{CLL}{}^{+}{}_{\beta}(p^*, s) = (2m)^{-1}[C^{-}R^{+}(i\not{p} - m)R^{-}C^{+}]_{\alpha\beta} \qquad \text{(H.245)}$$

where $\bar{u}_{CLL}{}^{+} = u_{CLL}{}^{+\dagger}\gamma^0$ and $\bar{v}_{CLL}{}^{+} = v_{CLL}{}^{+\dagger}\gamma^0$.

We now evaluate the canonical left-handed, light-front anti-commutation relation:

$$\{\psi_{CLL}{}^{+}{}_{a}(x), \psi_{CLL}{}^{+\dagger}{}_{b}(y)\} = \sum_{\pm s,s'} \int d^3p_i d^2p dp^+ \int d^3p_i' d^2p' dp'^+ N_{CLL}{}^{+}(p)\, N_{CLL}{}^{+}(p')\cdot$$
$$\cdot\theta(p^+)\theta(p'^+)\delta((p_i{}^3(p^+ - p^-)/\surd 2 + \mathbf{p}_{r\perp}\cdot\mathbf{p}_{i\perp})/m^2)\, \delta((p_i'^3(p'^+ - p'^-)/\surd 2 + \mathbf{p}'_{r\perp}\cdot\mathbf{p}'_{i\perp})/m^2)\cdot$$
$$\cdot[\{b_{CLL}{}^{+\dagger}(p'^*,s'),b_{CLL}{}^{+}(p,s)\}u_{CLL}{}^{+}{}_{a}(p,s)u_{CLL}{}^{+\dagger}{}_{b}(p'^*,s')e^{+i(p'^*\cdot y+p'\cdot y^*)2 - i(p\cdot x+p^*\cdot x^*)/2} +$$
$$+\{d_{CLL}{}^{+}(p'^*,s'),d_{CLL}{}^{+\dagger}(p,s)\}v_{CLL}{}^{+}{}_{a}(p,s)v_{CLL}{}^{+\dagger}{}_{b}(p'^*,s')e^{-i(p'^*\cdot y+p'\cdot y^*)/2 + i(p\cdot x + p^*\cdot x^*)/2}]$$

$$= 2^{-\frac{1}{2}}\sum_{\pm s} \int d^3p_i d^2p_r dp^+ [N_{CLL}{}^{+}(p)]^2\theta(p^+)[\delta((p_i{}^3(p^+ - p^-)/\surd 2 + \mathbf{p}_{r\perp}\cdot\mathbf{p}_{i\perp})/m^2)]^2 \cdot$$
$$\cdot[u_{CLL}{}^{+}{}_{a}(p,s)u_{CLL}{}^{+\dagger}{}_{b}(p^*,s)e^{+i(p^*\cdot(y-x)+p\cdot(y^*-x^*))/2} + v_{CLL}{}^{+}{}_{a}(p,s)v_{CLL}{}^{+\dagger}{}_{b}(p^*,s)e^{-i(p^*\cdot(y-x)+p\cdot(y^*-x^*))/2}]$$

$$= -2^{-3/2}\int d^3p_i d^2p dp^+\theta(p^+)[N_{CLL}{}^{+}(p)]^2\delta'((p_i{}^3(p^+ - p^-)/\surd 2 + \mathbf{p}_{r\perp}\cdot\mathbf{p}_{i\perp})/m^2)(2m)^{-1}\cdot$$
$$\cdot\{[C^{-}R^{+}(i\not{p} + m)\gamma^0R^{+}C^{-}]_{ab}e^{+i(p^*\cdot(y-x)+p\cdot(y^*-x^*))/2} +$$
$$+[C^{-}R^{+}(i\not{p} - m)\gamma^0R^{+}C^{-}]_{ab}e^{-i(p^*\cdot(y-x)+p\cdot(y^*-x^*))/2}\}$$

$$= -(1/2)C^{-}R^{+}\delta_{ab} \int d^3p_i\, d^2p_\perp \int_0^\infty dp^+\, \delta'((p_i{}^3(p^+ - p^-)/\surd 2 + \mathbf{p}_\perp\cdot\mathbf{p}_{i\perp})/m^2)(2\pi)^{-6}\cdot$$
$$\cdot\{ e^{+i\{p^+(y^- - x^-) - \mathbf{p}_{r\perp}\cdot(\mathbf{y}_{r\perp} - \mathbf{x}_{r\perp}) + \mathbf{p}_i\cdot(\mathbf{y}_i - \mathbf{x}_i)\}} + e^{-i\{p^+(y^- - x^-) - \mathbf{p}_{r\perp}\cdot(\mathbf{y}_{r\perp} - \mathbf{x}_{r\perp}) + \mathbf{p}_i\cdot(\mathbf{y}_i - \mathbf{x}_i)\}}\}$$

$$= -C^{-}R^{+}\delta_{ab}(4\pi)^{-1}\int_0^\infty dp^+\delta'(\nabla_r\cdot\nabla_i/m^2)\delta^3(y_i - x_i)\, \delta^2(y_r - x_r)\{e^{+ip^+(y^- - x^-)} + e^{-ip^+(y^- - x^-)}\}$$

whereupon we revert back to the original form of the constraint: $\delta(\nabla_r\cdot\nabla_i/m^2)$

$$\{\psi_{CLL}{}^{+}{}_{a}(x), \psi_{CLL}{}^{+\dagger}{}_{b}(y)\} = -(1/2)C^{-}R^{+}\delta_{ab}\,\delta'(\nabla_r\cdot\nabla_i/m^2)\delta(y^{-} - x^{-})\delta^2(y_r - x_r)\delta^3(y_i - x_i) \tag{H.246}$$

The result is the left-handed, light-front equivalent of the earlier non-tachyon result. Again the constraint is apparent in the anti-commutator. (The factor of 2 difference is due to light-front coordinate definitions.)

Therefore we have left-handed, light-front quantized tachyonic complexons with the equivalent of canonical anti-commutation relations, and with localized tachyonic complexons. As a result we have a canonical tachyonic complexon Quantum Field Theory.

H.5.3.2 Left-handed Case 4: Tachyonic Complexon Feynman Propagator

The light-front Feynman propagator for the left-handed $\psi_{CLL}{}^{+}$ *tachyonic* complexon field is

$$iS^{+}{}_{CLLF}(x,y) = \theta(x^{+} - y^{+})<0|\psi_{CLL}{}^{+}(x)\psi_{CLL}{}^{+\dagger}(y)\gamma^0|0> - \theta(y^{+} - x^{+})<0|\psi_{CLL}{}^{+\dagger}(y)\gamma^0\psi_{CLL}{}^{+}(x)|0> \tag{H.247}$$

$$= -\tfrac{1}{2}\int d^3p_i d^2p_r dp^{+}\theta(p^{+})N_{CLL}{}^{+2}\delta'((p_i{}^3(p^{+} - p^{-})/\sqrt{2} + \mathbf{p}_{r\perp}\cdot\mathbf{p}_{i\perp})/m^2)(2m)^{-1}C^{-}R^{+}\cdot$$
$$\cdot\{\theta(x^{+} - y^{+})[(i\not{p} + m)\gamma^0]e^{+i(p^*\cdot(y-x)+p\cdot(y^*-x^*))/2} +$$
$$+ \theta(y^{+} - x^{+})[(i\not{p} - m)\gamma^0]e^{-i(p^*\cdot(y-x)+p\cdot(y^*-x^*))/2}\}R^{+}C^{-}\gamma^0$$

If we define the on-shell momentum variables

$$p_0{}^{-} = (p_{r0}{}^1 p_{r0}{}^1 + p_{r0}{}^2 p_{r0}{}^2 - \mathbf{p}_{i0}\cdot\mathbf{p}_{i0} - m^2)/(2p_0{}^{+})$$
$$p_0{}^{+} = p^{+},\ p_{r0}{}^j = p_r{}^j \quad (\text{for } j = 1, 2),$$
$$\mathbf{p}_{i0} = \mathbf{p}_i,\ p_{r\perp0}{}^2 = p_{r0}{}^j p_{r0}{}^j$$
$$\not{p}_0 = p_0\cdot\gamma$$

with $p_0 = (p^0, \mathbf{p}_{r0} + \mathbf{ip}_{r0})$ then the above equation can be rewritten as

$$iS^{+}{}_{CLLF}(x,y) = -\tfrac{1}{2}C^{-}R^{+}\int d^4p d^3p_i N_{CLL}{}^{+2}\delta'((p_{i0}{}^3(p_0{}^+ - p_0{}^-)/\surd 2 + \mathbf{p}_{r\perp 0}\cdot\mathbf{p}_{i\perp 0})/m^2)(4\pi m)^{-1}e^{+i(p^*\cdot(y-x)+p\cdot(y^*-x^*))/2}\,\cdot$$
$$\cdot\{\theta(p^+)(i\not{p}_0 + m)\gamma^0]/[p^- - p_0{}^- + i\varepsilon] + \theta(-p^+)(i\not{p}_0 - m)\gamma^0]/[p^- + p_0{}^- - i\varepsilon]\}R^{+}C\gamma^0$$

$$= -\tfrac{1}{2}\int d^4p_r d^3p_i\, N_{CLL}{}^{+2}\delta'((p_{i0}{}^3(p^+ - p^-)/\surd 2 + \mathbf{p}_{r\perp}\cdot\mathbf{p}_{i\perp})/m^2)(p^+/4\pi m)\, e^{+i(p^*\cdot(y-x)+p\cdot(y^*-x^*))/2}\,\cdot$$
$$\cdot[C^{-}R^{+}(i\not{p} + m)\gamma^0 R^{+}C^{-}\gamma^0][(p^2 + m^2 + i\varepsilon)]^{-1}$$

with $p_r = (p^0, \mathbf{p}_r)$ and $p = (p^0, \mathbf{p}_r + \mathbf{i}\mathbf{p}_r)$. Substituting for N_{CLL} and using $x\delta'(x) = -\delta(x)$ we obtain

$$= -\tfrac{1}{2}\int d^4p_r d^3p_i(2\pi)^{-7}\delta'(\mathbf{p}_r\cdot\mathbf{p}_i/m^2)\exp[ip^0(y^0 - x^0) - i\mathbf{p}_r\cdot(\mathbf{y}_r - \mathbf{x}_r) + i\mathbf{p}_i\cdot(\mathbf{y}_i - \mathbf{x}_i)]\cdot$$
$$\cdot\,[C^{-}R^{+}(i\not{p} + m)R^{-}C^{+}]/(p^2 + m^2 + i\varepsilon)$$

since $C^{-}R^{+}(i\not{p} + m)\gamma^0 R^{+}C^{-}\gamma^0 = C^{-}R^{+}(i\not{p} + m)R^{-}C^{+}$. The integral can then be written:

$$iS^{+}{}_{CLLF}(x,y) = \int d^4p_r d^3p_i\delta'(\mathbf{p}_r\cdot\mathbf{p}_i/m^2)C^{-}R^{+}(i\not{p} + m)R^{-}C^{+}\cdot$$
$$\cdot\exp[-ip^0(x^0-y^0) + i\mathbf{p}_r\cdot(\mathbf{x}_r-\mathbf{y}_r) - i\mathbf{p}_i\cdot(\mathbf{x}_i - \mathbf{y}_i)]/(p^2 + m^2 + i\varepsilon)$$

$$= \int d^4p_r dM^2\delta'(\nabla_r\cdot\nabla_i/m^2)C^{-}R^{+}(ip^{\mathbf{0}}\gamma^{\mathbf{0}} - (\nabla_r - i\nabla_i)\cdot\boldsymbol{\gamma} + m)R^{-}C^{+}\cdot$$
$$\cdot\exp[-ip^0(x^0-y^0) + i\mathbf{p}_r\cdot(\mathbf{x}_r-\mathbf{y}_r)]J_2(\mathbf{x}_i - \mathbf{y}_i, M^2)/(p_r{}^2 + M^2 + i\varepsilon)$$

where

$$J_2(\mathbf{x}_i - \mathbf{y}_i, M^2) = (2\pi)^{-3}\int d^3p_i\,\delta(M^2 - \mathbf{p}_i{}^2 - m^2)\exp[-i\mathbf{p}_i\cdot(\mathbf{x}_i - \mathbf{y}_i)] \qquad \text{(H.248)}$$
$$= (2\pi)^{-2}|\mathbf{x}_i - \mathbf{y}_i|^{-1}\theta(M^2 - m^2)\sin((M^2 - m^2)^{1/2}|\mathbf{x}_i - \mathbf{y}_i|)$$

This tachyonic complexon Feynman propagator can be rearranged into the form of a spectral integral:

$$iS^{+}{}_{CLLF}(x, y) = -\int dM\, C^{-}R^{+}(\gamma^{\mathbf{0}}\partial/\partial x^{\mathbf{0}} + (\nabla_r - i\nabla_i)\cdot\boldsymbol{\gamma} - m)R^{-}C^{+}\,\delta'(\nabla_r\cdot\nabla_i/m^2)\,\cdot$$
$$\cdot J_2(\mathbf{x}_i - \mathbf{y}_i, M^2)\Delta_{FT}(x - y, M) \qquad \text{(H.249)}$$

with ∇_r and ∇_i derivatives with respect to $\mathbf{x_r}$ and $\mathbf{x_i}$ and where

$$\Delta_{FT}(x - y, M) = (2\pi)^{-4}\int d^4p_r \exp[-ip^0(x^0 - y^0) + i\mathbf{p_r}\cdot(\mathbf{x_r} - \mathbf{y_r})]/(p_r^2 + M^2 + i\varepsilon) \tag{H.250}$$

H.5.3.3 Case 5: Right-Handed Tachyonic Complexons

The case of right-handed tachyonic complexons is similar to left-handed complexons with only one difference: a minus sign in the canonical right-handed equal-time commutation relations resulting in a minus sign in the creation and annihilation operator anti-commutation relations. The right-handed tachyonic complexon wave function light-front Fourier expansion is:

$$\psi_{CLR}^{+}(x_r, x_i) = \sum_{\pm s} \int d^2p_r dp^+ d^3p_i\, N_{CLR}^{+}(p)\theta(p^+)\delta((p_i^3(p^+ - p^-)/\surd 2 + \mathbf{p_{r\perp}}\cdot\mathbf{p_{i\perp}})/m^2)\cdot$$
$$\cdot[b_{CLR}^{+}(p, s)u_{CLR}^{+}(p, s)e^{-i(p\cdot x + p^*\cdot x^*)/2} + d_{CLR}^{+\dagger}(p, s)v_{CLR}^{+}(p, s)e^{+i(p\cdot x + p^*\cdot x^*)/2}] \tag{H.251}$$

where

$$N_{CLR}^{+}(p) = (2\pi)^{-3}(2m/p^+)^{1/2} \tag{H.252}$$

Its hermitean conjugate is

$$\psi_{CLR}^{+\dagger}(x_r, x_i) = \sum_{\pm s} \int d^2p_r dp^+ d^3p_i\, N_{CLR}^{+}(p)\theta(p^+)\delta((p_i^3(p^+ - p^-)/\surd 2 + \mathbf{p_{r\perp}}\cdot\mathbf{p_{i\perp}})/m^2)\cdot$$
$$\cdot[b_{CLR}^{\dagger}(p^*,s)u_{CLR}^{\dagger}(p^*,s)e^{+i(p^*\cdot x + p\cdot x^*)/2} + d_{CLR}(p^*,s)v_{CLR}^{\dagger}(p^*,s)e^{-i(p^*\cdot x + p\cdot x^*)/2}] \tag{H.253}$$

where $\mathbf{p} = \mathbf{p_r} + i\mathbf{p_i}$, $\mathbf{x} = \mathbf{x_r} - i\mathbf{x_i}$, $p\cdot x = p^0x^0 - \mathbf{p}\cdot\mathbf{x}$, and † indicates hermitean conjugate. The right-handed spinors are

$$u_{CLR}^{+}(p, s) = C^+ R^+ S_{CR}w^1(0)$$
$$u_{CLR}^{+}(p, -s) = C^+ R^+ S_{CR}w^2(0)$$
$$v_{CLR}^{+}(p, s) = C^+ R^+ S_{CR}w^3(0)$$
$$v_{CLR}^{+}(p, -s) = C^+ R^+ S_{CR}w^4(0) \tag{H.254}$$

$$u_{CLR}^{+\dagger}(p^*, s) = w^{1T}(0)S_{CR}R^+C^+$$
$$u_{CLR}^{+\dagger}(p^*, -s) = w^{2T}(0)S_{CR}R^+C^+$$
$$v_{CLR}^{+\dagger}(p^*, s) = w^{3T}(0)S_{CR}R^+C^+$$
$$v_{CLR}^{+\dagger}(p^*, -s) = w^{4T}(0)S_{CR}R^+C^+$$

where the superscript "T" indicates the transpose. The anti-commutation relations of the Fourier coefficient operators are

$$\begin{aligned}
&\{b_{CLR}(p,s), b_{CLR}^{\dagger}(p'^*,s')\} = -2^{-1/2}\delta_{ss'}\delta(p^+ - p'^+)\delta^2(\mathbf{p}_r - \mathbf{p'}_{r'})\delta^3(\mathbf{p}_i + \mathbf{p'}_{i'})\\
&\{d_{CLR}(p,s), d_{CLR}^{\dagger}(p'^*,s')\} = -2^{-1/2}\delta_{ss'}\delta(p^+ - p'^+)\delta^2(\mathbf{p}_r - \mathbf{p'}_{r'})\delta^3(\mathbf{p}_i + \mathbf{p'}_{i'})\\
&\{b_{CLR}(p,s), b_{CLR}(p'^*,s')\} = \{d_{CLR}(p,s), d_{CLR}(p'^*,s')\} = 0\\
&\{b_{CLR}^{\dagger}(p,s), b_{CLR}^{\dagger}(p'^*,s')\} = \{d_{CLR}^{\dagger}(p,s), d_{CLR}^{\dagger}(p'^*,s')\} = 0\\
&\{b_{CLR}(p,s), d_{CLR}^{\dagger}(p'^*,s')\} = \{d_{CLR}(p,s), b_{CLR}^{\dagger}(p'^*,s')\} = 0\\
&\{b_{CLR}^{\dagger}(p,s), d_{CLR}^{\dagger}(p'^*,s')\} = \{d_{CLR}(p,s), b_{CRR}(p'^*,s')\} = 0
\end{aligned} \qquad \text{(H.255)}$$

The spinors satisfy

$$\sum_{\pm s} u_{CLR\ \alpha}^{+}(p, s)\bar{u}_{CLR\ \beta}^{+}(p^*, s) = (2m)^{-1}[C^+R^+(-i\not{p} + m)R^-C^-]_{\alpha\beta} \qquad \text{(H.256)}$$
$$\sum_{\pm s} v_{CLR\ \alpha}^{+}(p, s)\bar{v}_{CLR\ \beta}^{+}(p^*, s) = (2m)^{-1}[C^+R^+(-i\not{p} - m)R^-C^-]_{\alpha\beta}$$

where $\bar{u}_{CLR}^{+} = u_{CLR}^{+\dagger}\gamma^0$ and $\bar{v}_{CLR}^{+} = v_{CLR}^{+\dagger}\gamma^0$.

The right-handed anti-commutation relation with a minus sign follows in particular because of the minus signs in eqs. H.255.

H.5.3.4 Right-handed Case 5: Tachyonic Complexon Feynman Propagator

The Feynman propagator for right-handed tachyonic complexons can be obtained from eqs. H.249 and H.250 by changing the parity projection operator and some numerator signs in the integral (basically $p \rightarrow -p$) resulting in

$$iS^{+}_{CLRF}(x, y) = \int dM\, C^{+}R^{+}(\gamma^{0}\partial/\partial x^{0} + (\nabla_r - i\nabla_i)\cdot\boldsymbol{\gamma} - m)R^{-}C^{-}\,\delta'(\nabla_r\cdot\nabla_i/m^2)\cdot$$
$$\cdot J_2(\mathbf{x_i} - \mathbf{y_i}, M^2)\Delta_{FT}(x - y, M) \qquad \text{(H.257)}$$

with $\nabla_r + i\nabla_i$ derivatives with respect to $\mathbf{x_r}$ and $\mathbf{x_i}$ and where

$$\Delta_{FT}(x - y, M) = (2\pi)^{-4}\int d^4p_r \exp[-ip^0(x^0 - y^0) + i\mathbf{p_r}\cdot(\mathbf{x_r} - \mathbf{y_r})]/(p_r^{\,2} + M^2 + i\varepsilon) \qquad \text{(H.258)}$$

H.5.3.5 Other Cases? No

The four cases considered above are the only cases having symmetry under the real Lorentz group L and a single real energy (with a corresponding single real time parameter) that is independent of the direction of the boost thus preserving (real) spatial rotation invariance. The reality of the time variable survives the breakdown to conventional Lorentz invariance.

One might think that using the other type of spinor boost operator.

$$S_{CR}(\Lambda_{CR}(\omega, \hat{\mathbf{w}})) = \exp(-i\omega_R\sigma_{0i}w_i/2) = \exp(-\omega_R\gamma^0\boldsymbol{\gamma}\cdot\hat{\mathbf{w}}/2) \qquad \text{(H.259)}$$
$$= \cosh(\omega_R/2)I + \sinh(\omega_R/2)\gamma^0\boldsymbol{\gamma}\cdot\hat{\mathbf{w}}$$

where $\omega_R = \omega - i\pi/2$ might lead to more possible forms of spin ½ wave equations and particles. In fact it merely leads to the same particle types but with the role of the left-handed and right-handed fields reversed. The result would be a "right-handed" Standard Model contrary to experiment.

H.6 Spinor Boosts Generate 4 Species of Particles: Leptons and Quarks

In this appendix we have found four types of fermions using complex Lorentz boosts that correspond in a natural way with the four general *species* (types) of known fermions: charged leptons, neutrinos, up-type color quarks and down-type color quarks.[223]

H.6.1 Charged lepton fermions

The conventional Dirac equation and solutions.

H.6.2 Neutrinos

Simple tachyons with real energy and 3-momentum. Their free field equation is:

$$(\gamma^\mu \partial/\partial x^\mu - m)\psi_T(x) = 0 \qquad \text{(H.260)}$$

and their left-handed $\psi_{TL}{}^{+}$ Feynman propagator is:

$$iS^{+}{}_{TLF}(x, y) = \tfrac{1}{2}C^{-}R^{+}\gamma^{0}\int d^4p(2\pi)^{-4}\; p^{+}e^{-ip\cdot(x-y)}/(p^2 + m^2 + i\varepsilon) \qquad \text{(H.261)}$$

Similarly the light-front Feynman propagator for the right-handed $\psi_{TR}{}^{+}$ tachyon field is

$$iS^{+}{}_{TRF}(x,y) = -\tfrac{1}{2}C^{+}R^{+}\gamma^{0}\int d^4p(2\pi)^{-4}\; p^{+}e^{-ip\cdot(x-y)}/(p^2 + m^2 + i\varepsilon) \qquad \text{(H.262)}$$

H.6.3 Up-type Color Quarks

Up-type quarks are assumed[224] to be fermions with complex 3-momenta - complexons, and an internal color SU(3) symmetry, that satisfy $p^2 = m^2$. Their field equation with a color SU(3) index, denoted a, inserted is

---

[223] We call each type of fermion a *species*. Each species has three known generations.

[224] The complexon theory that we develop and use for quark dynamics in the Standard Model is <u>not</u> required. Our Standard Model could use Dirac fermion dynamics for the up-type quarks and tachyon dynamics for down-type quarks. Then the (broken) Left-handed complex Lorentz boosts would have the basic space-time group rather than L_C. We choose to use complexon dynamics for quarks because they have an internal SU(3)-like structure suggestive of color SU(3). More importantly, their spin dynamics is different and thus may resolve the differences between theory and experiment for the deep inelastic parton spin-dependent structure functions.

$$[i\gamma^0\partial/\partial t + i\boldsymbol{\gamma}\cdot(\nabla_r + i\nabla_i) - m]\psi_C{}^a(t, \mathbf{x_r}, \mathbf{x_i}) = 0 \tag{H.263}$$

with the subsidiary condition

$$\nabla_r\cdot\nabla_i\, \psi_C{}^a(t, \mathbf{x_r}, \mathbf{x_i}) = 0 \tag{H.264}$$

The free field solution is:

$$\psi_C{}^a(x) = \sum_{\pm s} \int d^3p_r d^3p_i\, N_C(p)\delta(\mathbf{p_r}\cdot\mathbf{p_i}/m^2)[b_C(p,a,s)u_C{}^a(p, s)e^{-i(p\cdot x + p^*\cdot x^*)/2} + $$
$$+ d_C{}^\dagger(p,a,s)v_C{}^a(p, s)e^{+i(p\cdot x + p^*\cdot x^*)/2}] \tag{H.265}$$

The free Feynman propagator arranged into the form of a spectral integral is

$$iS_C{}^{ab}(x,y) = -\delta^{ab}\int dM\, (i\gamma^0\partial/\partial x^0 - i(\nabla_r - i\nabla_i)\cdot\boldsymbol{\gamma} + m)\delta'(\nabla_r\cdot\nabla_i/m^2)J(\mathbf{x_i} - \mathbf{y_i}, M^2)\Delta_F(x - y, M) \tag{H.266}$$

where

$$\Delta_F(x - y, M) = (2\pi)^{-4}\int d^4p_r\, \exp[-ip^0(x^0 - y^0) + i\mathbf{p_r}\cdot(\mathbf{x_r} - \mathbf{y_r})]/(p_r{}^2 - M^2 + i\varepsilon) \tag{H.267}$$

and

$$J(\mathbf{x_i}, M^2) = (2\pi)^{-3}\int d^3p_i\, \delta(M^2 + \mathbf{p_i}^2 - m^2)\exp[-i\mathbf{p_i}\cdot(\mathbf{x_i} - \mathbf{y_i})] \tag{H.268}$$
$$= (2\pi)^{-2}|\mathbf{x_i} - \mathbf{y_i}|^{-1}\theta(m^2 - M^2)\sin((m^2 - M^2)^{1/2}|\mathbf{x_i} - \mathbf{y_i}|)$$

H.6.4 Down-type Color Quarks

Tachyonic complexons with complex 3-momenta, and an internal global SU(3) symmetry, that have mass shell condition $p^2 = -m^2$. Their field equation with a color SU(3) index, denoted a, inserted is

$$[\gamma^0\partial/\partial t + \boldsymbol{\gamma}\cdot(\nabla_r + i\nabla_i) - m]\psi_{CL}{}^a(t, \mathbf{x_r}, \mathbf{x_i}) = 0 \tag{H.269}$$

with the subsidiary condition on the wave function

$$\nabla_r \cdot \nabla_i \, \psi_{CL}{}^{a}(t, \mathbf{x_r}, \mathbf{x_i}) = 0 \qquad \text{(H.270)}$$

Its free field left-handed solution is:

$$\psi_{CLL}{}^{+a}(x_r, x_i) = \sum_{\pm s} \int d^2p_r dp^+ d^3p_i \, N_{CLL}{}^{+}(p)\theta(p^+)\delta((p_i{}^3(p^+ - p^-)/\surd 2 + \mathbf{p}_{r\perp}\cdot\mathbf{p}_{i\perp})/m^2)\cdot$$
$$\cdot[b_{CLL}{}^{+}(p,a,s)u_{CLL}{}^{a}(p,a,s)e^{-i(p\cdot x + p^*\cdot x^*)/2} + d_{CLL}{}^{+\dagger}(p,a,s)v_{CLL}{}^{+a}(p,a,s)e^{+i(p\cdot x + p^*\cdot x^*)/2}] \qquad \text{(H.271)}$$

and its right-handed solution is

$$\psi_{CLR}{}^{+a}(x_r, x_i) = \sum_{\pm s} \int d^2p_r dp^+ d^3p_i \, N_{CLR}{}^{+}(p)\theta(p^+)\delta((p_i{}^3(p^+ - p^-)/\surd 2 + \mathbf{p}_{r\perp}\cdot\mathbf{p}_{i\perp})/m^2)\cdot$$
$$\cdot[b_{CLR}{}^{+}(p,a,s)u_{CLR}{}^{+a}(p,a,s)e^{-i(p\cdot x + p^*\cdot x^*)/2} + d_{CLR}{}^{+\dagger}(p,a,s)v_{CLR}{}^{+a}(p,a,s)e^{+i(p\cdot x + p^*\cdot x^*)/2}] \qquad \text{(H.272)}$$

The free left-handed Feynman propagator arranged into the form of a spectral integral is

$$iS^{+}{}_{CLLF}{}^{ab}(x,y) = -\delta^{ab}\int dM\, C^-R^+(\gamma^0\partial/\partial x^0 + (\nabla_r - i\nabla_i)\cdot\boldsymbol{\gamma} - m)R^-C^+\delta'(\nabla_r\cdot\nabla_i/m^2)\cdot$$
$$\cdot J_2(\mathbf{x_i} - \mathbf{y_i}, M^2)\Delta_{FT}(x - y, M) \qquad \text{(H.273)}$$

with ∇_r and ∇_i derivatives with respect to $\mathbf{x_r}$ and $\mathbf{x_i}$ and where

$$\Delta_{FT}(x - y, M) = (2\pi)^{-4}\int d^4p_r \exp[-ip^0(x^0 - y^0) + i\mathbf{p_r}\cdot(\mathbf{x_r} - \mathbf{y_r})]/(p_r{}^2 + M^2 + i\varepsilon) \qquad \text{(H.274)}$$

and

$$J_2(\mathbf{x_i}, M^2) = (2\pi)^{-3}\int d^3p_i\,\delta(M^2 - \mathbf{p_i}^2 - m^2)\exp[-i\mathbf{p_i}\cdot(\mathbf{x_i} - \mathbf{y_i})] \qquad \text{(H.275)}$$
$$= (2\pi)^{-2}|\mathbf{x_i} - \mathbf{y_i}|^{-1}\theta(M^2 - m^2)\sin((M^2 - m^2)^{1/2}|\mathbf{x_i} - \mathbf{y_i}|)$$

The free right-handed Feynman propagator arranged into the form of a spectral integral is

$$iS^{+}{}_{CLRF}{}^{ab}(x, y) = \delta^{ab}\int dM\, C^{+}R^{+}(\gamma^{0}\partial/\partial x^{0} + (\nabla_{r} - i\nabla_{i})\cdot\boldsymbol{\gamma} - m)R^{-}C^{-}\delta'(\nabla_{r}\cdot\nabla_{i}/m^{2})\cdot \cdot J_{2}(\mathbf{x_i} - \mathbf{y_i}, M^{2})\Delta_{FT}(x - y, M) \qquad (H.276)$$

with ∇_r and ∇_i derivatives with respect to $\mathbf{x_r}$ and $\mathbf{x_i}$, and where

$$\Delta_{FT}(x - y, M) = (2\pi)^{-4}\int d^{4}p_{r}\, \exp[-ip^{0}(x^{0} - y^{0}) + i\mathbf{p}_{r}\cdot(\mathbf{x_r} - \mathbf{y_r})]/(p_{r}^{2} + M^{2} + i\varepsilon) \qquad (H.277)$$

H.7 First Step Towards The Superstandard Model

Thus we have derived a set of four fermion species that corresponds to the known fermions of one fermion generation from the Complex Lorentz Group.[225] We derive the four fermion generation form of the model based on a U(4) group that we call the Generation group. We derive the Generation group from conservation laws for baryon and lepton number.

The overall pattern that begins to emerge from the developments in this appendix divides particles and interactions into two categories (as seen in Nature):

Particles with real 4-Momenta	Complexons (Complex 3-Momenta)
Leptons	Color quarks
SU(2)⊗U(1) Vector Bosons	Color SU(3) gluons
Higgs Particles	Possibly Higgs Particles

Basically the leptons, SU(2)⊗U(1) Vector Bosons and a set of Higgs particles appear to be primarily based on the Left-handed boosts. These particles have real energies and momenta although some are "normal" and some are tachyons.

[225] Complex Lorentz group boosts lead to tachyons.

Another category of particles, complexons, emerges from our study of L_C. These particles have real energies and complex 3-momenta. In perturbation theory the loop integrations of loops of these particles would consist of a 7-fold integration over energy and complex 3-momenta with corresponding 7-fold delta functions to enforce energy-momentum conservation. As pointed out earlier the complex 3-momenta of these types of fermions has an SU(3) symmetry that it is natural to generalize to local color SU(3). (The other category of fermions, leptons. lack global SU(3) symmetry.) Thus we see the beginnings of the structure of the SuperStandard Theory in this appendix on spin ½ particles.

H.8 Dirac-like Equations of Matter from 4-Valued Logic

In our derivation every truly fundamental particle of matter, whether quark or lepton, has spin ½. We have seen in chapter 10 of Blaha (2011c) that the basic algebra of Operator Logic eigenvalue operators, and that of its raising and lowering operators, is the same as the algebra of creation and annihilarion operators for free spin ½ particles. Our goal is to build our theory on the scaffolding of Operator Logic. We view a fermion particle as a qube core which is dressed in spatial coordinates (and internal symmetries):

$$\text{Qube core + coordinates} \rightarrow \text{fermion particle} \qquad \text{(H.278)}$$

The creation and annihilation operators b(p,s) and $d^\dagger$(p,s) (and their hermitean conjugates $b^\dagger$(p,s) and d(p,s)) are mathematically similar to the raising and lowering operators of Operator (Matrix) Logic. They satisfy the anticommutation relations

$$\begin{aligned} \{b(q,s),\, b^\dagger(p,s')\} &= \delta_{ss'}\delta^3(\mathbf{q}-\mathbf{p}) \\ \{d(q,s),\, d^\dagger(p,s')\} &= \delta_{ss'}\delta^3(\mathbf{q}-\mathbf{p}) \end{aligned} \qquad \text{(H.279)}$$

Thus we see spin ½ particle wave functions originating from the Dirac-like spinors, and raising and lowering operators of the spinor formulation of Operator Logic.

When particles interact, the quantum field theory interaction terms use fermion creation operators, b(q,s) and $d^\dagger$(q,s), *and annihilation operators,* $b^\dagger$(p,s′) and d(q,s), *to implement the transformations between the Qubes of the interacting particles.*[226] *Thus the mathematics of the embedded Qubes' logic values is automatically implemented within quantum field theoretic calculations.*

An interesting point that emerges from this discussion is the nature of spin ½ particle states such as

$$|p, s> = b^{\dagger}(p, s)|0> \quad \text{(H.280)}$$

This state is interpreted as a one particle state. It also has an analogous interpretation in Operator Logic as creating a one term universe of discourse – a construct which is in part linguistic and in part logic. Thus particles are embodiments of Logic values and particle interactions change the logic values of the initial particles to those of the emergent particles. All in all, our universe can be viewed as an extraordinarily intricate logic machine. Serendipitously we are now seeing the use of particles to create quantum computers, which, in a sense, is bringing us full circle. Particles are Logic; Logic machines emerge from particle interactions.

H.9 Why Second Quantization of Fields?

One might have argued that the fermion field types that we have found could be treated as ordinary c-number fields and not be second quantized. However, particles are discrete entities that can be enumerated with integers. Second quantization implements the discrete particle concept in the most direct way and thus by Leibiz's Principle as well as Ockham's Razor second quantization is the best solution to obtain particle discreteness.

Quantum Theory is required by the discreteness of particles.

H.10 Why lagrangians? For dynamic evolution

Lagrangians naturally emerge as the 'preferred' formalism for quantum field theory due to their intimate relation with the energy-momentum tensor (particularly the Hamiltonian) that provides the generators of time evolution and of spatial translation.

---

[226] See chapter 3.

H.11 Functional Expression for Each of the four Species of Fermions

We have derived the four species of fermions in this appendix. We have used a 'conventional' notation for quantum fields. In this section we will define these quantum fields as inner products of functionals and fourier coordinate expansions.

Dirac Quantum Field:

$$\psi(x) = ({}_1f, \text{Dirac\_fourier\_expansion})$$

Tachyon Quantum Field:

$$\psi_T(x) = ({}_2f, \text{Tachyon\_fourier\_expansion})$$

Complexon Quantum Field:

$$\psi_C(x) = ({}_3f, \text{Complexon\_fourier\_expansion})$$

Complexon Tachyon Quantum Field:

$$\psi_{CT}(x) = ({}_4f, \text{Tachyon\_Complexon\_fourier\_expansion})$$

The digit prefixes of ${}_kf$ for k = 1, 2, 3, 4 distinguish the functionals for each species.

In addition we can decompose the above quantum fields into left-handed and right-handed fields. The left-handed functional representations are:

Left Dirac Quantum Field:

$$\psi_L(x) = ({}_{1L}f, \text{left-handed\_Dirac\_fourier\_expansion})$$

Left Tachyon Quantum Field:

$$\psi_{TL}(x) = ({}_{2L}f, \text{left-handed\_Tachyon\_fourier\_expansion})$$

Left Complexon Quantum Field:

$$\psi_{CL}(x) = ({}_{3L}f,\ \text{left-handed\_Complexon\_fourier\_expansion})$$

Left Complexon Tachyon Quantum Field:

$$\psi_{CTL}(x) = ({}_{4L}f,\ \text{left-handed\_Tachyon\_Complexon\_fourier\_expansion})$$

The right-handed cases have analogous forms.

H.12 Leptonic Tachyon Spinors

The general form of the solutions of the free tachyon Dirac equation can be written

$$\psi_T{}^r(x) = e^{-i\chi_r p\cdot x} w^r(p) \tag{H.281}$$

where $\chi_r = +1$ for r = 1, 2 and $\chi_r = -1$ for r = 3, 4. Denoting the spinors $w^r(p) = w^r(0)$ for a particle is at rest in a frame (E = m) we see they can take the form

$$w^r(0) = \begin{bmatrix} \delta_{1r} \\ \delta_{2r} \\ \delta_{3r} \\ \delta_{4r} \end{bmatrix} \tag{H.282}$$

where Kronecker deltas appear in the brackets. From eq. H.30 we find

$$S_L(\Lambda_L(\omega, \mathbf{u}))w^r(0) = w_T{}^r(p) \tag{H.283}$$

Using eq. H.66 for $S_L(\Lambda_L(\omega, \mathbf{u}))$ and

$$\mathbf{p} = m\mathbf{v}\gamma_s \qquad\qquad E = m\gamma_s \tag{H.284}$$

we see that eq. H.283 implies the columns of the resulting $S_L(\Lambda_L(\omega, \mathbf{u}))$ matrix are

$$
S_L(\Lambda_L(\omega, \mathbf{u})) = \begin{array}{c} \\ \end{array}
\begin{array}{cccc}
\underline{w_T^{3}(p)} & \underline{w_T^{4}(p)} & \underline{w_T^{1}(p)} & \underline{w_T^{2}(p)} \\
\end{array}
$$

$$
S_L(\Lambda_L(\omega, \mathbf{u})) = \begin{bmatrix}
\cosh(\omega_L/2) & 0 & \sinh(\omega_L/2)p_z/p & \sinh(\omega_L/2)p_-/p \\
0 & \cosh(\omega_L/2) & \sinh(\omega_L/2)p_+/p & -\sinh(\omega_L/2)p_z/p \\
\sinh(\omega_L/2)p_z/p & \sinh(\omega_L/2)p_-/p & \cosh(\omega_L/2) & 0 \\
\sinh(\omega_L/2)p_+/p & -\sinh(\omega_L/2)p_z/p & 0 & \cosh(\omega_L/2)
\end{bmatrix}
$$

(H.285)

based on the superluminal transformation of positive energy states to negative energy states with $p_\pm = p_x \pm ip_y$ and where $p = |\mathbf{p}|$. It is easy to verify

$$(i\not{p} - \chi_r m)w_T^{r}(p) = 0 \qquad \text{(H.286)}$$

where $\chi_r = -1$ for r = 1, 2 and $\chi_r = +1$ for r = 3, 4.

The spinors that we defined earlier can be generalized in a manner similar to Dirac spinors. We will use a similar notation to the Dirac spinor notation:

$$
\begin{aligned}
u_T(p, s) &= w_T^{1}(p) \\
u_T(p, -s) &= w_T^{2}(p) \\
v_T(p, s) &= w_T^{3}(p) \\
v_T(p, -s) &= w_T^{4}(p)
\end{aligned} \qquad \text{(H.287)}
$$

We define "double dagger" spinors:

$$
\begin{aligned}
u_T^{\ddagger}(p, s) &= u_T^{\dagger}(p, s)i\boldsymbol{\gamma}\cdot\mathbf{p}/|\mathbf{p}| \\
u_T^{\ddagger}(p, -s) &= u_T^{\dagger}(p, -s)i\boldsymbol{\gamma}\cdot\mathbf{p}/|\mathbf{p}| \\
v_T^{\ddagger}(p, s) &= v_T^{\dagger}(p, s)i\boldsymbol{\gamma}\cdot\mathbf{p}/|\mathbf{p}| \\
v_T^{\ddagger}(p, -s) &= v_T^{\dagger}(p, -s)i\boldsymbol{\gamma}\cdot\mathbf{p}/|\mathbf{p}|
\end{aligned} \qquad \text{(H.288)}
$$

.

where † indicates hermitean conjugate, which appear in important spinor "completeness" sums:

$$\sum_{\pm s} u_{T\alpha}(p, s)u_{T}{}^{\ddagger}{}_{\beta}(p, s) = (2m)^{-1}(i\not{p} - m)_{\alpha\beta} \tag{H.289}$$

$$\sum_{\pm s} v_{T\alpha}(p, s)v_{T}{}^{\ddagger}{}_{\beta}(p, s) = (2m)^{-1}(i\not{p} + m)_{\alpha\beta} \tag{H.290}$$

or

$$\sum_{\pm s} u_{T\alpha}(p, s)u_{T}{}^{\dagger}{}_{\beta}(p, s) = -i(2m)^{-1}[(i\not{p} - m)\boldsymbol{\gamma}\cdot\mathbf{p}/|\mathbf{p}|]_{\alpha\beta} \tag{H.291}$$

$$\sum_{\pm s} v_{T\alpha}(p, s)v_{T}{}^{\dagger}{}_{\beta}(p, s) = -i(2m)^{-1}[(i\not{p} + m)\boldsymbol{\gamma}\cdot\mathbf{p}/|\mathbf{p}|]_{\alpha\beta} \tag{H.292}$$

Lastly we define light-front, left-handed tachyon spinors by

$$\begin{aligned}
u_{TL}{}^{+}(p, s) &= C^{-} R^{+} S_L(\Lambda_L(\omega, \mathbf{u}))w^{1}(0) \\
u_{TL}{}^{+}(p, -s) &= C^{-} R^{+} S_L(\Lambda_L(\omega, \mathbf{u}))w^{2}(0) \\
v_{TL}{}^{+}(p, s) &= C^{-} R^{+} S_L(\Lambda_L(\omega, \mathbf{u}))w^{3}(0) \\
v_{TL}{}^{+}(p, -s) &= C^{-} R^{+} S_L(\Lambda_L(\omega, \mathbf{u}))w^{4}(0)
\end{aligned} \tag{H.293}$$

$$\begin{aligned}
u_{TL}{}^{+\dagger}(p, s) &= w^{1T}(0)\, S_L{}^{\dagger}(\Lambda_L(\omega, \mathbf{u}))\, R^{+}C^{-} \\
u_{TL}{}^{+\dagger}(p, -s) &= w^{2T}(0)\, S_L{}^{\dagger}(\Lambda_L(\omega, \mathbf{u}))R^{+}C^{-} \\
v_{TL}{}^{+\dagger}(p, s) &= w^{3T}(0)\, S_L{}^{\dagger}(\Lambda_L(\omega, \mathbf{u}))R^{+}C^{-} \\
v_{TL}{}^{+\dagger}(p, -s) &= w^{4T}(0)\, S_L{}^{\dagger}(\Lambda_L(\omega, \mathbf{u}))R^{+}C^{-}
\end{aligned} \tag{H.294}$$

where the superscript "T" indicates the transpose and † indicates hermitean conjugate.

Appendix I. Experimental Evidence for Faster-Than-Light Particles & Physics

Among the key assumptions of our Unified Superstandard Model are 1) that the speed of light is the same in all inertial reference frames and 2) that some fundamental particles (neutrinos and down-type quarks) travel faster than the speed of light.

In this appendix we describe convincing evidence for faster than light physics.

Until 1907 physicists thought that there was no limit on the speed of a particle or lump of matter. In 1907 Einstein and Poincaré showed that there was an inherent limit on the speed of a massive object – the speed of light. For the past 100 odd years physicists have generally accepted the speed of light as the limiting speed for particles with mass. Several theoretical physicists in the 1960's (E. C. Sudarshan and Gerald Feinberg) investigated the possibility of faster than light particles. They found that faster than light particles were theoretically possible but their theories – particularly their quantum field theories – had numerous discrepancies from canonical quantum field theory. These differences were taken by many to indicate that faster than light particles (called tachyons) were not present in nature. This belief was further supported by the happenings at particle accelerators where it was impossible to accelerate normal charged particles such as protons faster than the speed of light.

In the past fifteen years this author[227] developed a satisfactory quantum field theory of faster than light particles and found that if neutrinos and down-type quarks were faster than light particles he could derive the form of The Standard Model of Elementary Particles in detail. This theoretical development seems to have stimulated experimental groups at the new Linear Hadron Collider (LHC) at the CERN laboratory in Switzerland and the Gran Sasso Laboratory in Italy to measure the speed of neutrinos emitted in LHC particle collisions. The results, described below, were mixed and one can fairly say they neither proved nor disproved that neutrinos were tachyons.

---

[227] See Blaha (2012b) and earlier books extending back nine years.

However there is other experimental data that strongly indicate that neutrinos are tachyons, and that quantum mechanics requires – not just faster than light behavior – but in some circumstances instantaneous effects at a distance – infinite speed of transmission!

In this appendix we will look at experimentally proven instantaneous Quantum Mechanical effects, at tritium decay experiments over the past 20 years that imply faster than light neutrinos, at neutrino speed measurements at the CERN LHC and Gran Sasso, at tachyonic particle behavior inside of Black Holes, and at the tachyonic behavior of Higgs particles, the "so-called God particle." *The cumulative result of these considerations is that faster than light particles, and physics, are a part of nature.*

I.1 Instantaneous Quantum Mechanical Effects

Quantum entanglement is a quantum phenomenon wherein parts of a physical system are in a certain quantum state but are separated by a space-like distance. If a change is made in part of a quantum entangled system then it is known theoretically, and experimentally, that other parts of the system change instantaneously.[228] Many experiments have shown that the change in other parts of a system is instantaneous and thus can be viewed as taking place at infinite speed – obviously beyond the speed of light.[229] The most recent experiment by Juan Yin et al[230] has shown directly that quantum mechanical effects travel faster than 10,000 times the speed of light. These experimental results are consistent with the instantaneous speed predicted by quantum mechanics. Thus faster than light behavior is implicit in quantum theory and is experimentally verified.

I.2 Tritium Decay Experiments Yielding Neutrinos

Fact: Particles with negative values for the square of their mass are tachyons – particles moving faster than light.

---

[228] Matson, John, "Quantum Teleportation Achieved Over Record Distances" *Nature* **13**, August 2012.
[229] Francis, Matthew, "Quantum Entanglement Shows that Reality Can't be Local", *Ars Technica*, 30 October 2012.
[230] Juan Yin et al, arXiv[quant-ph]: 1303.0614V1 (March 4, 2013).

A series of experiments by various groups over recent years imply that electron neutrinos produced in tritium decay have negative mass squared despite the best efforts of experimenters to obtain positive values for the neutrino mass squared.

Experiment	measured mass squared	Year
Mainz	$-1.6 \pm 2.5 \pm 2.1$	2000
Troitsk	$-1.0 \pm 3.0 \pm 2.1$	2000
Zürich	$-24 \pm 48 \pm 61$	1992
Tokyo INS	$-65 \pm 85 \pm 65$	1991
Los Alamos	$-147 \pm 68 \pm 41$	1991
Livermore	$-130 \pm 20 \pm 15$	1995
China	$-31 \pm 75 \pm 48$	1995
1998 Average	-27 ± 20	1998

Table I.1 Electron neutrino mass squared values found in various tritium decay experiments. (Masses are in units of eV.) The average mass squared is negative suggesting electron neutrinos are tachyons.

Table I.1 summarizes the measured electron mass squared in these experiments. These experiments strongly suggest that neutrinos have negative mass squared and are thus faster-than-light particles - tachyons. However their small masses indicate that they only exceed the speed of light by a small amount.

I.3 LHC/Gran Sasso Direct Measurements of Neutrino Speeds

Two groups performed experiments at Gran Sasso Laboratory in Italy. They detected neutrinos emitted in interactions at the CERN LHC in Switzerland. The LVD collaboration in an exhaustive study of neutrino velocities found that the question was still open according to their data. Their refereed Physical Review Letter Abstract stated:

> We report the measurement of the time of flight of ν_μ on the CNGS baseline (732 km) with the Large Volume Detector (LVD) at the Gran Sasso Laboratory. The CERN-SPS accelerator has been operated from May 10th to May 24th 2012, with a tightly bunched-beam structure to allow the velocity of neutrinos to be accurately measured on an event-by-event basis. LVD has detected 48 neutrino events, associated with the beam, with a high absolute time accuracy. These events allow us to establish the following limit on the difference between the neutrino speed and the light velocity: $-3.8\times10^{-6} < (v_\nu - c)/c < 3.1\times10^{-6}$ (at 99% C.L.). This value is an order of magnitude lower than previous direct measurements.[231]

These results (involving at least 35 neutrino detections) slightly favor, and do not rule out, faster-than-light neutrinos. Another experiment at the same locations by the ATLAS group stated that they found neutrino velocities (Five neutrinos were measured.) were below c. This group has not published their results as yet. We conclude that the published data appears to support faster than light neutrinos – consistent with our theory of The Standard Model.

A new project is in the planning stages to measure neutrino beams at larger distances. The hope is that the masses of the various neutrinos will be determined by the experiment. If the neutrino mass squared values turn out to be negative then it will constitute additional proof that neutrinos are tachyons (confirming tritium decay data), and thus support this author's formulation of The Standard Model of Elementary Particles.

I.4 Tachyonic Behavior Within Black Holes

Inside a black hole (such as the Schwarzschild solution of General Relativity) the time coordinate effectively becomes a spatial coordinate and the radius coordinate effectively becomes a time coordinate. An in-falling particle has a constantly decreasing

---

[231] N. Yu. Agafonova et al. (LVD Collaboration), "Measurement of the Velocity of Neutrinos from the CNGS Beam with the Large Volume Detector" Phys. Rev. Lett. **109**, 070801 (15 August 2012).

radial distance from the center of the black hole just as time always increases outside a black hole.

As a result of the interchange of the roles of time and radius the velocity of a particle descending radially inside a Black Hole has a speed faster than light and is tachyonic.

I.5 Higgs Fields are Tachyons

Recently groups at the LHC CERN laboratory have announced the discovery of Higgs particles. The dynamic equations for Higgs bosons in The Standard Model have a negative mass squared. The mass squared must be negative or the Higgs Mechanism could not generate particle masses. Having negative mass terms implies that Higgs fields are tachyonic – faster than light particles. Their tachyonic nature is masked by a quartic self-interaction that generates a condensate and thereby the masses of other particles.

I.6 Conclusion: Faster-Than-Light Particles – Tachyons Exist in Nature

The bulk of the experimental and theoretical evidence presented in previous sections strongly favors the existence of faster-than-light particles such as neutrinos. Tachyonic neutrinos are an important part of our form of The Standard Model. This form of the theory also strongly suggests that quarks are tachyonic in parallel with tachyonic neutrinos in order to obtain the symmetries of The Standard Model.

Appendix J. SuperStandard Axioms

J.1 Basis of SuperStandard Axioms

In the following sections[232] we present a revised set of 'primitive' terms and axioms for our theory. A comparison of this new set of axioms with those provided in earlier editions will show that they are equivalent except for a few new axioms. They are also more simply stated, have fewer overlaps between axioms, and cleanly lead to our theory of elementary particles.

The goal of this edition is to derive the Unified Superstandard Model in the manner of Euclid with a clear connection between the steps of the derivation just as Euclid developed geometry from a progression of theorems.

J.2 Primitive Terms and Axioms

Primitive terms can be as simple as those of Euclid or they can be more complex. The level of simplicity depends on the nature of the theory and the Physical Laws that emerge from it. In the case at hand, a fundamental unified theory, the constructs that emerge in the construction of the theory are mathematically complex. Consequently, the choice of primitive terms and axioms may be expected to be mathematically complex as well, unless one wishes to expand the primitive terms into a more detailed, term by term description in simpler, more basic primitives. We will not pursue that alternative here since the terms that we use are 'self-explanatory' to the Elementary Particle Physics theorist knowledgeable about quantum field theory and particle symmetries.

---

[232] From Blaha (2019b).

J.3 Mathematics and Conceptual Prerequisites

Due to the complexity of the Theory we have chosen to specify mathematics prerequisites and use them in the derivation rather than devoting parts of the derivation to mathematical preliminaries. Therefore we use complex variable theory, Riemannian coordinates, group theory, classical and Quantum Logic, functionals, Chomsky-like computational languages, and so on without bringing in unnecessary supporting details from them.

We also assume certain physical concepts such as distance, quantum features, second quantization, covariance under a group transformation, and spatial curvature.

The axioms use some of these prerequisite concepts treating them as primitive terms for the derivation.

J.4 Primitive Terms for the Unified Superstandard Model

The set of primitive terms of the theory are:

Qubits
Qubes
Qubas
Core
Grammar
Terminal and Nonterminal Symbols
Production Rules
Speed of Light
Spatial Dimensions
Space and Time Coordinates
Covariance under group transformations
Asynchronous processes
Parallel Processes
Reference Frame
Complex Lorentz Group
General Coordinate Transformations
Gravity
Universe
Particle Masses

Fermions
Bosons
Particle States
Particle Rest State
Particle Momenta
Spin
Canonical Quantization
Quantum Process
Quantum Entanglement
Second Quantization
Quantum Field Theory
Quantum States
Asymptotic Particle States
Internal Symmetries
Coupling Constants
Discrete Symmetries
Yang-Mills Local Gauge Theory
Functionals
Functional space

In choosing these primitives, we understand that they generally embody a significant theoretic description or body of knowledge. We do not include names used in the mapping to reality (such as quark) in the list of primitives since the mapping to reality is a separate issue in our view.

J.5 Axioms for the Unified Superstandard Model

The set of axioms that we list below is supplemented by the Decision Axioms of Appendix C.1.3 of Blaha (2018e). The 'new' physical axioms are

PARTICLE AXIOMS

1. All matter and energy is composed of particles.
2. Each fundamental particle has a physico-logic structure within it that we designate its core.

3. Particles form an alphabet with a finite number of characters and combine in ways specified by the quantum probabilistic production rules of a quantum computational grammar.[233]
4. A core is a particle functional that combines with a free field fourier coordinate expansion in an inner product to produce a free second quantized particle field.
5. There is a 4-dimensional space of particle functionals, called *particle functional space*, with the distance measure, eq. 2.1 below, specifying the transformation group of particle functionals.
6. Particle functional space consists of a single point.
7. The core of a fermion functional is called a *qube*. Fundamental bosons have a core consisting of a boson functional called a *quba*.
8. Qubes have a a bare mass. Qubas have zero mass.

SPACE AXIOMS

9. The dimensions of a coordinate space-time are determined by the number of fundamental[234] interactions, and the requirement that all parallel processes, with parts perhaps separated by distances, can occur synchronously.
10. Spatial coordinates are inherently complex-valued.
11. Space has one complex-valued component that plays the role of time. Physical phenomena dynamically evolve based on the time variable.
12. The infinitesimal distance ds between two space-time points is given by

$$ds^2 = dt^2 - d\mathbf{x}^2 \tag{J.1}$$

where d**x** is a vector of the spatial coordinates. Transformations between coordinate systems preserve the value of ds and define a transformation group. (The Complex Lorentz Group)
13. Physically acceptable reference frames have real-valued coordinates. These coordinates can be obtained by group transformations from complex-valued

---

[233] See Blaha (2005b).

[234] Interactions that would exist in the absence of fermion particles.

coordinate systems. Physical space-time measurements are made in a real-valued coordinate system.

14. The speed of light is the same in all reference frames.
15. Free fundamental leptons must have a real-valued energy.
16. Gravity may cause space-time to be curved. (Complex General Coordinate transformations[235])

DYNAMICS AXIOMS

17. The complete theory has a lagrangian formulation. If the lagrangian is truncated to quadratic form (interactions set to zero) then symmetries appear that are the source of particle symmetry groups that persist with broken symmetry after interactions are reintroduced. The lagrangian specifies a set of production rules of a type 0 Chomsky language generalized to include production rules for the generation of all strings of symbols (particles) from any strings of symbols (including the *head symbol*.)[236]
18. The lagrangian of the theory must be invariant under coordinate system transformations.
19. Dynamical particle equations must be covariant under group transformations.
20. All interactions have a local Yang-Mills gauge theory formulation.
21. The vector bosons, and the interactions among them, are determined by terms in complete lagrangian, some of whose parts are obtained from the Riemann-Christoffel Curvature Tensor.

---

[235] If the metric tensor of space-time is analogous to one of the metric tensors of the superfluid phases of ^{3}He, then space-time might have several metric tensors in 'various regions.' If the space-time metric tensor is analogous to the ^{3}He-B superfluid phase metric tensor, which has an effective gravity with a complex metric tensor, the space-time metric tensor would be the familiar one of General Relativity. However if the space-time metric tensor is analogous to the metric tensor of superfluid ^{3}He-A, which exists at higher pressure and temperature, then the space-time metric tensor might be similar to the Penrose twistor theory metric tensor. In this case the corresponding General Relativity may have a twistor-like metric tensor: perhaps in the early universe, and/or inside black holes, and/or in small universes with higher pressure and temperature than our universe. We will assume the conventional metric for Complex Special and General Relativity.

[236] Chapter 8 of Blaha (2018b) discusses computational languages for particles in detail.

QUANTIZATION AXIOMS

22. All fields must be canonically quantized.
23. Fermion and Boson vacua can be defined that are valid in all coordinate systems.
24. The number of particles in an asymptotic state of any given type is invariant in all reference frames.
25. Quantum processes starting in an initial quantum state, with parts separated by a distance after a time, can have the parts synchronously change each other instantaneously. (Quantum Entanglement)

J.6 The Derivation of the Unified Superstandard Model

The derivation of the Unified Superstandard Model has been a multi-year process undertaken by the author. Much of the derivation appears in Blaha (2015a), (2016f), (2017b), (2017c), (2017d), (2018a), (2018b), and (2018c). Earlier work, upon which these books are based, is referenced in these books and listed in the References in this book.

We now show a clear logical development of the Unified Superstandard Model from first principles in a manner reminiscent of the derivation of Euclidean geometry. This derivation will be seen to be primarily based on a 'simple' concept—the space-time in our universe. The derivation explains the construction process of the physical theory. The manner of the derivation embodies the mapping of the theory to physical reality. The question of the 'Unmoved Mover' (Appendix C of Blaha (2018e)) is necessarily beyond the scope of Physics. (See the God Theory section of Blaha (2018e).)

Appendix K. Some Physical Constants Used in Calculating Numerical Expressions

Some physical constants that we found to be of use in the evaluation of expressions are (assuming units with $c = \hbar = 1$):

$$G \equiv 7.39 \times 10^{-29}\ \text{gm}^{-1}\ \text{cm} \quad \text{(K.1a)}$$
$$G \equiv 2.6 \times 10^{-66}\ \text{cm}^2 \quad \text{(K.1b)}$$
$$G \equiv 2.91 \times 10^{-87}\ \text{s}^2 \quad \text{(K.1c)}$$
$$G = 6.723 \times 10^{-57}\ \text{eV}^{-2} \quad \text{(K.1d)}$$

$$H_0 \equiv 2.133 \times 10^{-33}\text{h ev} \quad \text{(K.2a)}$$
$$H_0 \equiv 1.08 \times 10^{-28}\text{h cm}^{-1} \quad \text{(K.2b)}$$
$$H_0 \equiv 3.24 \times 10^{-18}\text{h s}^{-1} \quad \text{(K.2c)}$$
$$h \ = 0.689 \quad \text{(K.2d)}$$
$$H_0 = 100\ \text{h km s}^{-1}\ \text{Mpc}^{-1} \quad \text{(K.2e)}$$
$$1\ \text{s}^{-1} = 3.0864 \times 10^{-20}\ \text{km s}^{-1}\ \text{Mpc}^{-1} \quad \text{(K.2f)}$$

$$GH_0 \equiv 7.98 \times 10^{-57}\text{h gm}^{-1} \quad \text{(K.3)}$$

$$\rho_{crit} \equiv 1.88 \times 10^{-29}\text{h}^2\ \text{gm cm}^{-3} \quad \text{(K.4)}$$

$$M_{Planck} \equiv 1.22 \times 10^{28}\ \text{ev} \quad \text{(K.5a)}$$
$$M_{Planck} \equiv 2.18 \times 10^{-5}\ \text{g} \quad \text{(K.5b)}$$
$$M_{Planck} \equiv 6.20 \times 10^{32}\ \text{cm}^{-1} \quad \text{(K.6a)}$$
$$\text{Planck Length} = M_{Planck}{}^{-1} \equiv 1.61 \times 10^{-33}\ \text{cm} \quad \text{(K.6b)}$$
$$M_{Planck} \equiv 1.85 \times 10^{43}\ \text{s}^{-1} \quad \text{(K.7a)}$$

$$\text{Planck time} = M_{\text{Planck}}^{-1} \equiv 5.41 \times 10^{-44}\ \text{s} \qquad \text{(K.7b)}$$
$$1\ \text{eV} \equiv 5.08 \times 10^{4}\ \text{cm}^{-1} \qquad \text{(K.8a)}$$
$$1\ \text{eV} \equiv 1.52 \times 10^{15}\ \text{s}^{-1} \qquad \text{(K.8b)}$$
$$1\ \text{eV/c}^2 \equiv 1.783 \times 10^{-33}\ \text{g} \qquad \text{(K.8c)}$$
$$1\ \text{g} \equiv 2.85 \times 10^{37}\ \text{cm}^{-1} \qquad \text{(K.8d)}$$

$$\kappa \equiv 4.38\ {}^\circ\text{K}^{-1}\ \text{cm}^{-1} \qquad \text{(K.9)}$$
$$\kappa \equiv 1.31 \times 10^{11}\ {}^\circ\text{K}^{-1}\ \text{s}^{-1} \qquad \text{(K.10)}$$
$$\kappa \equiv 8.62 \times 10^{-5}\ \text{ev}\ {}^\circ\text{K}^{-1} \qquad \text{(K.11)}$$

$$1\ \text{Gyr} = 3.16 \times 10^{16}\ \text{s} \qquad \text{(KJ.12)}$$

where κ is Boltzmann's constant.

Constants based on M. Tanabashi *et al op. cit.*

h = Hubble Constant = 0.678(9)

ρ_{crit} = Critical density = $1.87840(9)\ h^2 \times 10^{-29}$ g/cm^{-3}

Ω_Λ = Dark Energy density/ρ_{cr} = ρ_{de}/ρ_{cr} = 0.692 ± 0.012

$\Omega_d = \Omega_c$ = Cold Dark matter density/ρ_{cr} = ρ_c/ρ_{cr} = $0.1186(20)\ h^{-2}$

Ω_b = Baryon density/ρ_{cr} = ρ_b/ρ_{cr} = $0.02226\ h^{-2}$

$\Omega_m = \Omega_b + \Omega_c$

= pressureless Matter density/ρ_{cr} = ρ_m/ρ_{cr} = 0.308 ± 0.012

$\Omega_{mtot} = \Omega_m + \Omega_\Lambda = 1.000 \pm 0.024$

$t_0 = t_{now}$ = Age of universe = 13.80 ± 0.04 Gyr

$r_{universe}(t_{now})$ = visible radius of the universe = 4.314×10^{28} cm

Ω_γ = radiation density/ρ_{cr} = ρ_γ/ρ_{cr} = $2.473h^{-2} \times 10^{-5}\ (T/2.7255)^4 h^{-2}$

$= 5.38 \times 10^{-5}$

$k = 5.56 \times 10^{-57}\ \text{cm}^{-2}$

REFERENCES

Akhiezer, N. I., Frink, A. H. (tr), 1962, *The Calculus of Variations* (Blaisdell Publishing, New York, 1962).

Bjorken, J. D., Drell, S. D., 1964, *Relativistic Quantum Mechanics* (McGraw-Hill, New York, 1965).

Bjorken, J. D., Drell, S. D., 1965, *Relativistic Quantum Fields* (McGraw-Hill, New York, 1965).

Blaha, S., 1998, *Cosmos and Consciousness* (Pingree-Hill Publishing, Auburn, NH, 1998).

_______, 2002, *A Finite Unified Quantum Field Theory of the Elementary Particle Standard Model and Quantum Gravity Based on New Quantum Dimensions™ & a New Paradigm in the Calculus of Variations* (Pingree-Hill Publishing, Auburn, NH, 2002).

_______, 2003, *A Finite Unified Quantum Field Theory of the Elementary Particle Standard Model and Quantum Gravity Based on New Quantum Dimensions™ and a New Paradigm in the Calculus of Variations* (Pingree-Hill Publishing, Auburn, NH, 2003).

_______, 2004, *Quantum Big Bang Cosmology: Complex Space-time General Relativity, Quantum Coordinates™ Dodecahedral Universe, Inflation, and New Spin 0, ½, 1 & 2 Tachyons & Imagyons* (Pingree-Hill Publishing, Auburn, NH, 2004).

______, *2005a, Quantum Theory of the Third Kind: A New Type of Divergence-free Quantum Field Theory Supporting a Unified Standard Model of Elementary Particles and Quantum Gravity based on a New Method in the Calculus of Variations* (Pingree-Hill Publishing, Auburn, NH, 2005).

______, 2005b, *The Metatheory of Physics Theories, and the Theory of Everything as a Quantum Computer Language* (Pingree-Hill Publishing, Auburn, NH, 2005).

______, 2005c, *The Equivalence of Elementary Particle Theories and Computer Languages: Quantum Computers, Turing Machines, Standard Model, Superstring Theory, and a Proof that Gödel's Theorem Implies Nature Must Be Quantum* (Pingree-Hill Publishing, Auburn, NH, 2005).

______, 2006a, *The Foundation of the Forces of Nature* (Pingree-Hill Publishing, Auburn, NH, 2006).

______, 2006b, *A Derivation of ElectroWeak Theory based on an Extension of Special Relativity; Black Hole Tachyons; & Tachyons of Any Spin.* (Pingree-Hill Publishing, Auburn, NH, 2006).

_______, 2007a, *Physics Beyond the Light Barrier: The Source of Parity Violation, Tachyons, and A Derivation of Standard Model Features* (Pingree-Hill Publishing, Auburn, NH, 2007).

______, 2007b, *The Origin of the Standard Model: The Genesis of Four Quark and Lepton Species, Parity Violation, the ElectroWeak Sector, Color SU(3), Three Visible Generations of Fermions, and One Generation of Dark Matter with Dark Energy* (Pingree-Hill Publishing, Auburn, NH, 2007).

______*, 2008a, A Direct Derivation of the Form of the Standard Model From GL(16) (Pingree-Hill Publishing, Auburn, NH, 2008).*

______, 2008b, *A Complete Derivation of the Form of the Standard Model With a New Method to Generate Particle Masses Second Edition* (Pingree-Hill Publishing, Auburn, NH, 2008)

______, 2009, *The Algebra of Thought & Reality: The Mathematical Basis for Plato's Theory of Ideas, and Reality Extended to Include A Priori Observers and Space-Time Second Edition* (Pingree-Hill Publishing, Auburn, NH, 2009).

______, 2010a, *Operator Metaphysics: A New Metaphysics Based on a New Operator Logic and a New Quantum Operator Logic that Lead to a Mathematical Basis for Plato's Theory of Ideas and Reality* (Pingree-Hill Publishing, Auburn, NH, 2010).

______, 2010b, *The Standard Model's Form Derived from Operator Logic, Superluminal Transformations and GL(16)* (Pingree-Hill Publishing, Auburn, NH, 2010).

______, 2010c, *SuperCivilizations: Civilizations as Superorganisms* (McMann-Fisher Publishing, Auburn, NH, 2010).

______, 2011a, *21st Century Natural Philosophy Of Ultimate Physical Reality* (McMann-Fisher Publishing, Auburn, NH, 2011).

______, 2011b, *All the Universe! Faster Than Light Tachyon Quark Starships & Particle Accelerators with the LHC as a Prototype Starship Drive Scientific Edition* (Pingree-Hill Publishing, Auburn, NH, 2011).

______, 2011c, *From Asynchronous Logic to The Standard Model to Superflight to the Stars* (Blaha Research, Auburn, NH, 2011).

______, 2012a, *From Asynchronous Logic to The Standard Model to Superflight to the Stars volume 2: Superluminal CP and CPT, U(4) Complex General Relativity and The Standard Model, Complex Vierbein General Relativity, Kinetic Theory, Thermodynamics* (Blaha Research, Auburn, NH, 2012).

______, 2012b, *Standard Model Symmetries, And Four And Sixteen Dimension Complex Relativity; The Origin Of Higgs Mass Terms* (Blaha Reasearch, Auburn, NH, 2012).

______, 2013a, *Multi-Stage Space Guns, Micro-Pulse Nuclear Rockets, and Faster-Than-Light Quark-Gluon Ion Drive Starships* (Blaha Research, Auburn, NH, 2013).

______, 2013b, *The Bridge to Dark Matter; A New Sister Universe; Dark Energy; Inflatons; Quantum Big Bang; Superluminal Physics; An Extended Standard Model Based on Geometry* (Blaha Reasearch, Auburn, NH, 2013).

______, 2014a, *Universes and Megaverses: From a New Standard Model to a Physical Megaverse; The Big Bang; Our Sister Universe's Wormhole; Origin of the Cosmological Constant, Spatial Asymmetry of the Universe, and its Web of Galaxies; A Baryonic Field between Universes and Particles; Megaverse Extended Wheeler-DeWitt Equation* (Blaha Reasearch, Auburn, NH, 2014).

______, 2014b, *All the Megaverse! Starships Exploring the Endless Universes of the Cosmos Using the Baryonic Force* (Blaha Research, Auburn, NH, 2014).

______, 2014c, *All the Megaverse! II Between Megaverse Universes: Quantum Entanglement Explained by the Megaverse Coherent Baryonic Radiation Devices – PHASERs Neutron Star Megaverse Slingshot Dynamics Spiritual and UFO Events, and the Megaverse Microscopic Entry into the Megaverse* (Blaha Research, Auburn, NH, 2014).

______, 2015a, *PHYSICS IS LOGIC PAINTED ON THE VOID: Origin of Bare Masses and The Standard Model in Logic, U(4) Origin of the Generations, Normal and Dark Baryonic Forces, Dark Matter, Dark Energy, The Big Bang, Complex General Relativity, A Megaverse of Universe Particles* (Blaha Research, Auburn, NH, 2015).

______, 2015b, *PHYSICS IS LOGIC Part II: The Theory of Everything, The Megaverse Theory of Everything, U(4)⊗U(4) Grand Unified Theory (GUT), Inertial Mass = Gravitational Mass, Unified Extended Standard Model and a New Complex General Relativity with Higgs Particles, Generation Group Higgs Particles* (Blaha Research, Auburn, NH, 2015).

______, 2015c, *The Origin of Higgs ("God") Particles and the Higgs Mechanism: Physics is Logic III, Beyond Higgs – A Revamped Theory With a Local Arrow of Time, The Theory of Everything Enhanced, Why Inertial Frames are Special, Universes of the Mind* (Blaha Research, Auburn, NH, 2015).

______, 2015d, *The Origin of the Eight Coupling Constants of The Theory of Everything: U(8) Grand Unified Theory of Everything (GUTE), S^8 Coupling Constant Symmetry, Space-Time Dependent Coupling Constants, Big Bang Vacuum Coupling Constants, Physics is Logic IV* (Blaha Research, Auburn, NH, 2015).

______, 2016a, *New Types of Dark Matter, Big Bang Equipartition, and A New U(4) Symmetry in the Theory of Everything: Equipartition Principle for Fermions, Matter is 83.33% Dark,*

Penetrating the Veil of the Big Bang, Explicit QFT Quark Confinement and Charmonium, Physics is Logic V (Blaha Research, Auburn, NH, 2016).

______, 2016b, *The Periodic Table of the 192 Quarks and Leptons in The Theory of Everything: The U(4) Layer Group, Physics is Logic VI* (Blaha Research, Auburn, NH, 2016).

______, 2016c, *New Boson Quantum Field Theory, Dark Matter Dynamics, Dark Matter Fermion Layer Mixing, Genesis of Higgs Particles, New Layer Higgs Masses, Higgs Coupling Constants, Non-Abelian Higgs Gauge Fields, Physics is Logic VII* (Blaha Research, Auburn, NH, 2016).

______, 2016d, *Unification of the Strong Interactions and Gravitation: Quark Confinement Linked to Modified Short-Distance Gravity; Physics is Logic VIII* (Blaha Research, Auburn, NH, 2016).

______, 2016e, *MoND: Unification of the Strong Interactions and Gravitation II, Quark Confinement Linked to Large-Scale Gravity, Physics is Logic IX* (Blaha Research, Auburn, NH, 2016).

______, 2016f, *CQ Mechanics: A Unification of Quantum & Classical Mechanics, Quantum/Semi-Classical Entanglement, Quantum/Classical Path Integrals, Quantum/Classical Chaos* (Blaha Research, Auburn, NH, 2016).

______, 2016g, *GEMS: Unified Gravity, ElectroMagnetic and Strong Interactions: Manifest Quark Confinement, A Solution for the Proton Spin Puzzle, Modified Gravity on the Galactic Scale* (Pingree Hill Publishing, Auburn, NH, 2016).

______, 2016h, *Unification of the Seven Boson Interactions based on the Riemann-Christoffel Curvature Tensor* (Pingree Hill Publishing, Auburn, NH, 2016).

______, 2017a, *Unification of the Eleven Boson Interactions based on 'Rotations of Interactions'* (Pingree Hill Publishing, Auburn, NH, 2017).

______, 2017b, *The Origin of Fermions and Bosons, and Their Unification* (Pingree Hill Publishing, Auburn, NH, 2017).

______, 2017c, *Megaverse: The Universe of Universes* (Pingree Hill Publishing, Auburn, NH, 2017).

______, 2017d, *SuperSymmetry and the Unified SuperStandard Theory* (Pingree Hill Publishing, Auburn, NH, 2017).

______, 2017e, *From Qubits to the Unified SuperStandard Theory with Embedded SuperStrings: A Derivation* (Pingree Hill Publishing, Auburn, NH, 2017).

______, 2017f, *The Unified SuperStandard Theory in Our Universe and the Megaverse: Quarks, … ,* (Pingree Hill Publishing, Auburn, NH, 2017).

______, 2018a, *The Unified SuperStandard Theory and the Megaverse SECOND EDITION A Deeper Theory based on a New Particle Functional Space that Explicates Quantum Entanglement Spookiness (Volume 1)* (Pingree Hill Publishing, Auburn, NH, 2018).

______, 2018b, *Cosmos Creation: The Unified SuperStandard Theory, Volume 2, SECOND EDITION* (Pingree Hill Publishing, Auburn, NH, 2018).

______, 2018c, *God Theory (*Pingree Hill Publishing, Auburn, NH, 2018).

______, 2018d, *Immortal Eye: God Theory: Second Edition* (Pingree Hill Publishing, Auburn, NH, 2018).

_____, 2018e, *Unification of God Theory and Unified SuperStandard Theory THIRD EDITION* (Pingree Hill Publishing, Auburn, NH, 2018).

______, 2019a, *Calculation of: QED α = 1/137, and Other Coupling Constants of the Unified SuperStandard Theory* (Pingree Hill Publishing, Auburn, NH, 2019).

______, 2019b, *Coupling Constants of the Unified SuperStandard Theory SECOND EDITION* (Pingree Hill Publishing, Auburn, NH, 2019).

______, 2019c, *New Hybrid Quantum Big_Bang-Megaverse_Driven Universe with a Finite Big Bang and an Increasing Hubble Constant* (Pingree Hill Publishing, Auburn, NH, 2019).

______, 2019d, *The Universe, the Electron, and the Vacuum* (Pingree Hill Publishing, Auburn, NH, 2019).

______, 2019e, *Quantum Big Bang-Quantum Vacuum Universes (Particles)* (Pingree Hill Publishing, Auburn, NH, 2019).

______, 2019f, *The Exact QED Calculation of the Fine Structure Constant Implies ALL 4D Universes have the Same Physics/Life Prospects* (Pingree Hill Publishing, Auburn, NH, 2019).

Eddington, A. S., 1952, *The Mathematical Theory of Relativity* (Cambridge University Press, Cambridge, U.K., 1952).

Fant, Karl M., 2005, *Logically Determined Design: Clockless System Design With NULL Convention Logic* (John Wiley and Sons, Hoboken, NJ, 2005).

Feinberg, G. and Shapiro, R., 1980, *Life Beyond Earth: The Intelligent Earthlings Guide to Life in the Universe* (William Morrow and Company, New York, 1980).

Gelfand, I. M., Fomin, S. V., Silverman, R. A. (tr), 2000, *Calculus of Variations* (Dover Publications, Mineola, NY, 2000).

Giaquinta, M., Modica, G., Souchek, J., 1998, *Cartesian Coordinates in the Calculus of Variations* Volumes I and II (Springer-Verlag, New York, 1998).

Giaquinta, M., Hildebrandt, S., 1996, *Calculus of Variations* Volumes I and II (Springer-Verlag, New York, 1996).

Gradshteyn, I. S. and Ryzhik, I. M., 1965, *Table of Integrals, Series, and Products* (Academic Press, New York, 1965).

Heitler, W., 1954, *The Quantum Theory of Radiation* (Claendon Press, Oxford, UK, 1954).

Huang, Kerson, 1992, *Quarks, Leptons & Gauge Fields 2nd Edition* (World Scientific Publishing Company, Singapore, 1992).

Jost, J., Li-Jost, X., 1998, *Calculus of Variations* (Cambridge University Press, New York, 1998).

Kaku, Michio, 1993, *Quantum Field Theory*, (Oxford University Press, New York, 1993).

Kirk, G. S. and Raven, J. E., 1962, *The Presocratic Philosophers* (Cambridge University Press, New York, 1962).

Landau, L. D. and Lifshitz, E. M., 1987, *Fluid Mechanics 2nd Edition*, (Pergamon Press, Elmsford, NY, 1987).

Misner, C. W., Thorne, K. S., and Wheeler, J. A., 1973, *Gravitation* (W. H. Freeman, New York, 1973).

Rescher, N., 1967, *The Philosophy of Leibniz* (Prentice-Hall, Englewood Cliffs, NJ, 1967).

Rieffel, Eleanor and Polak, Wolfgang, 2014, *Quantum Computing* (MIT Press, Cambridge, MA, 2014).

Riesz, Frigyes and Sz.-Nagy, Béla, 1990, *Functional Analysis* (Dover Publications, New York, 1990).
Sagan, H., 1993, *Introduction to the Calculus of Variations* (Dover Publications, Mineola, NY, 1993).

Sakurai, J. J., 1964, *Invariance Principles and Elementary Particles* (Princeton University Press, Princeton, NJ, 1964).

Sorokin, Pitirim, 1941, *Social and Cultural Dynamics* (Porter Sargent Publishers, Boston, MA, 1941).

Streater, R. F. and Wightman, A. S., 2000, *PCT, Spin, Statistics, and All That* (Princeton University Press, Princeton, NJ 2000).

Weinberg, S., 1972, *Gravitation and Cosmology* (John Wiley and Sons, New York, 1972).

Weinberg, S., 1995, *The Quantum Theory of Fields Volume I* (Cambridge University Press, New York, 1995).

Weinberg, S., 2000, *The Quantum Theory of Fields Volume III Supersymmetry* (Cambridge University Press, New York, 2000).

Weyl, H., 1950, *Space, Time, Matter* (Dover, New York, 1950).

Weyl, H., (Tr. S. Pollard et al), 1987, *The Continuum* (Dover Publications, New York, 1987).

INDEX

About the Author

Stephen Blaha is a well known Physicist and Man of Letters with interests in Science, Society and civilization, the Arts, and Technology. He had an Alfred P. Sloan Foundation scholarship in college. He received his Ph.D. in Physics from Rockefeller University. He has served on the faculties of several major universities. He was also a Member of the Technical Staff at Bell Laboratories, a manager at the Boston Globe Newspaper, a Director at Wang Laboratories, and President of Blaha Software Inc and of Janus Associates Inc. (NH).

Among other achievements he was a co-discoverer of the "r potential" for heavy quark binding developing the first (and still the only demonstrable) non-abelian gauge theory with an "r" potential; first suggested the existence of topological structures in superfluid He-3; first proposed Yang-Mills theories would appear in condensed matter phenomena with non-scalar order parameters; first developed a grammar-based formalism for quantum computers and applied it to elementary particle theories; first developed a new form of quantum field theory without divergences (thus solving a major 60 year old

problem that enabled a unified theory of the Standard Model and Quantum Gravity without divergences to be developed); first developed a formulation of complex General Relativity based on analytic continuation from real space-time; first developed a generalized non-homogeneous Robertson-Walker metric that enabled a quantum theory of the Big Bang to be developed without singularities at t = 0; first generalized Cauchy's theorem and Gauss' theorem to complex, curved multi-dimensional spaces; received Honorable Mention in the Gravity Research Foundation Essay Competition in 1978; first developed a physically acceptable theory of faster-than-light particles; first derived a composition of extrema method in the Calculus of Variations; first quantitatively suggested that inflationary periods in the history of the universe were not needed; first proved Gödel's Theorem implies Nature must be quantum; provided a new alternative to the Higgs Mechanism, and Higgs particles, to generate masses; first showed how to resolve logical paradoxes including Gödel's Undecidability Theorem by developing Operator Logic and Quantum Operator Logic; first developed a quantitative harmonic oscillator-like model of the life cycle, and interactions, of civilizations; first showed how equations describing superorganisms also apply to civilizations. A recent book shows his theory applies successfully to the past 14 years of history and to *new* archaeological data on Andean and Mayan civilizations as well as Early Anatolian and Egyptian civilizations.

He first developed an axiomatic derivation of the form of The Standard Model from geometry – space-time properties – The Unified SuperStandard Theory. It unifies all the known forces of Nature. It also has a Dark Matter sector that includes a Dark ElectroWeak sector with Dark doublets and Dark gauge interactions. It uses quantum coordinates to remove infinities that crop up in most interacting quantum field theories and additionally to remove the infinities that appear in the Big Bang and generate inflationary growth of the universe. It shows gravity has a MOND-like form without sacrificing Newton's Laws. It relates the interactions of the MOND-like sector of gravity with the r-potential of Quark Confinement. The axioms of the theory lead to the question of their origin. We suggest in the preceding edition of this book it can be attributed to an entity with God-like properties. We explore these properties in "God Theory" and show they predict that the Cosmos exists forever although individual universes (or incarnations of our universe) "come and go." Several other important results emerge from God Theory such a a functionally triune God. The Unified Superstandard Model has many other important parts described in the Current Edition of *The Unified Superstandard Model* and expanded in subsequent volumes.

Blaha has had a major impact on a succession of elementary particle theories: his Ph.D. thesis (1970), and papers, showed that quantum field theory calculations to all orders in ladder approximations could not give scaling deep inelastic electron-nucleon scattering. He later showed the eigenvalue equation for the fine structure constant α in Johnson-Baker-Willey QED had a zero at $\alpha = 1$ not 1/137 by solving the Schwinger-Dyson equations to all orders in an approximation that agreed with exact results to 4th order in α thus ending interest in this theory. In 1979 at Prof. Ken Johnson's (MIT) suggestion he calculated the proton-neutron mass difference in the MIT bag model and found the result had the wrong sign reducing interest in the bag model. These results all appear in Physical Review papers. In the 2000's he repeatedly pointed out the shortcomings of SuperString theory and showed that The Standard Model's form could be derived from space-time geometry by an extension of Lorentz transformations to faster than light transformations. This deeper space-time basis greatly increases the possibility that it is part of THE fundamental theory.Recently, Blaha showed that the Weak interactions differed significantly from the Strong, electromagnetic and gravitation interactions in important respects while these interactions had similar features, and suggested that ElectroWeak theory, which is essentially a glued union of the Weak interactions and Electromagnetism, possibly modulo unknown Higgs particle features, be replaced by a unified theory of the other interactions combined with a stand-alone Weak interaction theory. Blaha also showed that, if Charmonium calculations are taken seriously, the Strong interaction coupling constant is only a factor of five larger than the electromagnetic coupling constant, and thus Strong interaction perturbation theory would make sense and yield physically meaningful results.

In graduate school (1965-71) he wrote substantial papers in elementary particles and group theory: The Inelastic E- P Structure Functions in a Gluon Model. Phys. Lett. B40:501-502,1972; Deep-Inelastic E-P Structure Functions In A Ladder Model With Spin 1/2 Nucleons, Phys.Rev. D3:510-523,1971; Continuum Contributions To The Pion Radius, Phys. Rev. 178:2167-2169,1969; Character Analysis of U(N) and SU(N), J. Math. Phys. 10, 2156 (1969); and The Calculation of the Irreducible Characters of the Symmetric Group in Terms of the Compound Characters, (Published as Blaha's Lemma in D. E. Knuth's book: *The Art of Computer Programming Vols. 1 – 4*).

In the early 1980's Blaha was also a pioneer in the development of UNIX for financial, scientific and Internet applications: benchmarked UNIX versions showing that block size was critical for UNIX performance, developing financial modeling software, starting database benchmarking comparison studies, developing Internet-like UNIX networking (1982) and developing a hybrid shell programming technique (1982) that was a precursor to the PERL programming language. He was also the manager of the AT&T ten-year future products development database. His work helped lead to commercial UNIX on computers such as Sun Micros, IBM AIX minis, and Apple computers.

In the 1980's he pioneered the development of PC Desktop Publishing on laser printers. and was nominated for three "Awards for Technical Excellence" in 1987 by PC Magazine for PC software products that he designed and developed.

He has developed a theory of Megaverses – actual universes of which our universe is one – with quantum particle-like properties based on the Wheeler-DeWitt equation of Quantum Gravity. He has developed a theory of a baryonic force, which had been conjectured many years ago, and estimated the strength of the force based on discrepancies in measurements of the gravitational constant G. This force, operative in D-dimensinal space, can be used to escape from our universe in "uniships" which are the equivalent of the faster-than-light starships proposed in the author's earlier books. Thus travel to other universes, as well as to other stars is possible.

Blaha also considered the complexified Wheeler-DeWitt equation and showed that its limitation to real-valued coordinates and metrics generated a Cosmological Constant in the Einstein equations.

Recently he calculated the QED Fine Structure Constant exactly to the experimentally known 13 places. He also used the same approach to approximately calculate the Weak interaction SU(2) and Strong interaction SU(3) coupling constants successfully. Based on the origin of all coupling constants in quantum field theoretic vacuum polarization effects he suggested all universes would have the same interactions and consequently the same Physics, Chemistry and Biology. Thus all universes would be trivially Anthropic and capable of Life. Going further he suggested universes are particles and that universe expansion is co9mpletely analogous to vacuum polarization of a universe particle due to a gauge vector universe interaction. Universe expansion as a function of time is the fourier transform of universe vacuum polarization. This feature was demonstrated by almost exact (when compared to the universe scale factor) calculation of the small times universe scale factor in perturbation theory.
The author has also recently written a series of books on the serious problems of the United States and their solution as well as a book on the decline of Mankind that will follow from current social and genetic trends in Mankind.

In the past twelve years Dr. Blaha has written over 40 books on a wide range of topics. Some recent major works are: *From Asynchronous Logic to The Standard Model to Superflight to the Stars*, *All the Universe!*, *SuperCivilizations: Civilizations as Superorganisms*, *America's Future: an Islamic Surge, ISIS, al Qaeda, World Epidemics, Ukraine, Russia-China Pact, US Leadership Crisis*, *The Rises and Falls of Man – Destiny – 3000 AD: New Support for a Superorganism MACRO-THEORY of CIVILIZATIONS From CURRENT WORLD TRENDS and NEW Peruvian, Pre-Mayan, Mayan, Anatolian, and Early Egyptian Data, with a Projection to 3000 AD*, and *Mankind in Decline: Genetic Disasters, Human-Animal Hybrids, Overpopulation, Pollution, Global Warming, Food and Water Shortages, Desertification, Poverty, Rising Violence, Genocide, Epidemics, Wars, Leadership Failure.*

He has taught approximately 4,000 students in undergraduate, graduate, and postgraduate corporate education courses primarily in major universities, and large companies and government agencies.

www.ingramcontent.com/pod-product-compliance
Lightning Source LLC
Chambersburg PA
CBHW082002190326
41458CB00010B/3044
9781732824591